THE NATURE OF DIFFERENCE

THE NATURE OF DIFFERENCE

Sciences of Race in the United States from Jefferson to Genomics

edited by Evelynn M. Hammonds and Rebecca M. Herzig

The MIT Press
Cambridge, Massachusetts
London, England

This book was set in Stone Serif and Stone Sans by Asco Typesetters, Hong Kong.

Library of Congress Cataloging-in-Publication Data

The nature of difference : sciences of race in the United States from Jefferson to genomics / edited by Evelynn M. Hammonds and Rebecca M. Herzig.
    p.   cm.
Includes bibliographical references and index.
ISBN 978-0-262-08375-1 (hardcover : alk. paper) — ISBN 978-0-262-58275-9 (pbk. : alk. paper)
1. Race—Study and teaching—United States—History. 2. Race—Study and teaching—United States—History—Sources. 3. Science—Social aspects—United States—History.
4. Science—Social aspects—United States—History—Sources. 5. Difference (Psychology)—Social aspects—United States—History. 6. Difference (Psychology)—Social aspects—United States—History—Sources. 7. Race awareness—United States—History. 8. Race awareness—United States—History—Sources. 9. United States—Race relations—History. 10. United States—Race relations—History—Sources. I. Hammonds, Evelynn Maxine. II. Herzig, Rebecca M.
HT1506.N38   2009
305.800973—dc22   2007032265

Contents

Introduction

Evelynn M. Hammonds and Rebecca M. Herzig

This book documents how distinctions between people have been generated in, through, and around the natural sciences. It presents no singular claim about *the* role of race in scientific investigation, but instead illustrates multiple, conflicting efforts to assess difference. It is, in short, a book about unresolved intellectual and political debates. The appearance of such conflict in the natural sciences should not surprise us: at stake in the studies republished here is no less a matter than who we are, where we came from, and what we can do—that is to say, of what it might mean to be human.

Because the stakes of scientific investigations of race are (and have always been) so high, such studies have received considerable attention from scholars and activists. These critical reflections are as varied in their assumptions, approaches, and conclusions as the scientific studies themselves. Some analysts have sought to reveal nefarious personal interests behind specific scientific findings. Others address the logical contradictions of particular claims. Still others recalculate data to challenge the results of influential papers. Several of the best-known studies, such as Stephen Jay Gould's classic *The Mismeasure of Man* (1981), ferret out instances of bias and falsehood in sciences of difference. These efforts might be seen as part of an emancipatory tradition of scholarly criticism: one striving to replace error with accuracy and prejudice with impartiality. Recent analyses of race and racism in science tend to be cast in this mold, seeking to cast off false knowledge in favor of the truth about human difference: that, for example, race doesn't actually exist.[1]

While relying on the groundbreaking research of these previous studies, the approach of *The Nature of Difference* is somewhat different. Our aim is not to sort the "good" science from the "bad" science, to unveil the "myths" of race at long last. We see the task before those of us occupied by scientific studies of human variation as somewhat more difficult than removing bias from an essentially apolitical, culture-less method of inquiry. Given the resurgence of studies of racial difference across burgeoning fields such as neuroscience and pharmacogenomics, a tactic of "debunking" scientific racism appears insufficient to the demands of the present moment.

We instead turn our attention to particular sites in which human variation— including relations of inequality, disenfranchisement, and violence—has come to be incarnated in blood samples and hair sections, aliquot blots and statistical tables. We approach science not as a single instrument or method that reveals (or obscures) the real truth about human difference but instead, like race, as a profoundly heterogeneous array of practices. The documents presented here illuminate just a few of the countless ways in which specific categories of people have been brought into being through laborious acts of observation, quantification, and experimentation. We view these labors neither as the passage to an ideology-free future nor as hopelessly infested with racist thought. Resisting visions of salvation as well as damnation, *The Nature of Difference* instead shows just how entangled such hopes and fears about the relationships between science and race can be.

The book is organized chronologically and thematically. Any effort to divide the past into discrete chunks inevitably generates dilemmas and debates, and we wrestled at length with how best to arrange the sections that follow. Ultimately, we decided to concentrate on the period immediately following the close of the Civil War, when the scientific professions (and, by extension, professional scientific publications) grew in both size and scope of influence. As a result, we devote relatively little attention to the most recent generation of scientific research. Our hope is that in presenting a glimpse of the complexity of earlier, less readily accessible scientific studies of race, the book will encourage readers to explore contemporary resonances more fully on their own. Interested readers will find current scientific source material generally available in public libraries and via the Internet; the website www.racesci.org is one useful clearinghouse for recent scientific studies of race in North America. We trust that the readings and conceptual frameworks included here might be rearranged in any number of ways, in accordance with readers' own distinctive questions and concerns.

The usual problems with chronological organization were multiplied by the inherent ambiguity and complexity of the book's central themes. Drawing clean lines between social and natural sciences or between scientific and popular commentary proves as difficult as articulating the beginning or ending of any historical period. Rather than striving for comprehensive inclusion—which would be impossible, given the volume of scientific writings on race—we have here provided but one map of these diverse fields. In doing so, we sought to move away from craniometry, eugenics, the Tuskegee syphilis study, and equally famous or infamous chapters in the history of science. We have instead highlighted individuals, events, or processes that have attracted somewhat less attention to date. Readers will therefore find some well-known writers (such as the eugenicist Charles Davenport) in uncommon contexts (such as a discussion of racial mixing).

While acknowledging the fundamentally transnational character of each of the scientific trends discussed here, we have focused on studies written and published in the United States, to allow us to discuss the particular cultural circumstances in which each set of ideas was generated and received. To further accentuate the importance of context, we reproduced texts in their entirety wherever possible, a goal that dictated our decisions about inclusion: we opted for a few whole pieces rather than many partial excerpts. Again, we hope that our reliance on complete primary documents allows readers to identify other points of continuity or discontinuity than those we have selected, and to restructure the sections according to their expertise and interests.

Of course, in most fields of science today (including the so-called social sciences), a "primary" document would bear little resemblance to the sorts of texts gathered here. Evidence might instead include a column of figures, a radiograph, an assay, or a satellite image. Yet the kinds of documents included below ought not simply be seen as "unscientific." Just as footnotes have not always been the sign of reliable historical fact, so, too, experimental apparatuses, grainy images, and obscure tables have not always been central to credible knowledge about nature. Moreover, scientists, like all of us, speak differently before different audiences: from the after-dinner speech honoring a retiring colleague to the peer-reviewed article in a reputable professional journal. Indeed, one of the primary emphases of *The Nature of Difference* is the tremendous range in what has counted as science (and "pseudoscience") over the past two centuries. We invite readers to ask with us: what, exactly, makes something scientific? Does the measure reside in the reputation of the professional journal (an article in *Nature* versus one in *Scientific American*)? Does it instead reside in the reproducibility of results (the solidity of DNA sampling versus the vagaries of cranial measurements)? Is science defined by the period in which the research was conducted (the year 2000 rather than the year 1860)? By the instruments used in its practice (experimental electron scanners or dusty chalkboards)? Or perhaps by the characteristics of the investigator (a Nobel Prize-winning botanist or a second-grader nurturing a tiny bean plant; a twenty-first century African American geneticist or a nineteenth-century slaveholding physician)? We have avoided scattering awkward quotation marks around "science" throughout the book and instead endeavored to make the definitions of *science, scientist*, and the *scientific* themes worthy of fuller exploration. In lieu of a simple insistence that some texts, some claims, or some people are more genuinely scientific than others, we here encourage readers to approach the very definition of science as mutable and reflective of differential relations of power.

If science appears an unstable and shifting subject, its fluidity is matched by that of *race*, the second key term of this collection. Since its earliest known usage in 1508, the English word *race* has contained seemingly contradictory elements: at once natural and social, fixed and mutable, inherited and acquired. From the outset, the matter of race has been the topic of debate, from the number of races to the criteria for identifying members of a given race. Our aim is to display the protean (and deeply political) character of both the word and concept. As the following selections illustrate, even when race is the object of explicit study, its definition is rarely plain or stable. Within a single document, readers can track the various ideas embedded in the term at a given historical moment, including lineage, language, nationality, character, class, culture, ethnicity, self-designation, experience, blood, appearance, species, subspecies, gene, and population. These usages, moreover, have always been formed relationally. Naturalist Louis Agassiz, for instance, defined race in 1850 only through a process of triangulation: "The indomitable, courageous, proud Indian,—in how different a light he stands by the side of the submissive, obsequious, imitative negro, or by the side of the tricky, cunning, and cowardly Mongolian!"[2] These relational definitions extend not only from one racial group or type to another but also to other categories of difference, particularly those of age and sex.[3]

Although, as with the term *science*, we avoid incessant quotation marks around words such as Indian, Mongol, African, Caucasian, or Jew, we hope that readers will avoid the temptation to read racial designations backward in history, anachronistically equating a 1786 reference to "Negroes" to a 1964 reference to "Blacks," for instance, or a nineteenth-century discussion of "the Chinese" with twenty-first century discussions of "Asian Americans." It is one of the primary contentions of the volume that such semantic changes reflect much more substantial material transformations: a reorganization of the lived experience of difference. In an electrophoresis tray containing something labeled "Japanese" DNA, in the differential mortality rates of white Americans and American Indians, in surgical operations to sculpt the perfect nose, scientific framings of human difference come to inhere in the material world, demolishing some bodies and creating others. While we insist on the historical mutability of ideas about race, we also maintain that such changes merit careful consideration precisely because race continues to be so painfully corporeal. In the words of Donna Haraway: "Race kills, liberally and unequally; and race privileges, unspeakably and abundantly."[4] Studying the histories of various sciences helps draw out the peculiar "palpability and intangibility" of race.[5]

Demonstrating the ways in which the sciences in America have at once denied and depended on attaching racial categories to particular bodies, this book suggests that no amount of "clear thinking" or well-intentioned inquiry alone can eradicate the injustices historically associated with race.[6] It further suggests that matters of justice remain at the heart of sciences of difference in the United States: whether in evolutionary analyses of violence against women, the introduction of brain imagining techniques in criminal courts, or new forms of prenatal testing and selective abortion.[7] As biomedical innovations extend their reach into every capillary of American life, scientific authority is often summoned to resolve disputes in law, education, commerce, and entertainment. Given the enormously consequential role given to the sciences in our age, we hope that these selections will provide useful tools for a number of different audiences.

Our primary intended audience is students of the history of science, the history of race, and the history and cultural studies of the United States. We imagine that this slim collection of primary documents would pair well with recent related textbooks such as John P. Jackson Jr. and Nadine M. Weidman's *Race, Racism, and Science: Social Impact and Interaction* (ABC-Clio, 2004); other edited documentary collections, such as Carroll Pursell's *A Hammer in their Hands: A Documentary History of Technology and the African-American Experience* (MIT Press, 2005) or Emmanuel Chukwudi Eze's *Race and the Enlightenment* (Blackwell, 1997); and/or with one of the numerous volumes of relevant secondary essays, such as Maralee Mayberry et al.'s *Feminist Science Studies* (Routledge, 2001) or Sandra Harding's *The "Racial" Economy of Science* (Indiana University Press, 1993). Finally, we also hope that at least a few working scientists might find that the book provides useful context for their current research. Placing contemporary inquiry in a larger historical frame may help avoid reinventing dangerously rickety wheels.

While the sections that follow reveal tremendous change in the sciences of race (few of us now look to cranial capacity as a measure of intelligence), they also reveal odd and unexpected resonances across the centuries. While we seek to arrest simple narratives of progress by reminding readers of the monotonous recurrence of some scientific arguments about race, we also believe that further reflection on our shared past can only enhance our capacities for thoughtful collective action. Genetics, new evolutionary theory, and other sciences of the twenty-first certainly will not end controversy over the nature of race; if we learn one thing from the histories recorded here, it is that we *are* these debates.

Acknowledgments

We gratefully acknowledge the contributions of Abigail Bass, who composed the first drafts of the introductions to chapters four and ten and recommended selections for several other chapters. We also appreciate the advice and assistance provided by Lundy Braun, Heidi Chirayath, Anne Fausto-Sterling, Michelle Murphy, Jenny Reardon, Susan Reverby, and Katie Rieder, as well as the invaluable comments of three anonymous reviewers.

Notes

1. See, for example, Joseph L Graves, *The Race Myth: Why We Pretend Race Exists in America* (New York: Dutton, 2004). David Skinner and Paul Rosen offer a similar perspective in their history of the radical science movement, "Opening the White Box: The Politics of Racialised Science and Technology," *Science as Culture* 10, no. 3 (September 2001): 285–300.

2. Louis Agassiz, "The Diversity of Origin of the Human Races," *Christian Examiner* 49 (1850): 144.

3. Patricia Hill Collins, "Moving Beyond Gender: Intersectionality and Scientific Knowledge," in Myra Marx Ferree, Judith Lorber, and Beth B. Hess, ed., *Revisioning Gender* (Thousand Oaks: Sage Publications, 1999), 261–284.

4. Donna Haraway, *Modest_Witness@Second_Millennium.FemaleMan_Meets_OncoMouse* (New York: Routledge, 1997), 213.

5. Ann Laura Stoler, *Race and the Education of Desire: Foucault's "History of Sexuality" and the Colonial Order of Things* (Durham: Duke University Press, 1995): 206.

6. Cf. Roger Sanjek's proposition that "Clear thinking may help to purge human kind of race" ("The Enduring Inequalities of Race," in *Race*, edited by Steven Gregory and Roger Sanjek [New Brunswick: Rutgers University Press, 1996], 11). George Stocking similarly suggests that conflicting uses of the concept of "race" are less an issue of contestation than of *confusion*: he names "the persistence of widespread confusion of usage as the most immediately evident characteristics of racial thought" in the late nineteenth century. See George W. Stocking, Jr., "The Turn-of-the-Century Concept of Race," *Modernism/Modernity* 1, no. 1 (1994): 7.

7. Some of the most vociferous debates continue to revolve around alleged racial differences in intelligence. For one recent example, see J. Phillipe Rushton and Arthur R. Jensen, "Thirty Years of Research on Race Differences in Cognitive Ability," *Psychology, Public Policy, and the Law* (June 2005).

1 DICTIONARY DEFINITIONS OF "RACE"

Introduction

Introduction

Evelynn M. Hammonds and Rebecca M. Herzig

The definitions of *race* reproduced here provide a point of entry for the themes explored in *The Nature of Difference*. They range over time and by intended audience, from the standard desk reference used by generations of twentieth-century physicians to a dictionary designed for twenty-first-century evolutionary theorists and population geneticists.

These definitions certainly are not the only understandings of race used by working researchers, nor the only ones to be formalized in popular dictionaries. But in their variability and outright contradiction they highlight the role of definition in larger problems of classification. Where Thomas Lathrop Stedman's 1911 version of his famous *Practical Medical Dictionary* defines *race* in three brief phrases, in the same year the U.S. Immigration Commission required nothing less than a book-length treatise to clarify the meaning of race, the *Dictionary of Races or Peoples* prepared by Dr. Daniel Folkmar and Dr. Elnora C. Folkmar. (The Folkmars' definition of "Caucasian race" alone stretched over three pages.) Race does not appear in the 1890 version of George M. Gould's prominent medical dictionary, but does appear in the 1895 version. More recent scientific and medical dictionaries—including several others available via the Internet—continue to offer dozens of competing frameworks for understanding human variation, some that refer explicitly to *race* and some that do not.

These divergent, contested definitions carried (and continue to carry) consequences for people's everyday lives. In 1911, for example, the Immigration Commission's dictionary—with its distinctions between northern Italians and southern Italians, modern Greeks and ancient Greeks—spelled the difference between legal and illegal residence in the United States. While today's federal bureaus have abandoned the divisions between races, stocks, groups, and peoples used in 1911, definitions of racial groups still inform both the collection of federal statistics and the distribution of public resources based on those statistics. One need only recall the massive debate over which racial categories would be included on the 2000 U.S. Census in order to appreciate the import of the task of definition.

We include these selections, then, not in an attempt to stabilize a single definition of race, but with precisely the opposite intent: to emphasize the tremendous dissent ongoing over the content of the category even within scientific communities. By juxtaposing contemporary definitions with older, now archaic frameworks, ideas that we now take for granted can begin to appear unfamiliar, shaped by the particularities of time and place. As you read these definitions, consider the lines of divergence you see within them. Which aspects of the definitions appear to fade, which new elements arise, and which features appear to persist unaltered?

Definitions of "Race"

1.1   *A Complete Pronouncing Medical Dictionary* (1886)

J. Thomas

race   [From the Italian *raz 'za*, which is probably from the Latin *ra ' diz*, a "root"]. Races of men are permanent varieties of the human species, characterized by certain prominent distinctive traits. Blumenbach distinguished the following races:

1. The *Caucasian Race*—Skin white, passing into flesh-color, occasionally brownish; hair wavy, of a light or dark tint; face oval, facial angle large,-viz., from 80° to 85°: comprising the Europeans (except the Laplanders and Finns), the inhabitants of Western Asia as far as the Obi, the Ganges, and the Caspian Sea, and most of the tribes of Northern Attica.
2. The *Mongolian Race*—Skin yellow; hair black, straight, scanty; face broad, flat; glabella flat and broad: comprising the Tartars and Chinese; also the Laplanders, the Finns, and the Esquimaux and Greenlanders.
3. The *American Race*—Skin brownish copper-colored; hair black, straight, and scanty, comprising all the aborigines of America, except those included in the preceding variety.
4. The *Ethiopian Race*—Skin black, or brownish black; hair black, coarse, short, woolly, or frizzly; skull narrow, long; facial angle from 70° to 75°: including all the Africans (except those of the Caucasian variety), also the Negroes of Australia, those of Papua, etc.
5. The *Malay Race*—Skin black; hair black, soft, curling, and abundant; cranium moderately narrow: comprising the brown islanders of the South Sea, the inhabitants of the Sunda Isles, the Moluccas, the Philippine and Marianne Isles, and the true Malays of Malacca.

The classification of Blumenbach, however, has lost not a little of its prestige because it does not rest on a strictly scientific basis. Thus, it groups together under one head nations whose language proves them to be entirely distinct: *e.g.*, the Semitic Jews and Arabs are classed with Aryan nations like the Persians and Greeks. To classify

J. Thomas, *A Complete Pronouncing Medical Dictionary: Embracing the Terminology of Medicine and the Kindred Sciences, with their Signification, Etymology, and Pronunciation* (Philadelphia: J. B. Lippincott Co., 1886), 593.

nations by their complexion is scarcely more reasonable or more scientific than it would be to classify fruits by their color.

Classified by their language, mankind may be divided into three or more principal races or families,-viz.: 1. The *Aryan* (81" e-\U1) [nom the Sanscrit *ar 'ya*, "well-born," or "noble," a term applied to the high-cast Hindoos]. This name, as understood in modem science, includes not only the Sanscrit-speaking people of India, but also the ancient Persians, Greeks, and most of the nations of modern Europe, these being proved by their languages to be related to the Aryas of India. 2. The *Semit'ic* (or *Shemit'ic*) [nom *Shem*, the son of Noah]. This race, consisting of the descendants of Shem, includes the Arabians, Hebrews, ancient Assyrians, and probably portions of other nations. 3. The *Tura 'nian* [nom *Turan*, the ancient Persian name of Tartary]. This race includes the Turks, Mongolians, and most of the nations of Northeastern Asia. But there are many nations or tribes of Asia, Africa, and America which as yet, with our present imperfect knowledge, cannot be satisfactorily classified.

Race, in Botany, is a marked variety which may be perpetuated from seed. Our different sorts of wheat and maize are familiar examples.

1.2   *A Practical Medical Dictionary* (1911)

T. L. Stedman

race   1. A division of mankind, by some regarded as a species, such as the Caucasian, Mongolian, and Negro.   2. A tribal division.   3. An artificial division of animals kept distinct only by careful selection in breeding.

1.3   *Gould's Medical Dictionary* (1928)

G. M. Gould

race   *(rils)*.   1. A genealogic, ethnic, or triba [sic] stock; a breed or variety of plants or animals made permanent by constant transmission of its characters through the off-spring.   2. A Root, especially of ginger. r.-ginger, ginger in the race or root. r. suicide, artificial prevention of conception.

T. L. Stedman, *A Practical Medical Dictionary* (New York: William Wood and Co., 1911), 736.

G. M. Gould, *Gould's Medical Dictionary: Containing All the Words and Phrases Generally Used in Medicine and the Allied Sciences, with Their Proper Pronunciation, Derivation, and Definition Based on Recent Medical Literature* (Philadelphia: P. Blakiston's Son, 1928), 1179.

1.4   *Dictionary of Anthropology* (1956)

C. Winick

race   A major division of mankind, with distinctive, hereditarily transmissible physical characteristics, e.g., the Negroid, Mongoloid, and Caucasoid races. It may also be defined as a breeding group with gene organization differing from that of other intraspecies groups. Thus, the morphological and metrical features which members of a race have in common derive from their common descent. Each race has a tremendous range of internal variability. Such identifying criteria as skin color, hair and eye color, prognathism, cephalic index, nasal index, skull capacity, hair texture, and the degree of hirsuteness and lip eversion are generally used. There is no completely pure race, and the criteria for a given race may not be manifested by all the members although anyone member will probably manifest most. Current criteria for determining whether characteristics are racial include: hereditary transmission, comparative unalterability, lack of variability from external causes, and comparative independence of age and sex.

Blumenbach first divided mankind into races: Ethiopian, American, Malayan, Caucasian, and Mongolian. In Von Eickstedt's classification, there are now usually said to be three main races, 29 subraces. Coon, Garn, and Birdsell distinguish 30 races.

race, composite   A race that is a stable blend of two or more primary races and represents combinations of features from the different racial stocks involved. There are often geographic areas in which the blend has been stabilized.

race, genetic theory of   The classification of races on the basis of gene structure rather than morphology. At the present time, not enough is known about genes to make such a theory workable.

race, great   One of the major division of human beings.

1.5   *A Dictionary of Scientific Terms* (1957)

I. F. Henderson and W. D. Henderson

race   A permanent variety; a particular breed; a microspecies.

C. Winick, *Dictionary of Anthropology* (New York: Philosophical Library, 1956), 448–449.

I. F. Henderson and W. D. Henderson, *A Dictionary of Scientific Terms* (Princeton: D. Van Nostrand Co., 1957), 411.

1.6   Dictionary of the Biological Sciences (1967)

P. Gary

race 1   (*see also* race 2). A geographic enclave of a species the gene pool of which differs from that of other similar enclaves of the same species. The term is also loosely used in the sense of a variety that breeds true.

adaptive race   One which is physiologically, not morphologically, distinguished between race the offspring of a cross that shows some new types, but that mostly resemble the parent biological race = adaptive race.
climatic race   One which is adapted to a different climatic environment from that in which its nearest relatives live.
ecological race   Variously used to mean geographic race or subspecies.
mid race   One which can be improved by artificial selection and breeding.

race 2   (*see also* race 1). A *bot.* taxonomic category ranking below forma.

1.7   A Dictionary of Anthropology (1972)

D. M. Davies

race   A distinct group of people sharing certain inherited physical characteristics, such as skin color, hair type, etc., which are transmitted to their children.

racial types   Most people belong to one of the four racial types of man: the Mongoloid (*q. v.*), the European or Caucasian (*q. v.*), the Negroid (*q. v.*) and the Australasian (*q. v.*); and the rest are a mixture of these four. However, this is rather an over-simplification, as the Malays (*q. v.*), Polynesians (*q. v.*) and Arabs do not really fit into this pattern, nor do the mountain peoples of New Guinea, quite different from the coastal peoples who are Melanesians (*q. v.*) of a negrito type (*q. v.*). Possibly environment is the cause of these differences.

P. Gray, *Dictionary of the Biological Sciences* (New York: Reinhold Publishing Co., 1967), 436.

D. M. Davies, *A Dictionary of Anthropology* (New York: Crane, Russak, 1972), 155.

1.8   *McGraw-Hill Dictionary of the Life Sciences* (1976)

race [ANTHRO].   1. A distinctive human type possessing characteristic traits that are transmissible by descent.   2. Descendants of a common ancestor. [BIOL].   1. An intra-specific taxonomic group of organisms, such as subspecies or micro species.   2. A fixed variety or breed.

1.9   *MacMillan Dictionary of Anthropology* (1986)

C. Seymour-Smith

race The common use of the word in English to refer to a group of persons who share common physical characteristics and form a discrete and separable population unit has no scientific validity, since evolutionary theory and PHYSICAL ANTHROPOLOGY have long since demonstrated that there are no fixed or discrete racial groups in human populations. Instead, human groups constantly change and interact, to such an extent that modern population genetics focuses on CLINES or patterns of the distribution of specific genes rather than on artificially created racial categories. However as a folk concept in Western and non-Western societies the concept of race is a powerful and important one, which is employed in order to classify and systematically exclude members of given groups from full participation in the social system controlled by the dominant group. As a folk concept, race is employed to attribute not only physical characteristics but also psychological and moral ones to members of given categories, thus justifying or naturalizing a discriminatory social system. (*See* RACISM.)

1.10   *International Dictionary of Medicine and Biology* (1986)

S. I. Landau, ed.

race [Middle French, from Italian *razza* race, king]. A subspecies or other division or subdivision of a species. Human races are generally defined in terms of original geographic range and common hereditary traits which may be morphological, serological,

McGraw-Hill Dictionary of the Life Sciences (New York: McGraw-Hill, 1976), 717.

C. Seymour-Smith, *Macmillan Dictionary of Anthropology* (New York: Macmillian, 1986), 238.

S. I. Landau, ed., *International Dictionary of Medicine and Biology* (New York: Wiley, 1986), 2376.

hematological, immunological, or biochemical. The traditional division of mankind into several well-recognized racial types such as Caucasoid (white), Negroid (black), and Mongoloid (yellow) leaves a residue of populations that are of problematical classification, and its focus on a limited range of visible characteristics tends to oversimplify and distort the picture of human variation.

1.11   *Dictionary of Genetics and Cell Biology* (1987)

N. Maclean

race   Subspecific group of population that can be distinguished by phenotypic character and/or geographical isolation from other groups, or populations within the species. The degree of differences between races is variable, and the word is used according to convenience. Races may or may not be recognized as SUBSPECIES, and are only so recognized when the racial differences are considerable and deemed to be of taxonomic importance.

1.12   *Glossary of Genetics: Classical and Molecular* (1991)

R. Rieger

race   An intraspecific category, primarily a population or aggregate of populations, with characteristic gene frequencies or features of chromosome structure that distinguish a particular group of individuals from other groups of the same kind within formally recognizable ~ subspecies or within ~ species. Race differences are relative, not absolute. The term subspecies is frequently used in the same sense as race. Any race is able to interbreed freely with any other r. of the same species. Whenever different races of a cross-fertilizing species occupy geographically separate territories, they are said to be allopathic; those occupying the same territory are sympatric. Races may become distinct species (~ speciation) by the production of reproductive ~ isolation (with respect to the other races of the same species) and thus the formation of isolated ~ gene pools. (1) Geographic races are ~ subspecies occupying a geographic subdivision of the range of a species. (2) Ecological races are local races owing their most conspicuous attributes to the selective effect of a specific environment (~ ecotype). (3) Physiological races are

N. Maclean, *Dictionary of Genetics and Cell Biology* (New York: New York University Press, 1987), 333.

R. Rieger, *Glossary of Genetics: Classical and Molecular* (New York: Springer-Verlag, 1991), 409.

races differing in respect to features of chromosome structure (cytotypes) or in chromosome number (polyplotypes).

1.13   Encyclopedia & Dictionary of Medicine, Nursing, & Allied Health (1992)

B. F. Miller

race   (ras). A class or breed of animals; a group of individuals having certain characteristics in common, owing to a common inheritance.

1.14   A Dictionary of Genetics (1997)

R. King

race   A phenotypically and/or geographically distinctive subspecific group, composed of individuals inhabiting a defined geographical and/or ecological region, and possessing characteristic phenotypic and gene frequencies that distinguish it from other such groups. The number of racial groups that one wishes to recognize within a species is usually arbitrary but suitable for the purposes under investigation. *See* ecotype, subspecies.

1.15   The Cambridge Dictionary of Human Biology and Evolution (2005)

L. L. Mai, M. Young Owl, and M. P. Kersting

race   1. In nonhuman biology, a group of intraspecific populations, often geographically isolated, sharing certain conspicuous traits that make them perceptively distinct from other groups or sub-populations within a species; a taxonomic rank considered equivalent to or just below the subspecies. The biological concept of race is very diffi-

B. F. Miller, *Encyclopedia & Dictionary of Medicine, Nursing, & Allied Health* (Philadelphia: Saunders, 1992), 1259.

R. King, *Dictionary of Genetics* (New York: Oxford University Press, 1997), 285.

L. L. Mai, M. Young Owl, and M. P. Kersting, *The Cambridge Dictionary of Human Biology and Evolution* (New York: Cambridge University Press, 2005), 448.

cult to apply to observed patterns of human variation; rather human varieties can be considered cultural constructs that are contingent facts of history. See biological race, ecological race, geographic races, and physiological race.   2. In law, any group of persons related by common descent, blood, or heredity; a known stock; any group distinguished by self-identification or commonly known to others by a group name; a population so identified in the literature of the legal profession.   3. Colloquially, any commonly recognized nation, strain, tribe, or ethnic stock of humans distinguished by physical traits as well as by codes of behavior and moral conduct that are different from, and usually perceived as less correct than one's own; aka the peasant's perspective. See folk taxonomy.

2 ANATOMICAL OBSERVATIONS

Introduction

Evelynn M. Hammonds and Rebecca M. Herzig

Questions concerning the nature of human difference have been at the center of American society since the founding of the nation. Were bodily differences fixed at birth, or were they formed through circumstances such as climate, childrearing, diet or labor? Were differences in temperament or intelligence unalterable or mutable through education and religious discipline? Did variation between individuals override any apparent similarity among groups? As the first selection by Thomas Jefferson makes clear, the answers to such questions had clear political repercussions. Now remembered primarily as a statesman and slaveholder, Jefferson was also a well-regarded natural historian. In this excerpt on "Laws" from Jefferson's influential *Notes on the State of Virginia* (penned in 1781–1782 in response to a set of queries from a foreign visitor), Jefferson articulates his understanding of the relationship between racial anatomy and the administration of justice. On what does he appear to base his observations? What kinds of explanation does he offer for the variations he perceived?

Because questions of difference carried profound political consequences, scarcely any assessment of human nature went unchallenged. Indeed in 1791, Jefferson received a letter contesting just such claims, explicitly naming the gap between the Declaration of Independence's proclamations of equality and the brutal realities of African enslavement. The letter's author, Benjamin Banneker, was born in Maryland in 1731. The descendent of enslaved men and women, Banneker was, like Jefferson, a polymath who excelled in mathematics, astronomy, agriculture, mechanics, and other key fields of his era. Banneker is perhaps most widely remembered today as the surveyor of the site for the nation's capitol in Washington. In 1791, Banneker published the first edition of his *Almanac*, which included a table of stars and tides along with other useful information, and sent a copy of the book to Jefferson along with his letter. Note the ways in which Banneker stresses the importance of Christianity, and concepts of divine endowment and providence. When news of the letter was made public, pro-slavery advocates shot at Banneker's Maryland cabin and vandalized his farm. Two days after Banneker's death in 1806, attackers burned his cabin to the ground.

Violent controversy over relationships between human nature, divine intention, and social justice only intensified during the turbulent political climate of the early nineteenth century. National expansion after 1830 brought the forced relocation of Native peoples, increasingly fierce conflict over the institution of slavery, and the beginnings of a backlash against immigration from Asia. The passage of the Indian Removal Act (1830), the emergence of the American Anti-Slavery Society (1833), the conclusion of a bloody war with Mexico (1848), and the first Anti-Chinese Riot (1849) all point to the tumult of the period. Historians have argued that such tensions colluded to spark new attention to comparative physiology. As social debates grew more complex and acrimonious, new criteria were sought to rationalize (or challenge) the competing theories of racial hierarchy and racial equality found in reigning Protestant theology. Both in the US and in Europe, commentators increasingly drew on the authority of scientific inquiry in order to justify or resist practices of disenfranchisement, displacement, and genocide.

Discerning the nature of bodies, however, proved no less controversial than determining the course of social policy. Efforts to assess individual or collective racial differences were plagued by contradictory findings. Skin color—long a key to racial taxonomy—proved to be an unreliable standard of race, as broadsheets circulated stories of "Negro" slaves far more pale-skinned and fair-haired than their "white" owners. The enigmas of human variation attracted some of the most influential natural philosophers of the period, now collectively remembered as practitioners of the "American School" of ethnology. An explicit focus of their ethnological work was, as for Jefferson, the relative mutability or immutability of racial difference: not simply whether variation existed, but whether it could be ameliorated through circumstance.

In this selection from one of his most famous addresses, Frederick Douglass, an escaped slave whose life spanned most of the nineteenth century, challenges much of the work of the American School ethnologists. In opposition to the ethnologists' claims of innate inequality born of separate lines of descent, he argues for the essential unity of the human species and the mutability of apparent differences in appearance or capacity. As you read Douglass, you might compare his observations with those presented by Jefferson. To what does Douglass attribute distinctions of complexion or musculature? How might he refute Jefferson's assertion that anatomical variations are not merely an effect of different "conditions of life"?

While the Emancipation brought by Northern victory in the Civil War fundamentally altered the terrain of debates about the nature of race, scientists continued to seek standards of human variation amenable to precise measurement. Hair, pelvises, skulls, genitalia and other body parts seemed especially useful in this regard, since they could be easily dissected, stored, counted, and circulated by interested scholars. The development and elaboration of racial differences, in other words, required a kind of traffic in anatomical parts—their removal, display, purchase, and exchange. (See figures 2.7 and 2.8, for example.) The turn toward these inert, extractable body parts reveals an effort to move beyond the confusing *surfaces* of bodies toward their (allegedly) more stable components, thought to house the essence of racial identity. It also highlights the importance of observation and quantification as scientific methods—a topic to which we will return in chapter 5.

Such discussions of anatomy were often explicitly intertwined with discussions of disease: whether some forms of embodiment were more or less susceptible to tumors, fevers, or other vulnerabilities. The excerpt from an address Thomas W. Murrell, an Arkansas physician best known for his writings on syphilis, addresses these relations explicitly. Writing in the context of broader debates about the post-

Emancipation future of black Americans, Murrell contemplates the relations between anatomy, disease, and health through the structure of the ear. The selection includes portions of the discussion held after Murrell's 1887 presentation of this work at the Ninth Session of the International Medical Congress. Such hierarchical assessments of difference led not only to diagnoses of pathology, but also to new types of therapeutic intervention. Perceptions of inequality helped generate efforts to ameliorate racial "deformities," as evident in John Orlando Roe's discussion of the surgical correction of Irish pug noses. Roe, a New York physician born in 1848, is now remembered as a leading figure in the development of modern aesthetic rhinoplasty.

While reading the selections included here, you might consider how and why certain parts of the body might receive scientific focus at a particular moment: in what larger political, economic, and social contexts might ears, noses, index fingers, or brain hemispheres attract attention as reliable signs of racial difference? How do the various authors here emphasize or disavow distinctions between the physical body and its environment? What economic, political, or cultural consequences are contained in these various assessments of lungs, noses, and muscles? Which kinds of attributes do these authors understand to be innate, and which do they understand to be immutable? How do they distinguish between individual and group variation? It may also prove fruitful to contemplate these authors' statements on the *function* of difference. How do they explain the various shapes and capabilities of bodies? How do they connect their studies of body parts to earlier analyses?

Bibliography

Bedini, Silvio A. *The Life of Benjamin Banneker*, second edition. Baltimore: Maryland Historical Society, 1999.

Bieder, Robert E. *Science Encounters the Indian, 1820–1880*. Norman: University of Oklahoma, 1986.

Cerami, Charles A. *Benjamin Banneker: Surveyor, Astronomer, Publisher, Patriot*. New York: John Wiley and Sons, 2002.

Fee, Elizabeth. "Nineteenth-Century Craniology: The Study of the Female Skull," *Bulletin of the History of Medicine* 53 (1979): 415–433.

Gould, Stephen Jay. "American Polygeny and Craniometry before Darwin: Blacks and Indians as Separate, Inferior Species." In Sandra Harding, ed., *The 'Racial' Economy of Science*. Bloomington: Indiana University Press, 1993, pp. 84–115.

Graham, Shirley Dredge. *Your Most Humble Servant*. New York: Messner, [1949].

Johnson, Joseph Taber. "Apparent Peculiarities of Partuition in the Negro Race." *American Journal of Obstetrics* 8 (January 1875): 88–123.

Lindqvist, Sven. *The Skull Measurer's Mistake*. Translated by Joan Tate. New York: New Press, 1997.

Morton, Samuel G. *Crania Americana; or, A Comparative View of the Skulls of Various Aboriginal Nations of North and South America*. Philadelphia: J. Dobson, 1839.

Rogers, B. O. "John Orlando Roe—not Jacques Joseph—the Father of Aesthetic Rhinoplasty." *Aesthetic Plastic Surgery* 10, no. 2 (1986): 63–88.

Wailoo, Keith. *Dying in the City of the Blues: Sickle Cell Anemia and the Politics of Race and Health*. Chapel Hill: University of North Carolina Press, 2001.

2.1   "Laws" (Query XIV) (1781–1782)

Thomas Jefferson

THE ADMINISTRATION OF JUSTICE AND DESCRIPTION OF THE LAWS?

The state is divided into counties. In every county are appointed magistrates, called justices of the peace, usually from eight to thirty or forty in number, in proportion to the size of the county, of the most discreet and honest inhabitants. They are nominated by their fellows, but commissioned by the governor, and act without reward. These magistrates have jurisdiction both criminal and civil. If the question before them be a question of law only, they decide on it themselves; but if it be of fact, or of fact and law combined, it must be referred to a jury. In the latter case, of a combination of law and fact, it is usual for the jurors to decide the fact, and to refer the law arising on it to the decision of the judges. But this division of the subject lies with their discretion only. And if the question relate to any point of public liberty, or if it be one of those in which the judges may be suspected of bias, the jury undertake to decide both law and fact. If they be mistaken, a decision against right, which is casual only, is less dangerous to the state, and less afflicting to the loser, than one which makes part of a regular and uniform system. In truth, it is better to toss up cross and pile in a cause, than to refer it to a judge whose mind is warped by any motive whatever, in that particular case. But the common sense of twelve honest men gives still a better chance of just decision, than the hazard of cross and pile. These judges execute their process by the sheriff or coroner of the county, or by constables of their own appointment. If any free person commit an offence against the commonwealth, if it be below the degree of felony, he is bound by a justice to appear before their court, to answer it on indictment or information. If it amount to felony, he is committed to jail, a court of these justices is called; if they on examination think him guilty, they send him to the jail of the general court, before which court he is to be tried first by a grand jury of 24, of whom 13 must concur in opinion: if they find him guilty, he is then tried by a jury of 12 men of the county where the offence was committed, and by their verdict, which must be unanimous, he is acquitted or condemned without appeal. If the criminal be a slave the trial by the county court is final. In every case however, except that of high treason, there resides in the governor a power of pardon. In high treason, the pardon can only flow from the general assembly. In civil matters these justices have jurisdiction in all cases of whatever value, not appertaining to the department of the admiralty. This jurisdiction is twofold. If the matter in dispute be of less value than $4\frac{1}{6}$ dollars, a

Thomas Jefferson, "Laws" (Query XIV), in William Peden, ed., *Notes on the State of Virginia* (Chapel Hill: University of North Carolina Press, 1982), 130–149.

single member may try it at any time and place within his county, and may award execution on the goods of the party cast. If it be of that or greater value, it is determinable before the county court, which consists of four at the least of those justices, and assembles at the court-house of the county on a certain day in every month. From their determination, if the matter be of the value of ten pounds sterling, or concern the title or bounds of lands, an appeal lies to one of the superior courts.

There are three superior courts, to wit, the high-court of chancery, the general court, and court of admiralty. The first and second of these receive appeals from the county courts, and also have original jurisdiction where the subject of controversy is of the value of ten pounds sterling, or where it concerns the title or bounds of land. The jurisdiction of the admiralty is original altogether. The high-court of chancery is composed of three judges, the general court of five, and the court of admiralty of three. The two first hold their sessions at Richmond at stated times, the chancery twice in the year, and the general court twice for business civil and criminal, and twice more for criminal only. The court of admiralty sits at Williamsburgh whenever a controversy arises.

There is one supreme court, called the court of appeals, composed of the judges of the three superior courts, assembling twice a year at stated times at Richmond. This court receives appeals in all civil cases from each of the superior courts, and determines them finally. But it has no original jurisdiction.

If a controversy arise between two foreigners of a nation in alliance with the United States, it is decided by the Consul for their State, or, if both parties chuse it, by the ordinary courts of justice. If one of the parties only be such a foreigner, it is triable before the courts of justice of the country. But if it shall have been instituted in a county court, the foreigner may remove it into the general court, or court of chancery, who are to determine it at their first sessions, as they must also do if it be originally commenced before them. In cases of life and death, such foreigners have a right to be tried by a jury, the one half foreigners, the other natives.

All public accounts are settled with a board of auditors, consisting of three members, appointed by the general assembly, any two of whom may act. But an individual, dissatisfied with the determination of that board, may carry his case into the proper superior court.

A description of the laws.

The general assembly was constituted, as has been already shewn, by letters-patent of March the 9th, 1607, in the 4th year of the reign of James the First. The laws of England seem to have been adopted by consent of the settlers, which might easily enough be done whilst they were few and living all together. Of such adoption however we have no other proof than their practice, till the year 1661, when they were expressly adopted by an act of the assembly, except so far as "a difference of condition" rendered them inapplicable. Under this adoption, the rule, in our courts of judicature was, that the common law of England, and the general statutes previous to the 4th of James, were in force here; but that no subsequent statutes were, *unless we were named in them*, said the judges and other partisans of the crown, but *named or not named*, said those who reflected freely. It will be unnecessary to attempt a description of the laws of England, as that may be found in English publications. To those which were established here, by the adoption of the legislature, have been since added a number of acts of assembly passed during the monarchy, and ordinances of convention and acts of assembly enacted since the establishment of the republic. The following variations from the British model are perhaps worthy of being specified.

Debtors unable to pay their debts, and making faithful delivery of their whole effects, are released from confinement, and their persons for ever discharged from restraint for such previous debts: but any property they may afterwards acquire will be subject to their creditors.

The poor, unable to support themselves, are maintained by an assessment on the titheable persons in their parish. This assessment is levied and administered by twelve persons in each parish, called vestrymen, originally chosen by the housekeepers of the parish, but afterwards filling vacancies in their own body by their own choice. These are usually the most discreet farmers, so distributed through their parish, that every part of it may be under the immediate eye of some one of them. They are well acquainted with the details and œconomy of private life, and they find sufficient inducements to execute their charge well, in their philanthropy, in the approbation of their neighbours, and the distinction which that gives them. The poor who have neither property, friends, nor strength to labour, are boarded in the houses of good farmers, to whom a stipulated sum is annually paid. To those who are able to help themselves a little, or have friends from whom they derive some succours, inadequate however to their full maintenance, supplementory aids are given, which enable them to live comfortably in their own houses, or in the houses of their friends. Vagabonds, without visible property or vocation, are placed in workhouses, where they are well cloathed, fed, lodged, and made to labour. Nearly the same method of providing for the poor prevails through all our states; and from Savannah to Portsmouth you will seldom meet a beggar. In the larger towns indeed they sometimes present themselves. These are usually foreigners, who have never obtained a settlement in any parish. I never yet saw a native American begging in the streets or highways. A subsistence is easily gained here: and if, by misfortunes, they are thrown on the charities of the world, those provided by their own country are so comfortable and so certain, that they never think of relinquishing them to become strolling beggars. Their situation too, when sick, in the family of a good farmer, where every member is emulous to do them kind offices, where they are visited by all the neighbours, who bring them the little rarities which their sickly appetites may crave, and who take by rotation the nightly watch over them, when their condition requires it, is without comparison better than in a general hospital, where the sick, the dying, and the dead are crammed together, in the same rooms, and often in the same beds. The disadvantages, inseparable from general hospitals, are such as can never be counterpoised by all the regularities of medicine and regimen. Nature and kind nursing save a much greater proportion in our plain way, at a smaller expence, and with less abuse. One branch only of hospital institution is wanting with us; that is, a general establishment for those labouring under difficult cases of chirurgery. The aids of this art are not equivocal. But an able chirurgeon cannot be had in every parish. Such a receptacle should therefore be provided for those patients: but no others should be admitted.

Marriages must be solemnized either on special licence, granted by the first magistrate of the county, on proof of the consent of the parent or guardian of either party under age, or after solemn publication, on three several Sundays, at some place of religious worship, in the parishes where the parties reside. The act of solemnization may be by the minister of any society of Christians, who shall have been previously licensed for this purpose by the court of the county. Quakers and Menonists however are exempted from all these conditions, and marriage among them is to be solemnized by the society itself.

A foreigner of any nation, not in open war with us, becomes naturalized by removing to the state to reside, and taking an oath of fidelity: and thereupon acquires every right of a native citizen: and citizens may divest themselves of that character, by declaring, by solemn deed, or in open court, that they mean to expatriate themselves, and no longer to be citizens of this state.

Conveyances of land must be registered in the court of the county wherein they lie, or in the general court, or they are void, as to creditors, and subsequent purchasers.

Slaves pass by descent and dower as lands do. Where the descent is from a parent, the heir is bound to pay an equal share of their value in money to each of his brothers and sisters.

Slaves, as well as lands, were entailable during the monarchy: but, by an act of the first republican assembly, all donees in tail, present and future, were vested with the absolute dominion of the entailed subject.

Bills of exchange, being protested, carry 10 per cent. interest from their date.

No person is allowed, in any other case, to take more than five per cent. per annum simple interest, for the loan of monies.

Gaming debts are made void, and monies actually paid to discharge such debts (if they exceeded 40 shillings) may be recovered by the payer within three months, or by any other person afterwards.

Tobacco, flour, beef, pork, tar, pitch, and turpentine, must be inspected by persons publicly appointed, before they can be exported.

The erecting iron-works and mills is encouraged by many privileges; with necessary cautions however to prevent their dams from obstructing the navigation of the watercourses. The general assembly have on several occasions shewn a great desire to encourage the opening the great falls of James and Patowmac rivers. As yet, however, neither of these have been effected.

The laws have also descended to the preservation and improvement of the races of useful animals, such as horses, cattle, deer; to the extirpation of those which are noxious, as wolves, squirrels, crows, blackbirds; and to the guarding our citizens against infectious disorders, by obliging suspected vessels coming into the state, to perform quarantine, and by regulating the conduct of persons having such disorders within the state.

The mode of acquiring lands, in the earliest times of our settlement, was by petition to the general assembly. If the lands prayed for were already cleared of the Indian title, and the assembly thought the prayer reasonable, they passed the property by their vote to the petitioner. But if they had not yet been ceded by the Indians, it was necessary that the petitioner should previously purchase their right. This purchase the assembly verified, by enquiries of the Indian proprietors; and being satisfied of its reality and fairness, proceeded further to examine the reasonableness of the petition, and its consistence with policy; and, according to the result, either granted or rejected the petition. The company also sometimes, though very rarely, granted lands, independantly of the general assembly. As the colony increased, and individual applications for land multiplied, it was found to give too much occupation to the general assembly to enquire into and execute the grant in every special case. They therefore thought it better to establish general rules, according to which all grants should be made, and to leave to the governor the execution of them, under these rules. This they did by what have been usually called the land laws, amending them from time to time, as their defects were developed. According to these laws, when an individual wished a portion of unappropriated land, he was to locate and survey it by a public officer, appointed for that purpose: its breadth was to bear a certain proportion to its length: the grant was to be executed by the governor: and the lands were to be improved in a certain manner, within a given time. From these regulations there resulted to the state a sole and exclusive power of taking conveyances of the Indian right of soil: since, according to them, an Indian conveyance alone could give no right to an individual, which the laws would acknowledge. The state, or the crown, thereafter, made general purchases of the Indians from time to time, and the governor parcelled them out by special grants, conformed to the rules before described, which it was not in his power, or in that of the crown, to dispense with. Grants, unaccompanied by their proper legal circumstances, were set aside regularly by *scire facias*, or by bill in Chancery. Since the establishment of our new government, this order of things is but little changed. An individual, wishing to appropriate to himself lands still unappropriated by any other, pays to the public treasurer a sum of money proportioned to the quantity he wants. He carries the treasurer's receipt to the auditors of public accompts, who thereupon debit the treasurer with the sum, and order the register of the land-office to give the party a warrant for his land. With this warrant from the register, he goes to the surveyor of the county where the land lies on which he has cast his eye. The surveyor lays it off for him, gives him its exact description, in the form of a certificate, which certificate he returns to the land-office, where a grant is made out, and is signed by the governor. This vests in him a perfect dominion in his lands, transmissible to whom he pleases by deed or will, or by descent to his heirs if he die intestate.

Many of the laws which were in force during the monarchy being relative merely to that form of government, or inculcating principles inconsistent with republicanism, the first assembly which met after the establishment of the common-wealth appointed a committee to revise the whole code, to reduce it into proper form and volume, and report it to the assembly. This work has been executed by three gentlemen, and reported; but probably will not be taken up till a restoration of peace shall leave to the legislature leisure to go through such a work.

The plan of the revisal was this. The common law of England, by which is meant, that part of the English law which was anterior to the date of the oldest statutes extant, is made the basis of the work. It was thought dangerous to attempt to reduce it to a text: it was therefore left to be collected from the usual monuments of it. Necessary alterations in that, and so much of the whole body of the British statutes, and of acts of assembly, as were thought proper to be retained, were digested into 126 new acts, in which simplicity of stile was aimed at, as far as was safe. The following are the most remarkable alterations proposed:

To change the rules of descent, so as that the lands of any person dying intes-tate shall be divisible equally among all his children, or other representatives, in equal degree.

To make slaves distributable among the next of kin, as other moveables.

To have all public expences, whether of the general treasury, or of a parish or county, (as for the maintenance of the poor, building bridges, court-houses, &c.) sup-plied by assessments on the citizens, in proportion to their property.

To hire undertakers for keeping the public roads in repair, and indemnify indi-viduals through whose lands new roads shall be opened.

To define with precision the rules whereby aliens should become citizens, and citizens make themselves aliens.

To establish religious freedom on the broadest bottom.

To emancipate all slaves born after passing the act. The bill reported by the revisors does not itself contain this proposition; but an amendment containing it was prepared, to be offered to the legislature whenever the bill should be taken up, and fur-ther directing, that they should continue with their parents to a certain age, then be brought up, at the public expence, to tillage, arts or sciences, according to their genius-ses, till the females should be eighteen, and the males twenty-one years of age, when they should be colonized to such place as the circumstances of the time should render most proper, sending them out with arms, implements of houshold and of the handi-craft arts, seeds, pairs of the useful domestic animals, &c. to declare them a free and independant people, and extend to them our alliance and protection, till they shall have acquired strength; and to send vessels at the same time to other parts of the world for an equal number of white inhabitants; to induce whom to migrate hither, proper encouragements were to be proposed. It will probably be asked, Why not retain and incorporate the blacks into the state, and thus save the expence of supplying, by im-portation of white settlers, the vacancies they will leave? Deep rooted prejudices enter-tained by the whites; ten thousand recollections, by the blacks, of the injuries they have sustained; new provocations; the real distinctions which nature has made; and many other circumstances, will divide us into parties, and produce convulsions which will probably never end but in the extermination of the one or the other race.—To these objections, which are political, may be added others, which are physical and moral. The first difference which strikes us is that of colour. Whether the black of the negro resides in the reticular membrane between the skin and scarf-skin, or in the scarf-skin itself; whether it proceeds from the colour of the blood, the colour of the bile, or from that of some other secretion, the difference is fixed in nature, and is as real as if its seat and cause were better known to us. And is this difference of no importance? Is it not the foundation of a greater or less share of beauty in the two races? Are not the fine mixtures of red and white, the expressions of every passion by greater or less suffusions of colour in the one, preferable to that eternal monotony, which reigns in the countenances, that immoveable veil of black which covers all the emotions of the other race? Add to these, flowing hair, a more elegant symmetry of form, their own judgment in favour of the whites, declared by their preference of them, as uniformly as is the preference of the Oran-ootan for the black women over those of his own species. The circumstance of superior beauty, is thought worthy attention in the propagation of our horses, dogs, and other domestic animals; why not in that of man? Besides those of colour, figure, and hair, there are other physical distinctions proving a difference of race. They have less hair on the face and body. They secrete less by the kidnies, and more by the glands of the skin, which gives them a very strong and disagreeable odour. This greater degree of transpiration renders them more tolerant of heat, and less so of cold, than the whites. Perhaps too a difference of structure in the pulmonary apparatus, which a late ingenious experimentalist has discovered to be the principal regulator of animal heat, may have disabled them from extricating, in the act of inspiration, so much of that fluid from the outer air, or obliged them in expiration, to part with more of it. They seem to require less sleep. A black, after hard labour through the day, will be induced by the slightest amusements to sit up till midnight, or later, though knowing he must be out with the first dawn of the morning. They are at least as brave, and more adventuresome. But this may perhaps proceed from a want of forethought, which prevents their seeing a danger till it be present. When present, they do not go through it with more coolness or steadiness than the whites. They are more ardent after their female: but love seems with them to be more an eager desire, than a tender delicate mixture of sentiment and sensation. Their griefs are transient. Those numberless afflictions, which render it doubtful whether heaven has given life to us in mercy or in wrath, are less felt, and sooner forgotten with them. In general, their existence appears to participate more of sensation than reflection. To this must be ascribed their disposition to sleep when abstracted from their diversions, and unemployed in labour. An animal whose body is at rest, and who does not reflect, must be disposed to sleep of course. Comparing them by their faculties of memory, reason, and imagination, it appears to me, that in memory they are equal to the whites; in reason much inferior, as I think one could scarcely be found capable of tracing and comprehending the investigations of Euclid; and that in imagination they are dull, tasteless, and anomalous. It would be unfair to follow them to Africa for this investigation. We will consider them here, on the same stage with the whites, and where the facts are not apocryphal on which a judgment is to be formed. It will be right to make great allowances for the difference of condition, of education, of conversation, of the sphere in which they move. Many millions of them have been brought to, and born in America. Most of them indeed have been confined to tillage, to their own homes, and their own society: yet many have been so situated, that they might have availed themselves of the conversation of their masters; many have been brought up to the handicraft arts, and from that circumstance have always been associated with the whites. Some have been liberally educated, and all have lived in countries where the arts and sciences are cultivated to a considerable degree, and have had before their eyes samples of the best works from abroad. The Indians, with no advantages of this kind, will often carve figures on their pipes not destitute of design and merit. They will crayon out an animal, a plant, or a country, so as to prove the existence of a germ in their minds which only wants cultivation. They astonish you with strokes of the most sublime oratory; such as prove their reason and sentiment strong, their imagination glowing and elevated. But never yet could I find that a black had uttered a thought above the level of plain narration; never see even an elementary trait of painting or sculpture. In music they are more generally gifted than the whites with accurate ears for tune and time, and they have been found capable of imagining a small catch. Whether they will be equal to the composition of a more extensive run of melody, or of complicated harmony, is yet to be proved. Misery is often the parent of the most affecting touches in poetry.— Among the blacks is misery enough, God knows, but no poetry. Love is the peculiar œstrum of the poet. Their love is ardent, but it kindles the senses only, not the imagination. Religion indeed has produced a Phyllis Whately; but it could not produce a poet. The compositions published under her name are below the dignity of criticism. The heroes of the Dunciad are to her, as Hercules to the author of that poem. Ignatius Sancho has approached nearer to merit in composition; yet his letters do more honour to the heart than the head. They breathe the purest effusions of friendship and general philanthropy, and shew how great a degree of the latter may be compounded with strong religious zeal. He is often happy in the turn of his compliments, and his stile is easy and familiar, except when he affects a Shandean fabrication of words. But his imagination is wild and extravagant, escapes incessantly from every restraint of reason and taste, and, in the course of its vagaries, leaves a tract of thought as incoherent and eccentric, as is the course of a meteor through the sky. His subjects should often have led him to a process of sober reasoning: yet we find him always substituting sentiment for demonstration. Upon the whole, though we admit him to the first place among those of his own colour who have presented themselves to the public judgment, yet when we compare him with the writers of the race among whom he lived, and particularly with the epistolary class, in which he has taken his own stand, we are compelled to enroll him at the bottom of the column. This criticism supposes the letters published under his name to be genuine, and to have received amendment from no other hand; points which would not be of easy investigation. The improvement of the blacks in body and mind, in the first instance of their mixture with the whites, has been observed by every one, and proves that their inferiority is not the effect merely of their condition of life. We know that among the Romans, about the Augustan age especially, the condition of their slaves was much more deplorable than that of the blacks on the continent of America. The two sexes were confined in separate apartments, because to raise a child cost the master more than to buy one. Cato, for a very restricted indulgence to his slaves in this particular, took from them a certain price. But in this country the slaves multiply as fast as the free inhabitants. Their situation and manners place the commerce between the two sexes almost without restraint.—The same Cato, on a principle of œconomy, always sold his sick and superannuated slaves. He gives it as a standing precept to a master visiting his farm, to sell his old oxen, old waggons, old tools, old and diseased servants, and every thing else become useless. "Vendat boves vetulos, plaustrum vetus, ferramenta, vetera, servum senem, servum morbosum, & si quid aliud supersit vendat." The American slaves cannot enumerate this among the injuries and insults they receive. It was the common practice to expose in the island of Æsculapius, in the Tyber, diseased slaves, whose cure was like to become tedious. The Emperor Claudius, by an edict, gave freedom to such of them as should recover, and first declared, that if any person chose to kill rather than to expose them, it should be deemed homicide. The exposing them is a crime of which no instance has existed with us; and were it to be followed by death, it would be punished capitally. We are told of a certain Vedius Pollio, who, in the presence of Augustus, would have given a slave as food to his fish, for having broken a glass. With the Romans, the regular method of taking the evidence of their slaves was under torture. Here it has been thought better never to resort to their evidence. When a master was murdered, all his slaves, in the same house, or within hearing, were condemned to death. Here punishment falls on the guilty only, and as precise proof is required against him as against a freeman. Yet notwithstanding these and other discouraging circumstances among the Romans, their slaves were often their rarest artists. They excelled too in science, insomuch as to be usually employed as tutors to their master's children. Epictetus, ⟨Diogenes, Phaedon⟩, Terence, and Phædrus, were slaves. But they were of the race of whites. It is not their condition then, but nature, which has produced the distinction.—Whether further observation will or will not verify the conjecture, that nature has been less bountiful to them in the endowments of the head, I believe that in those of the heart she will be found to have done them justice. That disposition to theft with which they have been branded, must be ascribed to their situation, and not to any depravity of the moral sense. The man, in whose favour no laws of property exist, probably feels himself less bound to respect those made in favour of others. When arguing for ourselves, we lay it down as a fundamental, that laws, to be just, must give a reciprocation of right: that, without this, they are mere arbitrary rules of conduct, founded in force, and not in conscience: and it is a problem which I give to the master to solve, whether the religious precepts against the violation of property were not framed for him as well as his slave? And whether the slave may not as jusitfiably take a little from one, who has taken all from him, as he may slay one who would slay him?

That a change in the relations in which a man is placed should change his ideas of moral right and wrong, is neither new, nor peculiar to the colour of the blacks. Homer tells us it was so 2600 years ago.

'Ημισυ, γαζ τ' ἀρετῆς ἀποαίνυῖαι εὐρύθπα Ζεὺς
'Ανερος, εντ' ἄν μιν κατὰ δέλιον ἥμαζ ἔλησιν. Od. 17. 323.

Jove fix'd it certain, that whatever day
Makes man a slave, takes half his worth away.

But the slaves of which Homer speaks were whites. Notwithstanding these considerations which must weaken their respect for the laws of property, we find among them numerous instances of the most rigid integrity, and as many as among their better instructed masters, of benevolence, gratitude, and unshaken fidelity.—The opinion, that they are inferior in the faculties of reason and imagination, must be hazarded with great diffidence. To justify a general conclusion, requires many observations, even where the subject may be submitted to the Anatomical knife, to Optical glasses, to analysis by fire, or by solvents. How much more then where it is a faculty, not a substance, we are examining; where it eludes the research of all the senses; where the conditions of its existence are various and variously combined; where the effects of those which are present or absent bid defiance to calculation; let me add too, as a circumstance of great tenderness, where our conclusion would degrade a whole race of men from the rank in the scale of beings which their Creator may perhaps have given them. To our reproach it must be said, that though for a century and a half we have had under our eyes the races of black and of red men, they have never yet been viewed by us as subjects of natural history. I advance it therefore as a suspicion only, that the blacks, whether originally a distinct race, or made distinct by time and circumstances, are inferior to the whites in the endowments both of body and mind. It is not against experience to suppose, that different species of the same genus, or varieties of the same species, may posses different qualifications. Will not a lover of natural history then, one who views the gradations in all the races of animals with the ye of philosophy, excuse an effort to keep those in the department of man as distinct as nature has formed them? This unfortunate difference of colour, and perhaps of faculty, is a powerful obstacle to the emancipation of these people. Many of their advocates, while they wish to vindicate the liberty of human nature, are anxious also to preserve its dignity and beauty. Some of these, embarrassed by the question "What further is to be done with them?" join themselves in opposition with those who are actuated by sordid avarice only. Among the Romans emancipation required but one effort. The slave, when made free, might mix with, without staining the blood of his master. But with us a second is necessary, unknown to history. When freed, he is to be removed beyond the reach of mixture.

The revised code further proposes to proportion crimes and punishments. This is attempted on the following scale.

I. Crimes whose punishment extends to *Life*.
   1. High treason.   Death by hanging.
      Forfeiture of lands and goods to the commonwealth.
   2. Petty treason.   Death by hanging. Dissection.
      Forfeiture of half the lands and goods to the representatives of the
      party slain.
   3. Murder.   1. by poison.   Death by poison.
      Forfeiture of one-half as before.
      2. in Duel.   Death by hanging. Gibbeting, if the challenger.
      Forfeiture of one-half as before, unless it be the party
      challenged, then the forfeiture is to the commonwealth.
      3. in any other way.   Death by hanging.
      Forfeiture of one-half as before.
   4. Manslaughter.   The second offence is murder.
II. Crimes whose punishment goes to *Limb*.
   1. Rape,
   2. Sodomy,   } Dismemberment.
   3. Maiming,   } Retaliation, and the forfeiture of half the lands and goods to the
   4. Disfiguring,   } sufferer.
III. Crimes punishable by *Labour*.

1.	Manslaughter, 1st offence.	Labour VII. years for the public.	Forfeiture of half as in murder.
2.	Counterfeiting money.	Labour VI. years.	Forfeiture of lands and goods to the commonwealth.
3.	Arson — — — }	Labour V. years. — —	Reparation threefold.
4.	Asportation of vessels. }		
5.	Robbery. — — }	Labour IV. years. — —	Reparation double.
6.	Burglary — — }		
7.	Housebreaking. — }	Labour III. years. — —	Reparation.
8.	Horse-stealing. — — }		
9.	Grand Larceny. —	Labour II. years. — —	Reparation. — Pillory.
10.	Petty Larceny. —	Labour I. years. — — —	Reparation. — Pillory.
11.	Pretensions to witchcraft, &c.	Ducking. — — —	Stripes.
12.	Excusable homicide. — }		
13.	Suicide. }	to be pitied, not punished.	
14.	Apostacy. Heresy. }		

Pardon and privilege of clergy are proposed to be abolished; but if the verdict be against the defendant, the court in their discretion, may allow a new trial. No attainder to cause a corruption of blood, or forfeiture of dower. Slaves guilty of offences punishable in others by labour, to be transported to Africa, or elsewhere, as the circumstances of the time admit, there to be continued in slavery. A rigorous regimen proposed for those condemned to labour.

Another object of the revisal is, to diffuse knowledge more generally through the mass of the people. This bill proposes to lay off every county into small districts of five or six miles square, called hundreds, and in each of them to establish a school for teaching reading, writing, and arithmetic. The tutor to be supported by the hundred, and every person in it entitled to send their children three years gratis, and as much longer as they please, paying for it. These schools to be under a visitor, who is annually to chuse the boy, of best genius in the school, of those whose parents are too poor to give them further education, and to send him forward to one of the grammar schools, of which twenty are proposed to be erected in different parts of the country, for teaching Greek, Latin, geography, and the higher branches of numerical arithmetic. Of the boys thus sent in any one year, trial is to be made at the grammar schools one or two years, and the best genius of the whole selected, and continued six years, and the residue dismissed. By this means twenty of the best geniusses will be raked from the rubbish annually, and be instructed, at the public expence, so far as the grammar schools go. At the end of six years instruction, one half are to be discontinued (from among whom the grammar schools will probably be supplied with future masters); and the other half, who are to be chosen for the superiority of their parts and disposition, are to be sent and continued three years in the study of such sciences as they shall chuse, at William and Mary college, the plan of which is proposed to be enlarged, as will be hereafter explained, and extended to all the useful sciences. The ultimate result of the whole scheme of education would be the teaching all children of the state reading, writing, and common arithmetic: turning out ten annually of superior genius, well taught in Greek, Latin, geography, and the higher branches of arithmetic: turning out ten others annually, of still superior parts, who, to those branches of learning, shall have added such of the sciences as their genius shall have led them to: the furnishing to the wealthier part of the people convenient schools, at which their children may be educated, at their own expence.—The general objects of this law are to provide an education adapted to the years, to the capacity, and the condition of every one, and directed to their freedom and happiness. Specific details were not proper for the law. These must be the business of the visitors entrusted with its execution. The first stage of this education being the schools of the hundreds, wherein the great mass of the people will receive their instruction, the principal foundations of future order will be laid here. Instead therefore of putting the Bible and Testament into the hands of the children, at an age when their judgments are not sufficiently matured for religious enquiries, their memories may here be stored with the most useful facts from Grecian, Roman, European and American history. The first elements of morality too may be instilled into their minds; such as, when further developed as their judgments advance in strength, may teach them how to work out their own greatest happiness, by shewing them that it does not depend on the condition of life in which chance has placed them, but is always the result of a good conscience, good health, occupation, and freedom in all just pursuits.—Those whom either the wealth of their parents or the adoption of the state shall destine to higher degrees of learning, will go on to the grammar schools, which constitute the next stage, there to be instructed in the languages. The learning Greek and Latin, I am told, is going into disuse in Europe. I know not what their manners and occupations may call for: but it would be very ill-judged in us to follow their example in this instance. There is a certain period of life, say from eight to fifteen or sixteen years of age, when the mind, like the body, is not yet firm enough for laborious and close operations. If applied to such, it falls an early victim to premature exertion; exhibiting indeed at first, in these young and tender subjects, the flattering appearance of their being men while they are yet children, but ending in reducing them to be children when they should be men. The memory is then most susceptible and tenacious of impressions; and the learning of languages being chiefly a work of memory, it seems precisely fitted to the powers of this period, which is long enough too for acquiring the most useful languages antient and modern. I do not pretend that language is science. It is only an instrument for the attainment of science. But that time is not lost which is employed in providing tools for future operation: more especially as in this case the books put into the hands of the youth for this purpose may be such as will at the same time impress their minds with useful facts and good principles. If this period be suffered to pass in idleness, the mind becomes lethargic and impotent, as would the body it inhabits if unexercised during the same time. The sympathy between body and mind during their rise, progress and decline, is too strict and obvious to endanger our being misled while we reason from the one to the other.—As soon as they are of sufficient age, it is supposed they will be sent on from the grammar schools to the university, which constitutes our third and last stage, there to study those sciences which may be adapted to their views.—By that part of our plan which prescribes the selection of the youths of genius from among the classes of the poor, we hope to avail the state of those talents which nature has sown as liberally among the poor as the rich, but which perish without use, if not sought for and cultivated.—But of all the views of this law none is more important, none more legitimate, than that of rendering the people the safe, as they are the ultimate, guardians of their own liberty. For this purpose the reading the first stage, where *they* will receive their whole education, is proposed, as has been said, to be chiefly historical. History by apprising them of the past will enable them to judge of the future; it will avail them of the experience of other times and other nations; it will qualify them as judges of the actions and designs of men; it will enable them to know ambition under every disguise it may assume; and knowing it, to defeat its views. In every government on earth is some trace of human weakness, some germ of corruption and degeneracy, which cunning will discover, and wickedness insensibly open, cultivate, and improve. Every government degenerates when trusted to the rulers of the people alone. The people themselves therefore are its only safe depositories. And to render even them safe their minds must be improved to a certain degree. This indeed is not all that is necessary, though it be essentially necessary. An amendment of our constitution must here come in aid of the public education. The influence over government must be shared among all the people. If every individual which composes their mass participates of the ultimate authority, the government will be safe; because the corrupting the whole mass will exceed any private resources of wealth: and public ones cannot be provided but by levies on the people. In this case every man would have to pay his own price. The government of Great-Britain has been corrupted, because but one man in ten has a right to vote for members of parliament. The sellers of the government therefore get nine-tenths of their price clear. It has been thought that corruption is restrained by confining the right of suffrage to a few of the wealthier of the people: but it would be more effectually restrained by an extension of that right to such numbers as would bid defiance to the means of corruption.

Lastly, it is proposed, by a bill in this revisal, to begin a public library and gallery, by laying out a certain sum annually in books, paintings, and statues.

2.2 "Letter to the Secretary of State" (1791)

Benjamin Banneker

Maryland, Baltimore County, August 19, 1791

Sir,

I am fully sensible of the greatness of that freedom, which I take with you on the present occasion; a liberty which seemed to me scarcely allowable, when I reflected on that distinguished and dignified station in which you stand, and the almost general prejudice and prepossession, which is so prevalent in the world against those of my complexion.

I suppose it is a truth too well attested to you, to need a proof here, that we are a race of beings, who have long labored under the abuse and censure of the world; that we have long been looked upon with an eye of contempt; and that we have long been considered rather as brutish than human, and scarcely capable of mental endowments.

Sir, I hope I may safely admit, in consequence of that report which hath reached me, that you are a man far less inflexible in sentiments of this nature, than many others; that you are measurably friendly, and well disposed towards us; and that you are willing and ready to lend your aid and assistance to our relief, from those many distresses, and numerous calamities, to which we are reduced. Now Sir, if this is founded in truth, I apprehend you will embrace every opportunity, to eradicate that train of absurd and false ideas and opinions, which so generally prevails with respect to us; and that your sentiments are concurrent with mine, which are, that one universal Father hath given being to us all; and that he hath not only made us all of one flesh, but that he hath also, without partiality, afforded us all the same sensations and endowed us all with the same faculties; and that however variable we may be in society or religion, however diversified in situation or color, we are all of the same family, and stand in the same relation to him.

Sir, if these are sentiments of which you are fully persuaded, I hope you cannot but acknowledge, that it is the indispensible duty of those, who maintain for themselves the rights of human nature, and who possess the obligations of Christianity, to extend their power and influence to the relief of every part of the human race, from whatever burden or oppression they may unjustly labor under; and this, I apprehend, a full conviction of the truth and obligation of these principles should lead all to. Sir, I have long been convinced, that if your love for yourselves, and for those inestimable laws, which preserved to you the rights of human nature, was founded on

Benjamin Banneker, "Letter to the Secretary of State" (August 19, 1791) (Philadelphia: Printed and Sold by Daniel Lawrence, no. 33; Early American imprints, first series, no. 24073).

sincerity, you could not but be solicitous, that every individual, of whatever rank or distinction, might with you equally enjoy the blessings thereof; neither could you rest satisfied short of the most active effusion of your exertions, in order to their promotion from any state of degradation, to which the unjustifiable cruelty and barbarism of men may have reduced them.

Sir, I freely and cheerfully acknowledge, that I am of the African race, and in that color which is natural to them of the deepest dye; and it is under a sense of the most profound gratitude to the Supreme Ruler of the Universe, that I now confess to you, that I am not under that state of tyrannical thraldom, and inhuman captivity, to which too many of my brethren are doomed, but that I have abundantly tasted of the fruition of those blessings, which proceed from that free and unequalled liberty with which you are favored; and which, I hope, you will willingly allow you have mercifully received, from the immediate hand of that Being, from whom proceedeth every good and perfect Gift.

Sir, suffer me to recal to your mind that time, in which the arms and tyranny of the British crown were exerted, with every powerful effort, in order to reduce you to a state of servitude: look back, I entreat you, on the variety of dangers to which you were exposed; reflect on that time, in which every human aid appeared unavailable, and in which even hope and fortitude wore the aspect of inability to the conflict, and you cannot but be led to a serious and grateful sense of your miraculous and providential preservation; you cannot but acknowledge, that the present freedom and tranquility which you enjoy you have mercifully received, and that it is the peculiar blessing of Heaven.

This, Sir, was a time when you clearly saw into the injustice of a state of slavery, and in which you had just apprehensions of the horrors of its condition. It was now that your abhorrence thereof was so excited, that you publicly held forth this true and invaluable doctrine, which is worthy to be recorded and remembered in all succeeding ages: "We hold these truths to be self-evident, that all men are created equal; that they are endowed by their Creator with certain unalienable rights, and that among these are, life, liberty, and the pursuit of happiness." Here was a time, in which your tender feelings for yourselves had engaged you thus to declare, you were then impressed with proper ideas of the great violation of liberty, and the free possession of those blessings, to which you were entitled by nature; but, Sir, how pitiable is it to reflect, that although you were so fully convinced of the benevolence of the Father of Mankind, and of his equal and impartial distribution of these rights and privileges, which he hath conferred upon them, that you should at the same time counteract his mercies, in detaining by fraud and violence so numerous a part of my brethren, under groaning captivity and cruel oppression, that you should at the same time be found guilty of that most criminal act, which you professedly detested in others, with respect to yourselves.

I suppose that your knowledge of the situation of my brethren, is too extensive to need a recital here; neither shall I presume to prescribe methods by which they may be relieved, otherwise than by recommending to you and all others, to wean yourselves from those narrow prejudices which you have imbibed with respect to them, and as Job proposed to his friends, "put your soul in their souls' stead;" thus shall your hearts be enlarged with kindness and benevolence towards them; and thus shall you need neither the direction of myself or others, in what manner to proceed herein. And now, Sir, although my sympathy and affection for my brethren hath caused my enlargement thus far, I ardently hope, that your candor and generosity will plead with you in my behalf, when I make known to you, that it was not originally my design; but having taken up my pen in order to direct to you, as a present, a copy of an Almanac, which I have calculated for the succeeding year, I was unexpectedly and unavoidably led thereto.

This calculation is the production of my arduous study, in this my advanced stage of life; for having long had unbounded desires to become acquainted with the secrets of nature, I have had to gratify my curiosity herein, through my own assiduous application to Astronomical Study, in which I need not recount to you the many difficulties and disadvantages, which I have had to encounter.

And although I had almost declined to make my calculation for the ensuing year, in consequence of that time which I had allotted therefor, being taken up at the Federal Territory, by the request of Mr. Andrew Ellicott, yet finding myself under several engagements to Printers of this state, to whom I had communicated my design, on my return to my place of residence, I industriously applied myself thereto, which I hope I have accomplished with correctness and accuracy; a copy of which I have taken the liberty to direct to you, and which I humbly request you will favorably receive; and although you may have the opportunity of perusing it after its publication, yet I choose to send it to you in manuscript previous thereto, that thereby you might not only have an earlier inspection, but that you might also view it in my own hand writing.

And now, Sir, I shall conclude, and subscribe myself, with the most profound respect,
Your most obedient humble servant,
Benjamin Banneker

2.3   Thomas Jefferson's Reply to Banneker (1791)

To Mr. Benjamin Banneker
Philadelphia, August 30, 1791

Sir,

I thank you, sincerely, for your letter of the 19th instant, and for the Almanac it contained. No body wishes more than I do, to see such proofs as you exhibit, that nature has given to our black brethren talents equal to those of the other colors of men; and that the appearance of the want of them, is owing merely to the degraded condition of their existence, both in Africa and America. I can add with truth, that no body wishes more ardently to see a good system commenced, for raising the condition, both of their body and mind, to what it ought to be, as far as the imbecility of their present existence, and other circumstances, which cannot be neglected, will admit.

I have taken the liberty of sending your Almanac to Monsieur de Condozett, Secretary of the Academy of Sciences at Paris, and Member of the Philanthropic Society, because I considered it as a document, to which your whole color had a right for their justification, against the doubts which have been entertained of them.

I am with great esteem, Sir,
Your most obedient Humble Servant,
Thomas Jefferson

Thomas Jefferson's reply to Banneker (August 30, 1791) (Philadelphia: Printed and Sold by Daniel Lawrence, no. 33; Early American imprints, first series, no. 24074).

2.4   "The Effect of Circumstances upon the Physical Man" (1854)

Frederick Douglass

I may remark, just here, that it is impossible, even were it desirable, in a discourse like this, to attend to the anatomical and physiological argument connected with this part of the subject. I am not equal to that, and if I were, the occasion does not require it. The form of the *negro*—(I use the term *negro*, precisely in the sense that you use the term Anglo-Saxon; and I believe, too, that the former will one day be as illustrious as the latter)—has often been the subject of remark. His flat feet, long arms, high cheek bones and retreating forehead are especially dwelt upon, to his disparagement, and just as if there were no white people with precisely the same peculiarities. I think it will ever be found, that the *well* or *ill* condition of any part of mankind, will leave its mark on the physical as well as on the intellectual part of man. A hundred instances might be cited, of whole families who have degenerated, and others who have improved in personal appearance, by a change of circumstances. A man is worked upon by what *he* works on. He may carve out his circumstances, but his circumstances will carve him out as well. I told a boot maker, in Newcastle-upon-Tyne, that I had been a plantation slave. He said I must pardon him; but he could not believe it; no plantation laborer ever had a high instep. He said he had noticed that the coal heavers and work people in low condition had, for the most part, flat feet, and that he could tell, by the shape of the feet, whether a man's parents were in high or low condition. The thing was worth a thought, and I have thought of it, and have looked around me for facts. There is some truth in it; though there are exceptions in individual cases.

The day I landed in Ireland, nine years ago, I addressed, (in company with Father Spratt[30] and that good man who has been recently made the subject of bitter attack; I allude to the philanthropic James Haughton, of Dublin), a large meeting of the common people of Ireland, on temperance. Never did human faces tell a sadder tale. More than five thousand were assembled; and I say, with no wish to would the feelings of any Irishman, that these people lacked only a black skin and wooly hair, to complete their likeness to the plantation negro. The open, uneducated mouth—the long, gaunt arm—the badly formed foot and ankle—the shuffling gait—the retreating forehead and vacant expression—and, their petty quarrels and fights—all reminded me of the plantation, and my own cruelly abused people. Yet, *that* is the land of Grattan, of Curran, of O'Connell, and of Sheridan.[31] Now, while what I have said is true of the common people, the fact is, there are no more really handsome people in the world,

Frederick Douglass, "The Effect of Circumstances Upon the Physical Man," from John W. Blassingame, ed., "The Claims of the Negro Ethnologically Considered," in *The Frederick Douglass Papers* (New Haven: Yale University Press, 1982 [1854]), 2:520–525.

than the educated Irish people. The Irishman educated, is a model gentleman; the Irishman ignorant and degraded, compares in form and feature with the negro!

I am stating facts. If you go into Southern Indiana, you will see what climate and habit can do, even in one generation. The man may have come from New England, but his hard features, sallow complexion, have left little of New England on his brow. The right arm of the blacksmith is said to be larger and stronger than his left. The ship carpenter is at forty round-shouldered. The shoemaker carries the marks of his trade. One locality becomes famous for one thing, another for another. Manchester and Lowell, in America, Manchester and Sheffield, in England, attest this. But what does it all prove? Why, nothing positively, as to the main point; still, it raises the inquiry—May not the condition of men explain their various appearances? Need we go behind the vicissitudes of barbarism for an explanation of the gaunt, wiry, ape like appearance of some of the genuine negroes? Need we look higher than a vertical sun, or lower than the damp, black soil of the Niger, the Gambia, the Senegal, with their heavy and enervating miasma, rising ever from the rank growing and decaying vegetation, for an explanation of the negro's color? If a cause, full and adequate, can be found here, *why seek further*?

The eminent Dr. Latham, already quoted, says that nine-tenths of the white population of the globe are found between 30 and 65 degrees North latitude. Only about one-fifth of all the inhabitants of the globe are white; and they are as far from the Adamic complexion as is the negro. The remainder are—*what?* Ranging all the way from the brunette to jet black. There are the red, the reddish copper color, the yellowish, the dark brown, the chocolate color, and so on, to the jet black. On the mountains on the North of Africa, where water freezes in winter at times, branches of the same people who are *black* in the valley are *white* on the mountains. The Nubian, with his beautiful curly hair, finds it becoming frizzled, crisped, and even woolly, as he approaches the great Sahara. The Portuguese, white in Europe, is brown in Asia. The Jews, who are to be found in all countries, never intermarrying, are white in Europe, brown in Asia, and black in Africa. Again, what does it all prove? Nothing, absolutely; nothing which places the question beyond dispute; but it *does* justify the conjecture before referred to, that outward circumstances *may* have something to do with modifying the various phases of humanity; and that color itself is at the control of the world's climate and its various concomitants. It is the sun that paints the peach—and may it not be, that he paints the *man* as well? My reading, on this point, however, as well as my own observation, have convinced me that from the beginning the Almighty, within certain limits, endowed mankind with organizations capable of countless variations in form, feature and color, without having it necessary to begin a new creation for every new variety.

A powerful argument in favor of the oneness of the human family, is afforded in the fact that nations, however dissimilar, may be united in one social state, not only without detriment to each other, but, most clearly, to the advancement of human welfare, happiness and perfection. While it is clearly proved, on the other hand, that those nations freest from foreign elements present the most evident marks of deterioration. Dr. James McCune Smith, himself a colored man, a gentleman and scholar, alleges—and not without excellent reason—that this, our own great nation, so distinguished for industry and enterprise, is largely indebted to its composite character.[32] We all know, at any rate, that now, what constitutes the very heart of the civilized world—(I allude to England)—has only risen from barbarism to its present lofty eminence, through successive invasions and alliances with her people. The Medes and Persians constituted one of the mightiest empires that ever rocked the globe. The most terrible nation which now threatens the peace of the world, to make its will the law of Europe, is a grand piece of Mosaic work, in which almost every nation has its characteristic feature, from the wild Tartar to the refined Pole.[33]

But, gentlemen, the time fails me, and I must bring these remarks to a close. My argument has swelled beyond its appointed measure. What I intended to make special, has become, in its progress, somewhat general. I meant to speak here to-day, for the lonely and the despised ones, with whom I was cradled, and with whom I have suffered; and now, gentlemen, in conclusion, what if all this reasoning be unsound? What if the negro may not be able to prove his relationship to Nubians, Abyssinians and Egyptians? What if ingenious men are able to find plausible objections to all arguments maintaining the oneness of the human race? What, after all, if they are able to show very good reasons for believing the negro to have been created precisely as we find him on the Gold Coast—along the Senegal and the Niger—I say, what of all this? *"A man's a man for a' that."*[34] I sincerely believe, that the weight of the argument is in favor of the unity of origin of the human race, or species—that the arguments on the other side are partial, superficial, utterly subversive of the happiness of man, and insulting to the wisdom of God. Yet, what if we grant they are not so? What, if we grant that the case, on our part, is not made out? Does it follow, that the negro should be held in contempt? Does it follow, that to enslave and imbrute him is either *just* or *wise?* I think not. Human rights stand upon a common basis; and by all the reason that they are supported, maintained and defended, for one variety of the human family, they are supported, maintained and defended for *all* the human family; because all mankind have the same wants, arising out of a common nature. A diverse origin does not disprove a common nature, nor does it disprove a united destiny. The essential characteristics of humanity are everywhere the same. In the language of the eloquent Curran, "No matter what complexion, whether an Indian or an African sun has burnt upon him," his title deed to freedom, his claim to life and to liberty, to knowledge and to civilization, to society and to Christianity, are just and perfect.[35] It is registered in the Courts of Heaven, and is enforced by the eloquence of the God of all the earth.

I have said that the negro and white man are likely ever to remain the principal inhabitants of this country. I repeat the statement now, to submit the reasons that support it. The blacks can disappear from the face of the country by three ways. They may be colonized,—they may be exterminated,—or, they may die out. Colonization is out of the question; for I know not what hardships the laws of the land can impose, which can induce the colored citizen to leave his native soil. He was here in its infancy; he is here in its age. Two hundred years have passed over him, his tears and blood have been mixed with the soil, and his attachment to the place of his birth is stronger than iron. It is not probable that he will be exterminated; two considerations must prevent a crime so stupendous as that—the influence of Christianity on the one hand, and the power of self interest on the other; and, in regard to their dying out, the statistics of the country afford no encouragement for such a conjecture. The history of the negro race proves them to be wonderfully adapted to all countries, all climates, and all conditions. Their tenacity of life, their powers of endurance, their malleable toughness, would almost imply especial interposition on their behalf. The ten thousand horrors of slavery, striking hard upon the sensitive soul, have bruised, and battered, and stung, but have not killed. The poor bondman lifts a smiling face above the surface of a sea of agonies, *hoping on, hoping ever*. His tawny brother, the Indian, dies, under the flashing glance of the Anglo-Saxon. *Not* so the negro; civilization cannot kill him. He accepts it—becomes a part of it. In the Church, he is an Uncle Tom; in the State, he is the most abused and least offensive. All the facts in his history mark out for him a destiny, united to America and Americans. Now, whether this population shall, by freedom, industry, virtue and intelligence, be made a blessing to the country and the world, or whether their multiplied wrongs shall kindle the vengeance of an offended God, will depend upon the conduct of no class of men so much as upon the Scholars of the country. The future public opinion of the land, whether anti-slavery or pro-slavery, whether just or unjust, whether magnanimous or mean, must redound to the honor of the Scholars of the country or cover them with shame. There is but one safe road for nations or for individuals. The fate of a wicked man and of a wicked nation is the same. The flaming sword of offended justice falls as certainly upon the nation as upon the man. God has no children whose rights may be safely trampled upon. The sparrow may not fall to the ground without the notice of his eye, and men are more than sparrows.

Now, gentlemen, I have done. The subject is before you. I shall not undertake to make the application. I speak as unto wise men. I stand in the presence of Scholars. We have met here to-day from vastly different points in the world's condition. I have reached here—if you will pardon the egotism—by little short of a miracle: at any rate, by dint of some application and perseverance. Born, as I was, in obscurity, a stranger to the halls of learning, environed by ignorance, degradation, and their concomitants, from birth to manhood, I do not feel at liberty to mark out, with any degree of confidence, or dogmatism, what is the precise vocation of the Scholar. Yet, this I *can* say, as a denizen of the world, and as a citizen of a country rolling in the sin and shame of Slavery, the most flagrant and scandalous that ever saw the sun, "Whatsoever things are true, whatsoever things are honest, whatsoever things are just, whatsoever things are pure, whatsoever things are lovely, whatsoever things are of good report, if there be any virtue, and if there be any praise, think on these things."[36]

Notes

30. John Spratt.

31. Henry Grattan, John Philpot Curran, Daniel O'Connell, Richard B. Sheridan.

32. Douglass may refer to the article "Civilization: Its Dependence on Physical Circumstances" by the prominent black physician, abolitionist, and writer, James McCune Smith (1813–65). In the article, which was later published in the first issue of the *Anglo-African Magazine*, Smith claimed that "civilization depends upon the frequent intercourse of men differing in physical and mental endowments." The son of a slave father and a self-emancipated bondswoman, Smith was born in New York City, where he attended the New York African Free School. Denied admission to Columbia, Geneva Medical College, and the New York Academy of Medicine, Smith set sail for Scotland in 1832, receiving his B.A. (1835), M.A. (1836), and M.D. (1837) from the University of Glasgow. Smith opened a pharmacy upon his return to New York City, set up a medical practice that catered to both blacks and whites, and devoted his efforts to abolitionist concerns. He briefly served as an associate editor of the *Colored American* in 1839 and contributed regularly to the *Anglo-African Magazine* and under the pseudonym "Communipaw," to the *North Star* and *Frederick Douglass' Paper*. A longtime opponent of black colonization and emigration, Smith helped finance the revival of the *Weekly Anglo-African* as an anti-emigrationist organ in 1861. In 1863 Smith was appointed professor of anthropology at Wilberforce College but illness kept him from his post. He was a prominent member of the New York City Young Men's Association, the American and Foreign Anti-Slavery Society, and the New York African Society for Mutual Relief. He was also the sole attending physician of the Society for the Promotion of Education Among Colored Children, a member and vestryman of St. Philip's Episcopal Church, and a trustee of the New York Society for the Promotion of Education Among Colored Children. *FDP*, 18 May 1855; *Lib.*, 1 June 1838, 1 February 1839; James McCune Smith, "Civilization: Its Dependence on Physical Circumstances," *Anglo-African Magazine*, 1:5–17 (January 1859); Freeman, "Free Negro in New York City," 40–42, 177, 186, 195, 200, 247, 276, 286, 325, 353, 394; Quarles, *Black Abolitionists*, 115, 134; Pease and Pease, *They Who Would Be Free*, 90–92, 103, 110; Miller, *Search for a Black Nationality*, 243; *DAB*, 27:288–89.

33. Douglass alludes to Russia and the events leading to the Crimean War.

34. Douglass quotes a line from Robert Burns's song "For a' That and a' That." Smith, *Works of Robert Burns*, 227.

35. Thomas Davis, ed., *The Speeches of the Right Honorable John Philpot Curran* (Dublin, 1845), 182.

36. Phil. 4:8.

2.5 "Peculiarities in the Structure and Diseases of the Ear of the Negro" (1887)

T. E. Murrell

The negro in this country is rapidly losing his African cast of features by miscegenation, so that the pure type is becoming comparatively rare, even in the Southern States. The mulatto partakes of many of the characteristics of his white progenitor, including that of a roving disposition. Hence he is met with in the cities and on the railroads in all parts of the country, both North and South. But the more distinct type of African, with his black skin, woolly hair, flat nose, small auricles, large mouth, thick lips, prognathous jaw and dolicocephalic head, is seen chiefly on the plantations and in the cities in the cotton and rice regions of the South, and is only occasionally met with in the Northern States.

One practicing medicine in a Southern region will have so many types of so-called negroes under his observation that be may overlook some of the characteristic peculiarities of this people. He will, most likely, enter on his record book "white" and "negro," without discriminating between grades of caste, and following the custom in the South, of classing as negro, or *colored*, to use a popular word, any person whose associates are negroes—a matter which inflexible social custom forces upon any one bearing the faintest suspicion of African lineage—he will have mulattoes, quadroons, octoroons and every imaginable shade of color and type of feature recorded in one category. The census returns are faulty in the same particular, in that they make only two general racial classes, and are, for his reason, next to worthless for our present purpose. Fully ninety per centum of the affections occurring among the colored, as ordinarily classed in the census returns, belong to the mixed types—mulattoes of various shades of color. In my own records I have entered as negroes all socially so regarded, but have carefully differentiated between the castes in my study of the peculiarities hereinafter mentioned.

In its strict sense, therefore, negro, as here used, is intended to mean the full blood, and not the mixed races; but the slight admixture of white blood is not considered where the general features are satisfactorily African.

For several years I have been impressed with certain peculiarities in the negro, both as to some features of anatomy and as to the class of diseases most and least frequent in his ears. Since medical literature hitherto has taken no cognizance of these differences, I was the more impressed on noticing them. Excepting Dr. C. H. Burnett's mention, in his "Treatise on the Ear," of the large and straight external auditory canal noticed in the negro, I have nowhere met with any special references to this race of

T. E. Murrell, "Peculiarities in the Structure and Diseases of the Ear of the Negro," *Transactions of the International Medical Congress*, Ninth Session, vol. 3 (Washington, D.C., 1887), 817–824.

people in otological literature. Formerly I considered the facial contour and racial features, on the one hand, and mode of life, occupation and hygienic surroundings, on the other, sufficient to account for all the differences observed; but a larger experience and more thorough study have convinced me that there exist anomalies in these people which cannot be accounted for except by some racial peculiarity. In the Southern States we know the negro is typically odd in many respects—normally, mentally, socially and physically, and that he can live in the enjoyment of health where the white man would soon succumb to mephitic influences.

It became apparent to me some years ago that ear, throat, and nose affections were very infrequently met with in a genuine negro, while eye diseases were, perhaps, more common than in the white race.

I set about to ascertain, if possible, some of the reasons for the exemption of black people from these affections, with the results which follow:—

Peculiarities in Anatomy

The structure of the pinna, in so far as its aid to hearing is concerned, is of minor importance, and the peculiarities in this appendage in the negro belong, therefore, rather to the study of the ethnologist than the otologist. Allowing for many exceptions, it is ordinarily quite small, with smaller and less graceful development of the sinuous elevations and depressions on its anterior surface, than in the Caucasian. The large, flat surface between the helix and anti-helix, so common in white people, is usually of more limited extent in blacks. Instead of standing off from the head, it most generally adheres quite closely to it, often lying quite flatly against it. One reason for this position of the auricle may be the small size of the mastoid process, which is ordinarily but little developed and quite insignificant in external appearance. This, with the small transverse cranial measurement, presents a great narrowness to this part of the head, viewed from behind. Another reason for the closely set auricle is the predominance of the face over the occiput in the negro, by which it is thrown backward on its posterior outline. The external auditory canal presents such striking features that they cannot well be overlooked. The canal is often abnormally large in a full-sized African, sometimes almost admitting the little finger, while the largest-sized speculum in Toynbee's set can, in such cases, be introduced nearly to the drum membrane. The canal is also very straight, so much so that a speculum is seldom required to view the drum membrane. In fact, this peculiar straightness of the external auditory canal in the negro is so strongly typical that it is one of the last to be lost in approximating the Caucasian type of man, and so is to be found in nearly all mulattoes. The drum membrane can often be inspected in every detail, to its very periphery, including the annulus tympanicus, by merely turning the outer end of the canal toward a window and looking down it. The canal is also of slightly shorter average depth than in a white person, as would be inferred from the smaller transverse measurement of the negro skull. The integument lining the canal is commonly black, or deeply pigmented, but occasionally it is free from pigment in the blackest individual. The membrana tympani offers but few peculiarities, if we except the distinctness and ease with which it can ordinarily be seen. It is of necessity large in its diameters, to correspond with the size of the outer canal. The angle it forms with the axis of the external auditory canal is somewhat less than with the tortuous canal seen in many white persons, while the angle it forms with the sagittal plane does not materially differ. As is well known, however, differences in these angles occur in the white races. But actually, the angle the membrana tympani makes with the axis of the external auditory canal is due more to the direction of the canal than to differences in the position of the membrane. If its angle be measured by a line drawn straight through the head from the centre of one membrane to the centre of the other there will be really but little, if any, difference in the angle thus formed and that similarly formed in the white person. The membrana tympani in the negro, therefore, by reason of its larger size, more direct position relative to the external auditory canal, and the large diameter, less depth, and almost perfect straightness of this canal, whereby it can sometimes be seen binocularly by viewing it at some distance, so as to admit the binocular pencil into the canal, presents a most beautiful picture of this structure, exhibiting in every detail its many interesting features with a pleasing distinctness.

No measurements of the middle ear have been made, nor have any well-marked peculiarities in this part presented themselves. The mastoid process, as has been already mentioned, is usually quite small and contains less cell surface within than is common in white adults, while the bony exterior is generally excessively thick. It is a well-known fact that all the sinuses about the face of the negro are of much smaller dimensions than in the white race, and are encased in extremely thick bone; in fact, all the bony structures of his cranium are unusually developed. The Eustachian tubes open into an extremely wide and capacious pharynx, opposite very broad and unobstructed inferior nasal meati. The pharynx is peculiarly large in the full-blooded negro, with large mouth and great width between his malar eminences, presenting an inter-faucial space of remarkable dimensions. His nose is flat and extremely broad, with flared alæ nasi and a very low septum, which is almost never deflected. The great breadth to the inferior meati gives a large area to the choanæ and allows of unobstructed respiration.

Diseases Common and Uncommon in the Negro

External Ear

The auricle is less liable to cutaneous affections than in whites, owing, perhaps, to the greater thickness and toughness of the skin covering it. Eczema, so common in the delicate integument covering the auricle in fair-skinned and light-haired children, is quite rare in black children.

Deformities are more common than in whites, as the result of traumatism, negro children being usually very rough in their plays, and disposed to quarreling and fighting, using their teeth as a weapon of defense and offense.

The external auditory canal, being large and straight, is a trap for foreign bodies, and especially for insects. Quite a large bug can find its way into the canal. I have removed such that could by no means penetrate the ordinary canal of a white person. On the contrary, impacted cerumen is very rare in the negro. I have had about ten cases in white persons to one case in a negro, in the same number of ear patients of each. It is quite common to find an excess of wax in the canal, but there is little tendency to its complete impaction.

It is probable that the straightness and more uniform size of the canal allow of easy egress of dried cerumen, and it rids itself of such accumulations.

I have never met a case of otitis parasitica in the negro. This may be due, in part, to the fact that they mostly live in small cabins, and have fires during summer as well as winter, to do their cooking, hence mould is rare in their dwellings. Nor have I ever seen a furuncle in the external auditory canal of the negro, an affection so frequent in white people, and so very painful that, in his great abhorrence of pain, the negro would be certain to seek relief.

Middle Ear

It is in this part, where the greater portion of all aural affections are located, that we would wish to know if these people enjoy an immunity from disease. Suppurative processes occur, both acute and chronic, but are confined almost exclusively to children. Otitis media suppurativa acuta is a not infrequent affection in children, particularly those exhibiting a strumous diathesis, in whom there is often associated a catarrhal rhinities, and rarely it goes into the chronic form, but most generally reparation takes place with little or no care. Even the chronic form in children disappears in process of time, as in the adult it is almost unknown. This is far from true in the other and more careful race, and hence there must be a reason for it. In my many examinations of the ears of negroes, I have not met an instance of chronic suppuration in the adult; but in mulattoes, who are in all respects more vulnerable than either of the pure races, it is quite common. I have never seen a case of mastoiditis in a pure negro. The rarity of suppurative processes in the middle ear, together with the small development of the mastoid process, will sufficiently account for this. The catarrhal affections are also infrequent in the typical negro. Subacute catarrh of the middle ear rarely occurs from a cold in the head, and the negro is but little subject to acute rhinitis. Otitis media catarrhalis acuta occurs in children in whom nasal and pharyngeal disease exist, but in a far less proportion than in mulattoes and whites.

Chronic aural catarrh is so extremely rare in this race of people, that they may be considered to enjoy almost a complete exemption from it. Out of 421 aural affections among negroes, not one of chronic aural catarrh has been met. Yet, no doubt, it does occur. Certainly they do not apply for treatment, and were it as common as in the white races, there would be a far greater proportion of dullness of hearing among them. We would seek for the cause of this immunity from affections of the middle ears in the condition of the nose and pharynx.

As has been mentioned, strumous children often suffer from chronic rhinitis, with free discharge from the anterior nares, but it invariably disappears with bodily development and is not seen in the adult. In the adult, nasal and pharyngeal diseases are quite infrequent. Out of a large number of negroes treated for eye, ear, throat and nose affections, I have treated only one adult for chronic rhinitis, which speedily disappeared under very simple remedies, a success I have never attained with a similar case in my white patients.

Frequent examinations of the naso-pharyngeal structures of negroes has rarely shown any sufficiently morbid condition to call for treatment. Hypertrophic rhinitis is extremely rare in these people, and a case of adenoid growths in the vault of the pharynx has not been met with by me. In fact, as a general rule, the naso-pharynx in the negro comes as near habitually presenting a true healthy condition of these parts as is ever to be seen. In how far peculiarities in anatomy lend an influence to this exemption from disease, and in how far vigor of body, out-of-door life and plain living conduce to such end, is a question not yet fully solved. It is known that persons who live in open houses are much less prone to head colds than are those living in close houses. The negro in the Southern States commonly lives in a very open house, and burns wood for fires, so that he does not have a sore throat, or an attack of sneezing, for every change in the weather. Moreover, his life is one almost unexceptionally out of doors and engaged in some active labor. The women as well as the men venture out in all kinds of weather, and seldom know what it is to be delicate, like white ladies. In a state of slavery their lives were very methodical, but now they are far from it. They are particularly gregarious individuals, and love almost any kind of a big gathering, and as it falls to the lot of the most of them to earn their bread by their daily labor, they choose nights for their social and religious gatherings, in which they display anything but prudence in the hours they keep. Nevertheless, they seldom suffer from such dissipations. Upon the whole their lives are simple, they are a strong, hearty people, and notwithstanding their readily succumbing to certain affections that the stronger willed white man bears up under, they are particularly free from catarrhal affections of the naso-pharyngeal, or tubal region. Not that they are altogether exempt from the consequences of exposure, far from it; for the negro is very subject to pneumonia and bronchitis; but his middle ears commonly escape, by reason of the greater exemption from invasion of his naso-pharynx by such processes. It is, therefore, upon the ground of the anatomical conformation of the naso-pharynx on the one hand, and the mode of life and physical training upon the other, that these people enjoy so much greater freedom from catarrhal processes in the middle ear, and consequent deafness, than their white fellow-citizens. It may be argued that the negro manifests an indifference about his physical well-being not seen in the white American citizen, but this is not the case when any of his special senses are involved. Besides, were the middle-ear affections as common in them as in white people, we would meet with more cases of partial deafness among them than we actually find to be the case. In almost any city of a few thousand population, one can pick out numbers of persons more or less disagreeably deaf from some chronic middle ear process, but rarely, indeed, is one of them a negro.

In very old people, in whom senile changes in their ears produce more or less deafness in white persons, there are seldom such cases met with in the negro. It is true a few old negroes are dull of hearing, but not in the proportion that the same prevails in aged white people. Excluding deaf-mutes, then, there are very few negroes who have not sufficient hearing power to understand any ordinary conversation, and I have never known one have to resort to an ear trumpet or conversation tube.

Internal Ear

Affections involving the terminal acoustic apparatus are inflammatory, central, or traumatic. The inflammatory may be primary, as in Menière's disease, or secondary by extension from an otitis media, commonly with necrosis of the petrous bone. The central causes are lesions involving the portio mollis at its origin or in some part of its course. The traumatic may be direct, or remote. As slight affections of the auditory nerve are more or less obscure, both in diagnosis and etiology, only the graver forms will here be considered, which means, generally, more or less entire deafness, commonly associated with mutism.

As previously mentioned, chronic suppuration of the middle ear is quite rare in the negro, and I have never had a case in which there was necrosis of the temporal bone; hence, I have never seen an exfoliation of the cochlea, and deafness from this cause, in this race of people. It may rarely occur, however, as I have seen no statistics on this subject. Menière's disease I have not met in the negro, but I see no reason why it should not occur, as negro children frequently suffer from convulsions. It is probable they fall into the hands of general practitioners, who overlook the nature of the disease. This is true of all the cases in white children I have had to treat, and when they have been brought to me for consultation for deafness; afterward I have made the diagnosis from the history of the case. No such case has ever been brought to me in the negro child. The only instance of very great deafness in a negro child I have had, it twelve years' special practice amid a dense negro population, was a little boy about seven years old, who was suffering from obstruction of the Eustachian tubes caused by enlarged tonsils and pharyngitis, of which he was readily cured. As to traumatic causes of deafness, the negro is on an equal footing with the white man. In affections of the auditory nerve, of central origin, however, he was many things in his favor. It is a well-known fact that nervous diseases are much more infrequent in these people than in those higher up the scale of cerebral development. There is hardly a portion of all this country that has not at some time or other been swept by an epidemic of cerebro-spinal meningitis, leaving in its wake a very large percentage of its surviving victims inheritors of blindness, or deafness, and sometimes of both. In the Southern States many such epidemics are recorded in various localities, with their thousands of victims, but so rare is the occurrence of this affection in full-blooded negroes, that physicians of large experience who have seen a single case are few indeed.

As the result of many inquiries on this subject of physicians who have practiced medicine in southern cities and on southern plantations for many years, dating back to slavery times, I have found only one who could cite a case of cerebro-spinal fever in the negro, and this was his only instance.

It has been observed in epidemics of this strange and fatal malady that those who are in a state of high nervous tension, over-anxiety, or great mental worry, are far more likely to be stricken than are the more equable-minded, and less solicitous, or care-burdened.

The negro is proverbially light-hearted and care-free, and seldom allows a responsibility to weigh upon his mind. His inferior cerebral development also doubtless protects him from hyperæmia, and acute congestions of his brain and meninges.

Deaf-mutism, however, occurs in the negro to some extent. Statistics are unsatisfactory on this subject. The only source of information outside of the testimony of individual observers, is to be hand from the United States census reports. A much larger percentage of deaf-mutism in the colored race is here given, than my observations and inquiries had led me to imagine. But the very great error undoubtedly exists here of reporting all shades of color as colored, or negro, when by far the majority are of the mixed types.

Inquiries of many intelligent and experienced persons, both white and black, have proven to me that a deaf-mute pure negro must be somewhat of a rarity. Mulatto deaf-mutes are not so rare. And as no distinctions are made in the census tables as to shades of color, or caste, it must therefore be taken for granted that all socially regarded as negroes are so classed, which, as has been shown, will include every shade from black up to almost pure white. This fact, then, for our purpose, deprives the census returns of a greater portion of their value. Examination of these reports for 1880 shows the following proportion of deaf-mutism in white and colored in eleven States, containing the greatest negro population.

Alabama, { 1 white in   1358 population.
          { 1 colored in 2083   "

Arkansas, { 1 white in   1418   "
          { 1 colored in 2926   "

Florida, { 1 white in   2593   "
          { 1 colored in 2011   "

Georgia, { 1 white in   1637   "
          { 1 colored in 2266   "

Louisiana, { 1 white in   1387   "
           { 1 colored in 2467   "

Wisconsin, { 1 white in   1328   "
           { 1 colored in 1938   "

North Carolina, { 1 white in   1197   "
                { 1 eolored in 1725   "

South Carolina, { 1 white in   1000   "
                { 1 colored in 2298   "

Tennessee, { 1 white in   1310   "
           { 1 colored in 1679   "

Texas,   { 1 white in   1949   "
          { 1 colored in 2505   "

Virginia, { 1 white in   1249   "
          { 1 colored in 2155   "

We here see a very much larger ratio of white deaf-mutes than colored; almost double. Florida is the only State showing a greater proportion of deaf-mutism among the colored than the white. To account for this I am unable to offer an explanation. By taking an average of the whole eleven States we find the ratio to be as follows:—

White..1 in 1484   population.
Colored...1 in 2186   "

or about three white to two colored, in a given population of each.

　　If we leave out Florida, the other ten States give a ratio as follows:—

White..1 in 1383   population.
Colored...1 in 2204   "

or nearly two for the white, to one for the colored.

　　Arkansas and South Carolina, both containing a large negro population, show more than two white to one colored in the same population of each.

　　The causes of deaf-mutism among negroes is not sufficiently well known to serve the purpose for which it would be very desirable here to use it. Another factor in ear affections, to which attention has of late been called, should be left out almost altogether in considering aural patients in the pure type of negro; that is, malaria. If malaria play any active rôle in the production of aural troubles, our well-favored black man could stalk the most miasmatic regions with next to perfect ears. In conclusion, I will say I have omitted tables of statistics and reports of cases from my records, in order to avoid unnecessary tediousness, but have endeavored to present facts as they have impressed themselves upon my mind from a careful study of all the cases in my private and dispensary practice, and gathered from other physicians, and from any and all sources available. The subject is a new one, and worthy of further investigation, for scientific reasons, if for no other. My investigations have been alone and single-handed; but it is to be hoped those engaged in special practice in the Southern States will give this subject careful attention in future, and make known their observations through the proper channels.

Discussion

Dr. C. M. Hobby, Iowa City, Iowa, thinks the excellent paper of Professor Murrell suggests the necessity of a more thorough investigation into the influence of race upon the production of disease. The census reports show fifty cases of mutism among colored people to sixty-six cases among white people. Is this difference due to race influence, or to the neglect or inability of the negro to receive observation and attention? The only reports in reference to causation of mutism among the negroes show a nearly equal proportion of mutism from cerebro-spinal fever with the white race, and the same is true of all kinds of alleged causes. Professor Murrell's objection to including under the term "colored" all possessing acknowledged African blood is undoubtedly good, but even his extensive experience among the so-called "colored" people cannot have brought to his attention any very great number of those of uncontaminated African blood.

Dr. R. Tilley, Chicago, Illinois.—I wish to express my high appreciation of the paper which has just been read, and I am sorry the author is not present to receive my compliments. The subject is so completely new to us who practice in the North that it is necessarily difficult to speak upon it. The subject is, however, peculiarly interesting to me in connection with the relative susceptibility of the negro to the ravages of syphilis. I refer to this subject in a paper which I hope to have the honor of presenting to the Section. The statistics presented to us are exceedingly interesting, but require further corroboration, and it would be important if we could bring any influence to bear upon the officers of the next census, to render its reports more exact on this question. The writer's theoretical explanation of the absence of impacted cerumen among the negroes, because of the large size of the external meatus, scarcely corresponds with my observation. For, although I have had very little experience with the negro, yet there is a marked difference in the size of individual canals, and it has been my observation that large canals are more associated with impacted cerumen than the smaller ones.

Dr. L. Turnbull, of Philadelphia, Pennsylvania, remarked that if he understood the paper, no mention has been made of fibrous tumors of the lobe of the ear in the negro. In his manual, "Diseases of the Ear," published in 1872, he had made special mention of the great tendency of the negro to this form of disease. He had seen several cases among negro girls of one, two, and in one instance three such tumors, which he removed by operation. In his clinic he has every year several cases, in the negro and mixed races, of acute otitis media; by negro he meant, not a brown or yellow, but a *black* man.

Dr. S. O. Richey, Washington, D. C.—Among charity patients of this country are to be found many Irish. Those of them over forty years of age who have ear troubles commonly have very large meati, often impacted cerumen, and persistently impaired hearing, associated together. No one of these symptoms is the cause of any other, but we have to look to a more central common cause in the nervous system. The torpid mental and nervous organization of the negro stands in strong contrast to his exaggerated emotional nature, the offspring of his ignorance and superstition. But in this torpor of his nervous system we may find some explanation of his comparative freedom from progressive deafness. Keloid of the lobe of the ear is not uncommon in the negro race.

Professor G. E. Frothingham, Ann Arbor, Michigan, said that though Professor Murrell's paper had already been quite thoroughly and ably discussed, he wished to call attention to the statements it contained that were important in connection with the report made yesterday by Dr. Bishop, of Chicago. He desired to do so because they tend to support the theory which he advocated in the discussion of Dr. Bishop's paper. Professor Murrell declares that middle ear inflammations seldom occur among the colored people, and that they nearly all live in dwellings thoroughly ventilated, not from architectural design, but from incomplete construction. They were thus freed from exposure to contaminated atmosphere, which has a tendency, through the germs which it contains, to produce naso-pharyngeal catarrhs, and acute and chronic inflammations of the middle ear, especially when the changes of temperature are sudden and extreme. The conditions under which the population lives from which Dr. Bishop has gathered his statistics, and, so far as inflammatory diseases of the naso-pharyngeal and tympanic cavities are concerned, the results, are just the reverse; Dr. Bishop finding these inflammations prevailing to a great extent; Professor Murrell finding them very infrequent. Professor Frothingham believes the infrequency of impacted cerumen can best be explained by the absence of chronic aural catarrh. He has found this to exist in a large proportion of the cases of impacted cerumen, to such an extent as to impair the hearing seriously, often leading to progressive deafness. In many of these cases, the ceruminous glands are stimulated, and in many other cases of over stimulation of glands, we get an excessive and pathological secretion. This, with the desquamated epithelium, from the same cause, also in excess, leads to impaction. He is more inclined to this view than the one offered by Dr. Richey.

Dr. H. B. Young, Burlington, Iowa, agrees with Dr. Tilley in the idea that large, straight canals do not, *per se*, tend to prevent accumulations of cerumen. Experience shows that large and straight canals are quite commonly affected in this way. It has also been noticed that in some cases there appears to be an hereditary tendency. If parents have it, adult sons and daughters may expect it.

2.6 "The Deformity Termed 'Pug Nose' and Its Correction by a Simple Operation" (1887)

John Orlando Roe

The nose is the central and most prominent feature of the face; and on its shape, size, and appearance, to a great degree, depends the relative facial beauty of the person.

Physiognomists emphasize the importance of the nose in the category of anatomical conformations that are indicative of special traits of character; and regard it as a measure of force in nations and individuals.

Says Wells: "A skillful dissembler may disguise, in a degree, the expression of the mouth; the hat may be slouched over the eyes; the chin may be hidden in an impenetrable thicket of beard; but the nose will stand out 'and make its sign' in spite of all precautions. It utterly refuses to be ignored, and we are, as it were, compelled to give it our attention."

Even in ancient times much attention was given to its shape and appearance. Among the ancient Persians no man who had a crooked or deformed nose was allowed to sit upon the throne. Cyrus, it is said, had an asymmetrical nose, which was made a thing of beauty through the kind assistance of his emasculated attendants. In order to secure symmetrical and handsomely formed noses, in the children of the royal blood, the eunuchs who had charge of the royal offspring were accustomed to mould their noses into perfect shape (Mackenzie).

Considered from the profile point of view alone, noses are classified according to their shape by students of physiognomy into five main classes:

1. The Roman noses;
2. The Greek noses;
3. The Jewish noses;
4. The Snub or Pug noses; and
5. The Celestial noses.

These classes of noses, considered in the light of the characteristics of the race or class to which they are peculiar, are observed to indicate prominent traits of character, as follows:

The Roman indicates executiveness or strength; the Greek, refinement; the Jewish, commercialism or desire for gain; the Snub or Pug, weakness and lack of development; the Celestial, weakness, lack of development, and inquisitiveness.

John Orlando Roe, "The Deformity Termed 'Pug Nose' and Its Correction by a Simple Operation" (1887), reprinted in Frank McDowell, *Source Book of Plastic Surgery* (Baltimore: Williams and Wilkins Co., 1977), 114–118.

"Le nez retroussé" of the French is applied to the Celestial nose, which is simply the pug lengthened and turned upward so as to form a gentle curve from the root to the tip.

The fact that the deductions of physiognomists almost completely harmonize with the anatomical and physiological facts in the case of the last two classes becomes striking, when we consider that those deductions have been made from observation alone.

Mr. Warwick says: "A snub-nose is to us a subject of most melancholy interest. We behold in it a proof of a degeneracy of the human race."

Tristram Shandy's father, regretting his son's misfortunes, remarked: "No family, however high, could stand against a succession of short noses;" and his grandfather "when tendering his hand and heart to the lady who afterward consented to 'make him the happiest of men,' was forced to capitulate to her terms owing to the brevity of his nose."

There are three conditions that may occur during the development of the nose that give it the appearance called snub or pug. They are: (1) excessive development of the alae and cartilaginous portions on the end of the nose; (2) a lack of sufficient development, or a sunken or flattened condition of the base and bridge of the nose, while the end of the nose may be but normally developed; (3) the combination, to a greater or less degree, of the conditions just mentioned. This last condition is the one most frequently found.

During development the nose and parts comprising the central portion of the face, as the ethmoid and sphenoid bones, and parts adjacent, are late in developing, and are also the last portions of the face to undergo ossification. At birth the nose, at its base and central portions, is flat and nearly level with the face, but later this depressed line is replaced by a more prominent one as the nose becomes developed. From this it will be seen that anything interfering with the proper development of these parts so as to cause them to remain in their infantile condition, while the end of the nose undergoes due development, will give the nose a snubbed and unsightly shape.

The best developed and most beautiful noses are one-third the length of the face. But noses often vary from this proportion, and in some instances an ill-formed nose is inherited, it being a special family mark. Ribot says, "that of all the features, the nose is the one which heredity preserves the best" (*Hereditary Traits*, Richard A. Proctor). But in other instances it is the result of diseased conditions affecting its growth and proper development during infancy and early childhood.

There are many conditions that operate to produce this result. The principal one is obstruction of the nasal passages which cuts off nasal respiration. During the inspiratory act of respiration and deglutition, when the nasal passages are obstructed, a partial vacuum is produced in the naso-pharynx. This suction force, being exerted on the inner side of the yielding cartilaginous nasal tissues, tends thereby to draw them inward, and thus in a corresponding degree retards or prevents their normal expansion and development. This obstruction of the nasal passages may also cause an enlargement or undue development of the portion of the nose below and beyond the obstruction, especially if this obstruction is composed of firm tissues, through interfering with the return circulation. The end of the nose thus becomes engorged, the vessels distended, and a marked thickening of the tissues takes place. The importance of attention to obstructed nostrils in infants, commonly called snuffles, is thus clearly demonstrated.

All chronic affections of the nose, even when unattended by obstruction of the passages, tend to produce by sympathetic irritation more or less congestion of the vessels of the end of the nose, and, by reason of these vessels having less power of resistance, an undue distention of them takes place; the surrounding tissues become thickened, and an enlargement of the end of the nose occurs. This is very commonly observed during the treatment of nasal diseases.

This diminished resistance of the peripheral vessels explains the effect of alcohol in the coloration and enlargement of the end of the nose in "old topers," which is so often observed. Alcohol produces congestion of, or sends the blood into, the capillaries and terminal blood-vessels. Since the capillaries in the nose have less resistance than the other superficial vessels of the face or other parts, the effect of imbibition is first shown in the end of the nose.

A crooked or wrinkled septum will have the effect to lower the contour of the nose, as well as to cause an undue arching of the palatine vault.

A snubbed appearance may be given to the nose by injuries to its bridge or base, and also by ulceration and necrosis of the bones of the nasal chambers, especially the vomer, resulting in the removal of the support to the centre of the nose, which then falls inward.

The operation for the correction of the deformity under consideration is easily performed, although I can find no record of it, and have no knowledge of its having been proposed or performed.

It may be classed about the same as the operation for strabismus, and, like many other operations, is mainly to improve the personal appearance of the individual.

The operation consists in the removal from the end of the nose that tissue which is in excess, or which is disproportionate in amount to the other portions of the nose. In other words, we are to make the nose symmetrical from one end to the other.

In cases where the bridge is low and undeveloped, if the end is lowered, made smaller, and brought down so that the top of the nose forms a straight line from its base, or junction of the frontal and nasal bones, to the end, the nose ceases to be unduly noticeable or unsightly; and, although the nose will be smaller, it will appear much larger than before by reason of its being symmetrical and proportionate throughout.

The nose does not appear ugly by reason of the fact that its size is disproportionate to that of the face (for noses vary greatly in this respect), but by reason of the disproportionate relations to one another of the different parts of the nose itself.

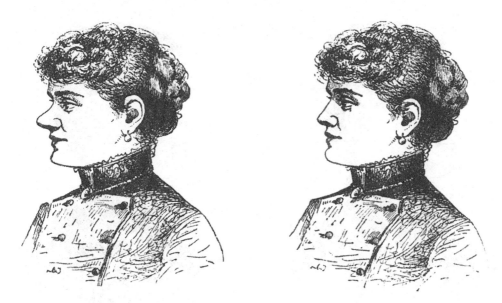

Figures 2.6.1–2.6.2
(Courtesy of National Library of Medicine, Bethesda, Md.)

In those cases in which the deformity consists in an undue enlargement of the end or cartilaginous portion of the nose, while the bony framework is normally developed, it will be seen that the main portion of the nose is straight until we come to the enlarged end, which suddenly tilts upward. Even in this class of cases it is not the end of the nose that really appears too large, but it is the base or bridge that appears too small or depressed.

This fact is shown by the foregoing illustrations, made from photographs taken in each case shortly before and shortly after the operation.

In Fig. 2.6.1 the bridge of the nose appears much lower than it does in Fig. 2.6.2, although it is of the same height. This is also true of Figs. 2.6.3 and 2.6.4.

The expression of the face in Fig. 2.6.2 is decidedly different from that of the face in Fig. 2.6.1. The same is also true of Figs. 2.6.4 and 2.6.3. This change is due entirely to the alteration made by removing the excessive tissue in the end of the nose. It will also be observed that in the case of the boy, excepting the nose, the features are the same in both pictures; and so in the case of the girl. This can be very easily demonstrated in the foregoing illustrations by comparing the shapes of the noses before and after the operation.

The operation is performed as follows: We first deaden the sensibility of the interior of the end of the nose by cocaine (general anaesthesia being unnecessary) and then brightly illuminate this part.

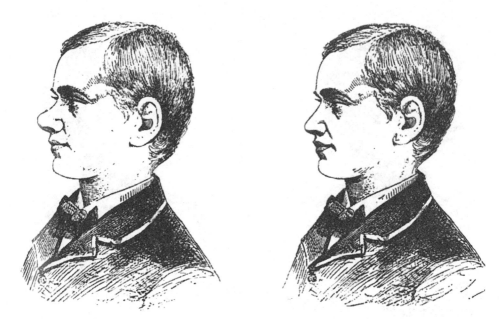

Figures 2.6.3–2.6.4
(Courtesy of National Library of Medicine, Bethesda, Md.)

If the tissue is to be removed from that portion where the mucous membrane is not too firmly adherent, the membrane should be dissected back, to be replaced after the operation.

The end of the nose is turned upward and backward, and held with a retractor by an assistant; then sufficient of the superfluous tissue is removed or dissected out to allow the nose to conform to the shape that we desire. Great care must, however, be exercised not to remove too much tissue, and also not to cut through into the skin, lest we may have afterward a scar or a dent in the external surface of the nose.

In some cases no after-treatment is required, but in others it is advisable to mould a saddle or splint, as it were, to the top of the nose, so as to make it, while healing, assume the shape we wish to obtain.

In some instances the large and unsightly end of the nose is not due to an excessive tissue but to a malformation of the cartilages of the alae, bulging outward with a corresponding concavity on the inside.

These noses can be very readily moulded into a handsome shape by cutting, with a small tenotomy knife, through these cartilages, in different places, sufficiently to destroy their elasticity. Then by inserting a silver or hard-rubber tube, of the proper size and shape, into the nostril, and conforming the saddle to the outside of the nose, we have it encased in an outside and inside splint that compels it to conform to the exact shape we desire.

While performing this operation and moulding the nose into shape, we must not neglect to preserve the nasal passages free and unobstructed.

Thus far I have performed this operation on five persons. I have, however, illustrations of but the last two, although all have been successful. With the first two patients it did not occur to me to have photographs of the nose taken before the operation, and the third patient would not permit it.

John O. Roe, M.D.
Rochester, N.Y.

SCALE OF DIAMETERS OF PILE TO EXPLAIN THE CLASSIFICATION OF MANKIND
BY THE HAIR AND WOOL OF THEIR HEADS.

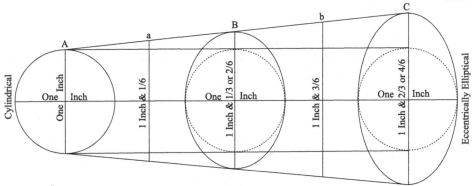

REFERENCES:–
A. –Cylindrical ; both diameters the same.
B. –Oval ; one diameter 1-3 more than the other.
C. –Eccentrically elliptical ; one diameter 2-3 more than the other.
  Pile which exceed the cylindrical by less than 1-6, may be called *cylindroidal*.
  Those which exceed 1-6, but not 2-6, are *lesser ovoidal*.
  Those that exceed 2-6, but are not 3-6, are *ovoidal*.
  Those which exceed 3-6, but do not reach 4-6, are *eccentrically elliptoidal*.

Figure from "The Hair and Wool of the Different Species of Man," *United States Magazine and Democratic Review* 27 (November 1850): 455.

2.8 "On Some of the Apparent Peculiarities of Partuition in the Negro Races" (1875)

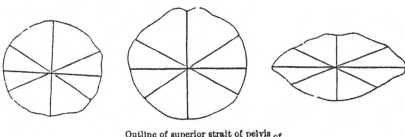

Outline of superior strait of pelvis of

African Negress. American Negress (Mulatto ?). White Female.

Joseph Taber Johnson. Figure from "On Some of the Apparent Peculiarities of Partuition in the Negro Races" *American Journal of Obstetrics* 8 (January 1875): 117.

3 IMMUNITY AND CONTAGION

Introduction

Introduction

Evelynn M. Hammonds and Rebecca M. Herzig

Differential experiences of disease have long been central to efforts to understand human variation. American Indian tribes suffering the blights of new plagues brought by Europeans were compelled to contend with the relationships between foreign peoples and susceptibility to disease. Russian, Spanish, French, and English travelers all remarked on the connections between bodily illness and America's unfamiliar landscapes and customs. Early settlers investigated the "bad airs" of their new homes, and debated whether proper "seasoning" (hardening through experience in a specific location) might increase resistance to contagion.

By the mid-nineteenth century, ongoing processes of migration and immigration shifted scientists' focus from European experiences of arrival and dislocation and the dangerous airs of the New World to instead address disease as an external threat, a threat carried *to* America by others. Scientists, physicians, and intellectuals began to consider the incidence of disease to be symptomatic of more immutable, underlying distinctions between peoples. In 1856, physician Josiah Nott spoke for many when he suggested that "the races of men" differed not only in morphology (differences in cranial capacity or musculature) but also in disease morbidity and mortality. Other medical and scientific commentators debated whether Irishmen were particularly prone to rheumatic fever, whether Jews were impervious to the skin condition called plica polonica, or whether the presence of "negro blood" might increase resistance to yellow fever.

The first selection in this section, published in 1851, was originally presented by the physician Samuel A. Cartwright before the Medical Association of Lousiana. At the time a man widely esteemed in medical communities, Cartwright echoed dozens of similar Southern writers offering advice for slaveholders on topics such as slave nutrition and discipline. Cartwright's work, with its discovery of diseases such as "drapetomania" (an illness causing enslaved men and women to wish to run away), has since been widely discussed by historians as a cogent example of antebellum scientific racism. You might note how concepts of anatomical and physiological differences inform the physician's understanding of disease susceptibility, and vice versa. How would you describe the relations between body and mind, physique and temperament that Cartwright presumes here?

As racial character was considered an index of disease susceptibility, so, too, vulnerability to disease was thought to reveal something about intrinsic racial characteristics. For instance, white women's purported sensitivity to puerperal fever and other frailties of childbirth (compared to the alleged hardiness of American Indian and Black women) was trumpeted as an indication of a refined racial character. Depending on the particular disease in question, vulnerability might be ennobling or debasing, a sign of civilization or degeneracy.

By the last half of the nineteenth century, these questions tended to revolve around the "Chinese question." Immigration from China closely followed the discovery of gold in the mountains of California in 1848. From approximately 50 people in the late 1840s, the state's Chinese population grew to more than 25,000 by 1852. By 1880, the Chinese population in the mainland U.S. had grown to more than 100,000, and by 1890 Chinese men and women were living in every state and territory. Many of these immigrants arrived under the "credit-ticket system," in which they were indentured to repay—with heavy interest—the costs associated with trans-Pacific relocation. Most immigrants were able-bodied men, and so their influence on wage work was disproportionate to their numbers, particularly in Western states and territories. In California, for instance, Chinese immigrants comprised only a tenth of the state's population, but 25% of its physical labor force. Managers and owners used the threat of displacement by Chinese laborers to threaten white strikers. Antagonism toward Chinese workers grew, as did the contradictions between demands for profitably cheap labor and for an efficiently healthy workforce.

In the context of economic and social upheaval, arguments about disease flickered between the realms of science, medicine, and public policy, and commentators often came to dress their anti-immigrant sentiments in the language of contagion. These interconnections are laid bare in Dr. Arthur B. Stout's *Chinese Immigration and the Physiological Causes of the Decay of a Nation*. Stout's treatise, published in San Francisco in 1862, demonstrates the relations between debates over immigration and investigations into the etiology and treatment of diphtheria, syphilis, typhoid, and other contagious diseases. It also highlights the mutability of the category of contagion itself; Stout, for example, includes "mental alienation" as a communicable disease conveyed by immigrants.

Other physicians and scientific investigators similarly intertwined physiology and economy in their writings, contemplating the moral and physical "decay" of disease. In this selection by Mary P. Sawtelle, the author addresses the question of differential susceptibility to disease not only by race but also by sex. Writing for a largely female readership (the short-lived *Medico-Literary Journal* was dedicated to the "diffusion of medical knowledge among women"), Sawtelle's piece appeared just three years before the passage of the first federal Chinese exclusion act in 1882, which banned the entry of Chinese workers and barred Chinese people from being naturalized as U.S. citizens.

The debates between Washington Matthews, a U.S. Army surgeon, and Thomas Mays, a Philadelphia physician, focuses on racial variation in susceptibility to consumption. Consumption, so named because of the characteristic wasting experienced by disease sufferers, is generally considered the archaic name for the modern disease tuberculosis. Debates over relative racial susceptibility to consumption or tuberculosis (and about the relative lung vitality of various races) continued well into the twentieth century. Here the two authors debate the affects of "civilizing influences" on rates of consumption, discussing the effects of diet, work habits, outdoor life-style, and climate on comparative rates of disease. What does *race* appear to entail for these two investigators? And how is *civilization* related to race?

Bibliography

Arnold, David, ed. *Imperial Medicine and Indigenous Societies*. Manchester: Manchester University Press, 1988.

Gardner, Martha Mabie. "Working on White Womanhood: White Working Women in the San Francisco Anti-Chinese Movement, 1877–1890." *Journal of Social History* 33, no. 1 (1999): 73–95.

Guillory, James D. "The Pro-slavery Arguments of Dr. Samuel Cartwright." *Louisiana History* 9 (1968): 209–227.

Hammonds, Evelynn M. *Childhood's Deadly Scourge: The Campaign to Control Diphtheria in New York City, 1880–1930*. Baltimore: Johns Hopkins University Press, 1999.

Ileto, Reynaldo C. "Cholera and the Origins of the American Sanitary Order in the Philippines." In David Arnold, ed., *Imperial Medicine and Indigenous Societies*. Manchester: Manchester University Press, 1988, 125–148.

Miller, Stuart Creighton. *The Unwelcome Immigrant: The American Image of the Chinese, 1785–1882*. Berkeley: University of California Press, 1969.

Nott, Josiah. "Thoughts on Acclimation and Adaptation of Races to Climate." *American Journal of the Medical Sciences* 32 (October 1856): 320–334.

Shah, Nayan. *Contagious Divides: Epidemics and Race in San Francisco's Chinatown*. Berkeley: University of California Press, 2001.

Stout, Arthur B. *Chinese Immigration and the Physiological Causes of the Decay of a Nation*. San Francisco: Agnew and Deffebach, 1862.

Wailoo, Keith. *Dying in the City of the Blues: Sickle Cell Anemia and the Politics of Race and Health*. Chapel Hill: University of North Carolina Press, 2001.

3.1 "Report on the Diseases and Physical Peculiarities of the Negro Race" (1851)

Samuel A. Cartwright

Gentlemen:—On the part of the Committee, consisting of Doctors Copes, Williamson, Browning and myself, to investigate the diseases and physical peculiarities of our negro population, we beg leave to report—

That, although the African race constitutes nearly a moiety of our southern population, it has not been made the subject of much scientific investigation, and is almost entirely unnoticed in medical books and schools. It is only very lately, that it has, in large masses, dwelt in juxta position with science and mental progress. On the Niger and in the wilds of Africa, it has existed for thousands of years, excluded from the observation of the scientific world. It is only since the revival of learning, that the people of that race have been introduced on this continent. They are located in those parts of it, not prolific in books and medical authors. No medical school was ever established near them until a few years ago; hence, their diseases and physical peculiarities are almost unknown to the learned. The little knowledge that Southern physicians have acquired concerning them, has not been derived from books or medical lectures, but from facts learned from their own observation in the field of experience, or picked up here and there from others.

Before going into the peculiarities of their diseases, it is necessary to glance at the anatomical and physiological differences between the negro and the white man; otherwise their diseases cannot be understood. It is commonly taken for granted, that the color of the skin constitutes the main and essential difference between the black and the white race; but there are other differences more deep, durable and indelible, in their anatomy and physiology, than that of mere color. In the albino the skin is white, yet the organization is that of the negro. Besides, it is not only in the skin, that a difference of color exists between the negro and white man, but in the membranes, the muscles, the tendons and in all the fluids and secretions. Even the negro's brain and nerves, the chyle and all the humors, are tinctured with a shade of the pervading darkness. His bile is of a deeper color and his blood is blacker than the white man's. There is the same difference in the flesh of the white and black man, in regard to color, that exists between the flesh of the rabbit and the hare. His bones are whiter and harder than those of the white race, owing to their containing more phosphate of lime and less gelatine. His head is hung on the atlas differently from the white man; the face is thrown more upwards and the neck is shorter and less oblique; the spine more inwards, and the pelvis more obliquely outwards; the thigh-bones larger and flattened from before backwards; the bones more bent; the legs curved outwards or bowed;

Samuel A. Cartwright, "Report on the Diseases and Physical Peculiarities of the Negro Race," *New Orleans Medical and Surgical Journal* 7 (1851): 691–715.

the feet flat; the gastrocnemii muscles smaller; the heel so long, as to make the ankle appear as if planted in the middle of the foot; the gait, hopper-hipped, or what the French call *l'allure déhanchée*, not unlike that of a person carrying a burden. The projecting mouth, the retreating forehead, the broad, flat nose, thick lips and wooly hair, are peculiarities that strike every beholder. According to Sœmmerring and other anatomists, who have dissected the negro, his brain is a ninth or tenth less than in other races of men, his facial angle smaller, and all the nerves going from the brain, as also the ganglionic system of nerves, are larger in proportion than in the white man. The nerves distributed to the muscles are an exception, being smaller than in the white race. Sœmmerring remarks, that the negro's brain has in a great measure run into nerves. One of the most striking differences is found in the much greater size of the *foramen magnum* in the negro than the white man. The foramen, or orifice between the brain and the spinal marrow, is not only larger, but the medulla oblongata, and particularly the nerves supplying the abdominal and pelvic viscera. Although the nose is flat, the turbinated bones are more developed, and the pituitary membrane, lining the internal cavities of the nose, more extensive than in the white man, and causing the sense of smell to be more acute. The negro's hearing is better, his sight is stronger, and he seldom needs spectacles.

The field of vision is not so large in the negro's eye as in the white man's. He bears the rays of the sun better, because he is provided with an anatomical peculiarity in the inner canthus, contracting the field of vision, and excluding the sun's rays,— something like the membrana nictitans, formed by a preternatural development of the plica lunaris, like that which is observed in apes. His imitative powers are very great, and he can agitate every part of the body at the same time, or what he calls *dancing all over*. From the diffusion of the brain, as it were, into the various organs of the body, in the shape of nerves to minister to the senses, everything, from the necessity of such a conformation, partakes of sensuality, at the expense of intellectuality. Thus, music is a mere sensual pleasure with the negro. There is nothing in his music addressing the understanding; it has melody, but no harmony; his songs are mere sounds, without sense or meaning—pleasing the ear, without conveying a single idea to the mind; his ear is gratified by sound, as his stomach is by food. The great development of the nervous system, and the profuse distribution of nervous matter to the stomach, liver and genital organs, would make the Ethiopian race entirely unmanageable, if it were not that this excessive nervous development is associated with a deficiency of red blood in the pulmonary and arterial systems, from a defective atmospherization or arterialization of the blood in the lungs—constituting the best type of what is called the lymphatic temperament, in which lymph, phlegm, mucus, and other humors, predominate over the red blood. It is this defective hematosis, or atmospherization of the blood, conjoined with a deficiency of cerebral matter in the cranium, and an excess of nervous matter distributed to the organs of sensation and assimilation, that is the true cause of that debasement of mind, which has rendered the people of Africa unable to take care of themselves. It is the true cause of their indolence and apathy, and why they have chosen, through countless ages, idleness, misery and barbarism, to industry and frugality,—why social industry, or associated labor, so essential to all progress in civilisation and improvement, has never made any progress among them, or the arts and sciences taken root on any portion of African soil inhabited by them; as is proved by the fact that no letters, or even hieroglyphics—no buildings, roads or improvements, or monuments of any kind, are any where found, to indicate that they have ever been awakened from their apathy and sleepy indolence, to physical or mental exertion. To the same physiological causes, deeply rooted in the organization, we must look for an explanation of the strange facts, why none of the languages of the native tribes of Africa, as proved by ethnographical researches, have risen above common names, standing for things and actions, to abstract terms or generalizations;—why no form of government on abstract principles, with divisions of power into separate departments, has ever been instituted by them;—why they have always preferred, as more congenial to their nature, a government combining the legislative, judicial and executive powers in the same individual, in the person of a petty king, a chieftain or master;—why, in America, if let alone, they always prefer the same kind of government, which we call slavery, but which is actually an improvement on the government of their forefathers, as it gives them more tranquility and sensual enjoyment, expands the mind and improves the morals, by arousing them from that natural indolence so fatal to mental and moral progress. Even if they did not prefer slavery, tranquility and sensual enjoyment, to liberty, yet their organization of mind is such, that if they had their liberty, they have not the industry, the moral virtue, the courage and vigilance to maintain it, but would relapse into barbarism, or into slavery, as they have done in Hayti. The reason of this is founded in unalterable physiological laws. Under the compulsive power of the white man, they are made to labor or exercise, which makes the lungs perform the duty of vitalizing the blood more perfectly than is done when they are left free to indulge in idleness. It is the red, vital blood, sent to the brain, that liberates their mind when under the white man's control; and it is the want of a sufficiency of red, vital blood, that chains their mind to ignorance and barbarism, when in freedom.

The excess of organic nervous matter, and the deficiency of cerebral—the predominance of the humors over the red blood, from defective atmospherization of the blood in the lungs, impart to the negro a nature not unlike that of a new-born infant of the white race. In children, the nervous system predominates, and the temperament is lymphatic. The liver, and the rest of the glandular system, is out of proportion to the sanguineous and respiratory systems, the white fluids predominating over the red; the lungs consume less oxygen, and the liver separates more carbon, than in the adult age. This constitution, so well marked in infancy, is the type of the Ethiopian constitution, of all ages and sexes. It is well known, that in infancy, full and free respiration of pure fresh air in repose, so far from being required, is hurtful and prejudicial. Half smothered by its mother's bosom, or the cold external air carefully excluded by a warm room or external covering over the face, the infant reposes—re-breathing its own breath, warmed to the same temperature as that of its body, and loaded with carbonic acid and aqueous vapor. The natural effect of this kind of respiration is, imperfect atmospherization of the blood in the lungs, and a hebetude of intellect, from the defective vitalization of the blood distributed to the brain. But it has heretofore escaped the attention of the scientific world, that the defective atmospherization of the blood, known to occur during sleep in infancy, and to be the most congenial to their constitutions, is the identical kind of respiration most congenial to the negro constitution, of all ages and sexes, when in repose. This is proved by the fact of the universal practice among them of covering their head and faces, during sleep, with a blanket, or any kind of covering that they can get hold of. If they have only a part of a blanket, they will cover their faces when about to go to sleep. If they have no covering, they will throw their hands or arms across the mouth and nose, and turn on their faces, as if with an instinctive design to obstruct the entrance of the free external air into the lungs during sleep. As in the case with infants, the air that negroes breathe, with their faces thus smothered with blankets or other covering, is not so much the external air as their own breath, warmed to the same temperature as that of their bodies, by confinement and reïnspiration. This instinctive and universal method of breathing, during sleep, proves the similarity of organization and physiological laws existing between negroes and infants, as far as the important function of respiration is concerned. Both are alike in re-breathing their own breath, and in requiring it to be warmed to their own temperature, by confinement which would be insupportable to the white race after passing the age of infancy. The inevitable effect of breathing a heated air, loaded with carbonic acid and aqueous vapor, is defective hematosis and hebetude of intellect.

Negroes, moreover, resemble children in the activity of the liver and in their strong assimilating powers, and in the predominance of the other systems over the sanguineous; hence they are difficult to bleed, owing to the smallness of their veins. On cording the arm of the stoutest negro, the veins will be found scarcely as large as a white boy's of ten years of age. They are liable to all the convulsive diseases, cramps, spasms, colics, etc., that children are so subject to.

Although their skin is very thick, it is as sensitive, when they are in perfect health, as that of children, and like them they fear the rod. They resemble children in another very important particular; they are very easily governed by love combined with fear, and are ungovernable, vicious and rude under any form of government whatever, not resting on love and fear as a basis. Like children, it is not necessary that they be kept under the fear of the lash; it is sufficient that they be kept under the fear of offending those who have authority over them. Like children, they are constrained by unalterable physiological laws, to love those in authority over them, who minister to their wants and immediate necessities, and are not cruel or unmerciful. The defective hematosis, in both cases, and the want of courage and energy of mind as a consequence thereof, produces in both an instinctive feeling of dependence on others, to direct them and to take care of them. Hence, from a law of his nature, the negro can no more help loving a kind master, than the child can help loving her who gives it suck.

Like children, they require government in every thing; food, clothing, exercise, sleep—all require to be prescribed by rule, or they will run into excesses. Like children, they are apt to over-eat themselves or to confine their diet too much to one favorite article, unless restrained from doing so. They often gorge themselves with fat meat, as children to with sugar.

One of the greatest mysteries to those unacquainted with the negro character, is the facility with which an hundred, even two or three hundred, able-bodied and vigorous negroes are kept in subjection by one white man, who sleeps in perfect security among them, generally, in warm weather, with doors and windows open, with all his people, called slaves, at large around him. But a still greater mystery is the undoubted fact of the love they bear to their masters, similar in all respects to the love that children bear to their parents, which nothing but severity or cruelty in either case can alienate. The physiological laws, on which this instinctive and most mysterious love is founded in the one case, are applicable to the other. Like children, when well-behaved and disposed to do their duty, it is not the arbitrary authority over them that they dread, but the petty tyranny and imposition of one another. The overseer among them, like the school-master among children, has only to be impartial, and to preserve order by strict justice to all, to gain their good will and affections, and to be viewed, not as an object of terror, but as a friend and protector to quiet their fears of one another.

There is a difference between infant negroes and infant white children; the former are born with heads like gourds, the fontinelles being nearly closed and the sutures between the various bones of the head united,—not open and permitting of overlapping, as in white children. There is no necessity for the overlapping of the bones of the head in infant negroes, as they are smaller, and the pelvis of their mothers larger than in the white race. All negroes are not equally black—the blacker, the healthier and stronger; any deviation from the black color, in the pure race, is a mark of feebleness or ill health. When heated from exercise, the negro's skin is covered with an oily exudation that gives a dark color to white linen, and has a very strong odor. The odor is strongest in the most robust; children and the aged have very little of it.

I have thus hastily and imperfectly noticed some of the more striking anatomical and physiological peculiarities of the negro race. The question may be asked, Does he belong to the same race as the white man? Is he a son of Adam? Does his peculiar physical conformation stand in opposition to the Bible, or does it prove its truth? These are important questions, both in a medical, historical and theological point of view. They can better be answered by a comparison of the facts derived from anatomy, physiology, history and theology, to see if they sustain one another. We learn from the Book of Genesis, that Noah had three sons, Shem, Ham and Japheth, and that Canaan, the son of Ham, was doomed to be servant of servants unto his brethren. From history, we learn, that the descendants of Canaan settled in Africa, and are the present Ethiopians, or black race of men; that Shem occupied Asia, and Japheth the north of Europe.

In the 9th chapter and 27th verse of Genesis, one of the most authentic books of the Bible, is this remarkable prophecy: "God shall enlarge Japheth, and he shall dwell in the tents of Shem; and Canaan *shall be* his servant." Japheth has been greatly enlarged by the discovery of a new world, the continent of America. He found in it the Indians, whom natural history declares to be of Asiatic origin, in other words, the descendants of Shem: he drove out Shem, and occupied his tents: and now the remaining part of the prophecy is in the process of fulfilment, from the facts every where before us, of Canaan having become his servant. The question arises, Is the Canaanite, or Ethiopian, qualified for the trying duties of servitude, and unfitted for the enjoyment of freedom? If he be, there is both wisdom, mercy and justice in the decree dooming him to be servant of servants, as the decree is in conformity to his nature. Anatomy and physiology have been interrogated, and the response is, that the Ethiopian, or Canaanite, is unfitted, from his organization and the physiological laws predicated on that organization, for the responsible duties of a free man, but, like the child, is only fitted for a state of dependence and subordination. When history is interrogated, the response is, that the only government under which the negro has made any improvement in mind, morals, religion, and the only government under which he has led a happy, quiet and contented life, is that under which he is subjected to the arbitrary power of Japheth, in obedience to the Divine decree. When the original Hebrew of the Bible is interrogated, we find, in the significant meaning of the original name of the negro, the identical fact set forth, which the knife of the anatomist at the dissecting table has made appear; as if the revelations of anatomy, physiology and history, were a mere re-writing of what Moses wrote. In the Hebrew word "Canaan," the original name of the Ethiopian, the word *slave by nature*, or language to the same effect, is written by the inspired penman. Hence there is no conflict between the revelations of the science of medicine, history, and the inductions drawn from the Baconian philosophy, and the authority of the Bible; one supports the other.

As an illustration, it is known that all the Hebrew names are derived from verbs, and are significant. The Hebrew verb *Canah*, from which the original name of the negro is derived, literally means *to submit himself—to bend the knee*. Gesenius, the best Hebrew scholar of modern times, renders both the Kal, Hiphil and Niphal form of the verb from which Canaan, the original name of the negro is derived, in the following Latin: *Genu flexit*—he bends the knee; *in genua procidet*—he falls on his knees; *depressus est animus*—his mind is depressed; *submisse se gessit*—he deports himself submissively; *fractus est*—he is crouched or broken; or in other words, *slave by nature*, the same thing which anatomy, physiology, history, and the inductions drawn from philosophical observations, prove him to be.

A knowledge of the great primary truth, that the negro is a slave by nature, and can never be happy, industrious, moral or religious, in any other condition than the one he was intended to fill, is of great importance to the theologian, the statesman, and to all those who are at heart seeking to promote his temporal and future welfare. This great truth, if better known and understood, would go far to prevent the East India Company and British government from indulging in any expectation of seeing their immense possessions in Asia enhanced in value, by the overthrow of slave labor in America, through the instrumentality of northern fanaticism; or of seeing the Union divided into two or more fragments, hostile to each other; or of gaining any advantages, that civil commotion on this side of the Atlantic would give to the tottering monarchies of Europe. With the subject under this aspect, the science of Medicine has nothing to do, further than to uncover its light, to show truth from error.

Without a knowledge of the physical differences between the Ethiopian and the Caucasian, the Queen of England's medical advisers would not be much better qualified to prescribe for a negro, than her parliament to legislate for him; or her subjects to dictate to us what position he should occupy in our republican Union of Sovereign States.

The Diseases of Negroes—Pulmonary Congestions, Pneumonia, etc.

One of the most formidable complaints among negroes, and which is more fatal than any other, is congestion of the lungs, or what European writers would call false pleurisy, or peri-pneumonia notha. It is often called cold plague, typhus pneumonia, bilious pleurisy, etc., according to its particular type, and the circumstances attending it; sometimes the head complains more than any other part, and it then bears the misnomer, "head pleurisy." It occurs, mostly, in winter and spring, but is met with at every season of the year, when cold nights succeed to warm days. It is more common among those who sleep in open houses, without sufficient fires to keep them warm and comfortable. It is seldom observed among negroes who inhabit log cabins, with cemented or clay floors, or warm houses made of brick, or any material to exclude the cold wind and air. The frame houses, with open weather-boarding and loose floors, admitting air both at the sides and from below, are buildings formed in ignorance of the peculiar physiological laws of the negro's organization, and are the fruitful sources of many of his most dangerous diseases.

Want of sufficient fires and warm blankets, is also another cause of thoracic complaints. The negro's lungs, except when the body is warmed by exercise, are very sensitive to the impressions of cold air. When not working or taking exercise, they always crowd around a fire, even in comparatively warm weather, and seem to take a positive pleasure in breathing heated air and warm smoke. In cold weather, instead of sleeping with their feet to the fire, as all other kinds of people do, whether civilized or savage, they turn their head to the fire—evidently for the satisfaction of inhaling warm air, as congenial to their lungs, in repose, as it is to infants. In bed, when disposing themselves for sleep, the young and old, male and female, instinctively cover their heads and faces, as if to insure the inhalation of warm, impure air, loaded with carbonic acid and aqueous vapor. The natural effect of this practice is imperfect atmospherization of the blood—one of the heaviest chains that binds the negro to slavery. In treating, therefore, their pulmonary affections, the important fact should be taken into consideration, that cold air is inimical to the lungs of healthy negroes, when the body is in repose, and not heated by exercise, and consequently more prejudicial in the diseases of those organs. A small, steady fire, a close room, and plenty of thick blanket covering, aided with hot stimulating teas, are very essential means in the treatment of the pulmonary congestions to which their lungs are so prone. An accurate diagnosis, whether the complaint be a mere congestion, pleuritis or pneumonia, is not of much practical importance in the first instance, because, whether it be one or the other, warm air is equally essential, and warm stimulating teas, to determine to the surface. It is proper, first to warm the body by external means and stimulating drinks, after which, an emetic, followed by a purgative of a mild kind, will be beneficial. When there is pain in taking a full inspiration, a moderate blood-letting from the arm, followed by half grain or grain doses of tartar emetic, repeated at intervals of an hour or two, and combined with a little anodyne, to prevent its running off by the bowels, will be found a very effectual remedy in subduing inflammation and promoting expectoration. In the typhoid forms of pneumonia, the quinine, in efficient doses, combined with camphor, aromatics and calomel, is generally the best practice. Bleeding is not admissible in this form of pneumonia, otherwise they bear blood-letting in chest complaints much better than any others. But even in these, they will not bear repeated blood-letting, as the white race do.

Bilious and Adynamic Fevers—Remittents and Intermittents

The next class of complaints to which they are mostly liable, are bilious and adynamic fevers—remittents and intermittents. Evacuating the stomach and bowels by a mild emetico-cathartic, combined with a weak anodyne carminative, to prevent its excessive action, is generally the best medicine to begin with; for, whatever be the type of the fever, as negroes are hearty eaters, it will be an advantage, in the after treatment of the case, to have the *primæ viæ* cleared of their load of undigested food, and the superabundant mucosities poured out into the alimentary canal of a people so phleghmatic, when attacked with a fever suspending digestion, and interrupting absorption.

For this purpose, a combination of ipecacuanha, rhubarb and cream of tartar, each half a drachm, and a tea-spoonful of paregoric, in ginger or pepper tea, is a very safe and effectual medicine. It will vomit, if there be bile or much mucosity, and will afterwards act on the bowels, promote secretion of urine, and determine to the surface; after which, a dose or two of quinine will generally effect a cure. Calomel is used too indiscriminately in the treatment of their diseases; nevertheless, in obstinate cases, it is not to be dispensed with. Negroes are very liable to become comatose, particularly after watery operations, or in torpid states of the liver. Such cases are best treated by a combination of calomel, camphor, capsicum, quinine and laudanum, and a blister to the back of the neck. Cold water to the head is dangerous. Nearly all their complaints bear stimulating, aromatic substances much better than similar affections among white people, and will not tolerate evacuations so well. The pure anti-phlogistic treatment by evacuations, cold air, starvation and gum water, so effectual in the inflammatory complaints of the hematose white man, will soon sink them into hopeless collapse. Even under the use of anti-phlogistics in their inflammatory complaints, pepper or ginger tea, or some stimulant, is necessary to support the vital actions, which would soon fail under such insipid drinks as gum water. The reason of this is, that the fluids and all the secretions are more acrid than those of the white man. In the latter, the lungs consume more oxygen; the blood is redder and more stimulating, and all the fluids more bland and sweet: whereas, in the negro, the deficient hematosis renders the blood less stimulating, and requires acrid and piquant substances addressed to the digestive system, to supply the stimulus that would otherwise be derived from the air in the lungs. Although they are so liable to congestive and bilious fevers—remittents and intermittents—they are not liable to the dreaded *el vomito*, or yellow fever. At least, they have it so lightly, that I have never seen a negro die with black vomit, although I have witnessed a number of yellow fever epidemics. This is a strong proof against the identity of yellow fever, and the other fevers just named.

Scrofula, etc.

Like children, negroes are very liable to colics, cramps, convulsions, worms, glandular and nervous affections, sores, biles, warts, and other diseases of the skin. Scrofula is very common among them. Rickets, diseases of the spine and hip joint, and white swellings, are not uncommon. They are also subject to the goitre. All very fat negroes, except women who have passed the prime of life, are unhealthy and scrofulous. The great remedy for the whole tribe of their scrofulous affections, without which all other remedies do very little good, is *sunshine*. The solar rays is one of the most efficient therapeutic agents in the treatment of many other affections to which they are liable. A good, wholesome, mixed diet, warm clothing, warm, dry lodgings, and inunction of the skin, with oleaginous substances, and occasional tepid baths of salt and water, are also very necessary remedies. The limits of this report will not permit me to go into details of familiar treatment, as the use of iodine, and the usual remedies.

Frambæsia, Pian, or Yaws

The Frambæsia, Pian, or Yaws, is a disease thought to be peculiar to negroes. I have seen it in its worst forms, in the West Indies. I have occasionally met with it in its modified form, in the States of Mississippi and Louisiana, where it is commonly mistaken for syphilis. It is a contagious disease, communicable by contact among those who greatly neglect cleanliness. Children are liable to it, as well as adults. It is supposed to be communicable, in a modified form, to the white race, among whom it resembles pseudo-syphilis, or some disease of the nose, throat or larynx. Further observations are wanting in regard to it. It is said to be vary prevalent in Tamaulipas, in Mexico. Attacking the nose and throat, in the first instance, very similar to secondary syphilitic affections, without ever having appeared on the genital organs at all, except in the shape of a slight herpes preputialis. According to my experience, no other remedies have been found to make the least impression upon it but the dento-chloride of mercury, combined with guaiacum and dulcamara. Our planters do not go to the North or to Europe to learn the art of making sugar, cotton, rice, and tobacco, but they send their sons there to study medicine in the hospitals, where nearly all the diseases they see arise from causes unknown on our plantations—want of food, fire, and the common necessaries of life. Very good physicians they might be, if they staid there; but, on returning home, they have to study Medicine over again, in the school of experience, before they can practice with success, particularly among negroes. It would be very strange, that among the whole multitude of medical schools in the Unitd States, there is not one that has made any special provision for instruction in regard to three millions of people in the Southern States, and representing half the value of Southern property, differently organized in mind and body from any other people, and having diseases requiring peculiar treatment,—if it were not for the well-known fact of the predominance of a most erroneous hypothesis among statesmen, divines, and other classes of people nearly everywhere, "That there are no radical or physical differences in mankind, other than those produced by external circumstances, and that the treatment applicable to the white man would be just as good, under similar external circumstances, for the negro." This false hypothesis is at the root of the doctrine that the liberty and political institutions so beneficial to the white man, would be equally beneficial to the negro—that there is no internal or physical difference between the two races. The every-day experience of the Southern people, where the two races dwell together, prove this hypothesis to be unfounded; whereas its fallacy is not so apparent to the people of the North and of Europe, where only one race of mankind is found in numbers sufficient to make comparisons between the two. Hence they have not the data to arrive at the truth, and nothing to correct the erroneous views that a false dogma has given them in regard to negro slavery. But it is most strange that our institutions for medical learning, South, should be doing nothing, with such ample materials around them, to overturn an hypothesis, founded in gross ignorance of the anatomy and physiology of the African race—an hypothesis threatening to cause a disruption of our federal government, and that could be disproved and put down forever at the dissecting table; as it also could be by contrasting the phenomena, drawn from daily observations taken among three millions of negroes, in health and disease, with the phenomena already drawn from observations of the white race; and thereby proving the difference of organization in mind and body between the two races. Stranger still, that our Southern schools in Medicine should be content to linger behind those of the North, without even hope of rivaling them in the numbers of their students, when a provision for including, in their course of instruction, the three millions of people in our midst, not cared for by any school, would, in time, put them far a-head, by attracting the current of students South, who have heretofore been attracted to the North. Some provision in our schools especially devoted to the anatomy and physiology of our negroes,—to the treatment of their diseases,—to the best means to prevent sickness among them,—to improve their condition, and at the same time to make them more valuable to their owners, and governed with more ease and safety,—would be sending Science into a new and wide field of usefulness, to reap immense benefits for the millions of both races inhabiting the South.

Negro Consumption

Negro consumption is a disease almost unknown to medical men of the Northern States and Europe. A few Southern physicians have acquired some valuable information concerning it, from personal experience and observation; but this knowledge is scattered in fragments about, and has never been condensed in a form to make it of much practical utility. It is hoped that Dr. Fenner's Southern Reports will collect the experience of our physicians, and make that experience more available than it has heretofore been; some physicians, looking upon negro consumption through Northern books, suppose it to be a variety of phthisis pulmonalis—but it has no form or resemblance to the phthisis of the white race, except in the emaciation, or when it is complicated with the relics of pneumonia, or a badly-cured pleurisy. Others regard it as a dyspepsia, or some disease of the liver or stomach; the French call it *mal d'estomac*. But dyspepsia is not a disease of the negro; it is, *par excellence*, a disease of the Anglo-Saxon race; I have never seen a well-marked case of dyspepsia among the blacks. It is a disease that selects its victims from the most intellectual of mankind, passing by the ignorant and unreflecting.

The popular opinion is, that negro consumption is caused by *dirt-eating*. The eating of dirt is not the cause, but only one of the effects—a mere symptom, which may or may not attend it. As in pica, there is often a depraved appetite for substances not nutritious, as earth, chalk, lime, etc.; but oftener, as in malacia, a depraved appetite, for nutritious substances to a greater degree, than for non-nutritious. In negro consumption, the patients are generally hearty eaters of all kinds of food; but there are exceptions.

The disease may be detected, at a very early stage of its existence, by the pale, whitish color of the mucous membrane lining the gums and the inside of the mouth, lips and cheeks: so white are the mucous surfaces, that some overseers call it the paper-gum disease. It can be detected, however, in its incipient state, by making the patient ascend a flight of stairs; the pulse will be accelerated from eighty or ninety beats, to an hundred and thirty or forty. All kinds of active exercise will greatly accelerate the pulse, that of walking up hill or up stairs more than any other. The skin is ashy, pale and dry; the veins of the head are distended, and show more than in health; occasionally during the day, there is some heat of the skin, and febrile excitement; the blood is poor, pale and thin, in the advanced stages, containing very few red globules; but the pathognomonic symptoms of the complaint are the acceleration of the pulse on exercise, and the whiteness of the lining membrane of the cheeks, lips and gums; the lining membrane of the eye-lids is also pale and whitish. It is of importance to know the pathognomic signs in its early stages, not only in regard to its treatment, but to detect impositions, as negroes afflicted with the complaint are often for sale; the acceleration of the pulse on exercise incapacitates them for labor, as they quickly give out and have to leave their work. This induces their owners to sell them, although they may not know the cause of their inability to labor. Many of the negroes brought south for sale are in the incipient stage of the disease; they are found to be inefficient laborers, and are sold in consequence thereof.

In order to be able to prevent or cure any malady, it is necessary to know its cause, and its seat. The seat of negro consumption is not in the lungs, stomach, liver or any organ of the body, but in the mind, and its cause is generally mismanagement or bad government on the part of the master, and superstition on the part of the negro. The patients themselves believe that they are poisoned; they are right, but it is not the body, but the mind that is poisoned. Negroes are very jealous and suspicious; hence, if they are slighted or imposed on in any way, or over-tasked, or do not get what they call their rights, they are apt to fall into a morbid state of mind, with sulkiness and dissatisfaction very plainly depicted in their countenances. It is bad government to let them remain in this sulky, dissatisfied mood, without inquiring into its causes, and removing them; otherwise, its long continuance leads to the disease under consideration. They fancy, that their fellow-servants are against them, that their master or overseer cares nothing for them, or is prejudiced against them, and that some enemy on the plantation or in the neighborhood has tricked them, that is, laid poison for them to walk over, or given it to them in their food or drinks. On almost every large plantation, there are one or more negroes, who are ambitious of being considered in the character of conjurers—in order to gain influence, and to make the others fear and obey them. The influence that these pretended conjurers exercise over their fellow servants, would not be credited by persons unacquainted with the superstitious mind of the negro. Nearly all, particularly those who have passed the age of puberty, are at times kept in constant dread and terror by the conjurers. These impostors, like all other impostors, take advantage of circumstances to swell their importance, and to inculcate a belief in their miraculous powers to bring good or evil upon those they like or dislike. It may be thought that the old superstitition about conjuration has passed away with the old stock of native Africans; but it is too deeply radicated in the negro intellect to pass away: intelligent negroes believe in it, who are ashamed to acknowledge it. The effect of such a superstition—a firm belief that he is poisoned or conjured—upon the patient's mind, already in a morbid state, and his health affected from hard usage, over-tasking or exposure, want of wholesome food, good clothing, warm comfortable lodging, and the distressing idea, that he is an object of hatred or dislike, both to his master and fellow servants, and has no one to befriend him, tends directly to generate that erythism of mind, which is the essential cause of negro consumption. This erythism of mind, like the erythism of the gravid uterus in delicate females, often causes a depraved appetite for earth, chalk, lime, and such indigestible substances. The digestive passages, in both cases, become coated with acescent mucosities, or clogged with saburricious matters. Natural instinct leads such patients to absorbents, to correct the state of the stomach.

In the depraved appetite caused by pregnancy, or in young women afflicted with leuchorrhœa, true art improves upon instinct, or the natural medication of the patients themselves, by substituting magnesia, cathartics, bitters and tonics. But for the same morbid appetite in negro consumption, the natural medication, resorted to by the instinctive wants of the patient, is mistaken for the cause of the disease. It is not only earth or clay that the patients have an appetite for, but, like chlorotic girls, they desire vinegar, pepper, salt, and stimulants. Their skins are dry, proving want of cutaneous exhalation; very little aqueous vapor is thrown off from the lungs, owing to their inability to take exercise. Consequently, defluxions occur on the mucous coat of the digestive passages, from want of action of the skin and lungs; the mucosity, lining the intestinal canal, interrupts the absorption of chyle—the blood becomes impoverished, and the body wastes away from interstitial absorption and want of nutriment.

As far as medication is concerned, I have found a combination of tartar emetic half grain, capsicum five grains, a tea-spoonful of charcoal, a table-spoonful of tincture of gum guaiacum, three times a-day, a good remedy; also, rubbing the whole surface of the body over with some oily substance. But these, as various other remedies, as purgatives, tonics, etc., should be assisted by removing the original cause of the dissatisfaction or trouble of mind, and by using every means to make the patient comfortable, satisfied and happy.

Drapetomania, or the Disease Causing Slaves to Run Away

Drapetomania is from δζαπέτης, a runaway salve, and μανια, *mad or crazy*. It is unknown to our medical authorities, although its diagnostic symptom, the absconding from service, is well known to our planters and overseers, as it was to the ancient Greeks, who expressed by the single word δζαπέτης the fact of the absconding, and the relation that the fugitive held to the person he fled from. I have added to the word meaning runaway slave, another Greek term, to express the disease of the mind causing him to abscond. In noticing a disease not heretofore classed among the long list of maladies that man is subject to, it was necessary to have a new term to express it. The cause, in the most of cases, that induces the negro to run away from service, is as much a disease of the mind as any other species of mental alienation, and much more curable, as a general rule. With the advantages of proper medical advice, strictly followed, this troublesome practice that many negroes have of running away, can be almost entirely prevented, although the slaves be located on the borders of a free State, within a stone's throw of the abolitionists. I was born in Virginia, east of the Blue Ridge, where negroes are numerous, and studied medicine some years in Maryland, a slave State, separated from Pennsylvania, a free State, by Mason & Dixon's line—a mere air line, without wall or guard. I long ago observed that some persons, considered as very good, and others as very bad masters, often lost their negroes by their absconding from service; while the slaves of another class of persons, remarkable for order and good discipline, but not praised or blamed as either good or bad masters, never ran away, although no guard or forcible means were used to prevent them. The same management which prevented them from walking over a mere nominal, unguarded line, will prevent them from running away anywhere.

To ascertain the true method of governing negroes, so as to cure and prevent the disease under consideration, we must go back to the Pentateuch, and learn the true meaning of the untranslated term that represents the negro race. In the name there given to that race, is locked up the true art of governing negroes in such a manner that they cannot run away. The correct translation of that term declares the Creator's will in regard to the negro; it declares him to be the submissive knee-bender. In the anatomical conformation of his knees, we see *"genu flexit"* written in the physical structure of his knees, being more flexed or bent, than any other kind of man. If the white man attempts to oppose the Deity's will, by trying to make the negro anything else than *"the submissive knee-bender,"* (which the Almighty declared he should be,) by trying to raise him to a level with himself, or by putting himself on an equality with the negro; or if he abuses the power which God has given him over his fellowman, by being cruel to him or punishing him in anger, or by neglecting to protect him from the wanton abuses of his fellow-servants and all others, or by denying him the usual comforts and necessaries of life, the negro will run away: but if he keeps him in the position that we learn from the Scriptures he was intended to occupy, that is, the position of submission, and if his master or overseer be kind and gracious in his bearing towards him, without condescension, and at the same time ministers to his physical wants and protects him from abuses, the negro is spell-bound, and cannot runaway. *"He shall serve Japheth*; he shall be his servant of servants;"—on the conditions above mentioned— conditions that are clearly implied, though not directly expressed. According to my experience, the "genu flexit"—the awe and reverence, must be exacted from them, or they will despise their masters, become rude and ungovernable and run away. On Mason and Dixon's line, two classes of persons were apt to lose their negroes; those who made themselves too familiar with them, treating them as equals, and making little or no distinction in regard to color; and, on the other hand, those who treated them cruelly, denied them the common necessaries of life, neglected to protect them against the abuses of others, or frightened them by a blustering manner of approach, when about to punish them for misdemeanors. Before negroes run away, unless they are frightened or panic-struck, they become sulky and dissatisfied. The cause of this sulkiness and dissatisfaction should be inquired into and removed, or they are apt to run away or fall into the negro consumption. When sulky and dissatisfied without cause, the experience of those on the line and elsewhere was decidedly in favor of whipping them out of it, as a preventive measure against absconding or other bad conduct. It was called whipping the devil out of them.

If treated kindly, well fed and clothed, with fuel enough to keep a small fire burning all night, separated into families, each family having its own house—not permitted to run about at night, or to visit their neighbors, or to receive visits, or to use intoxicating liquors, and not overworked or exposed too much to the weather, they are very easily governed—more so than any other people in the world. When all this is done, if any one or more of them, at any time, are inclined to raise their heads to a level with their master or overseer, humanity and their own good require that they should be punished until they fall into that submissive state which it was intended for them to occupy in all after time, when their progenitor received the name of Canaan, or "submissive knee-bender." They have only to be kept in that state, and treated like children, with care, kindness, attention and humanity, to prevent and cure them from running away.

Dysæsthesta Æthiopis, or Hebetude of Mind and Obtuse Sensibility of Body—A Disease Peculiar to Negroes—Called by Overseers, "Rascality"

Dysæsthesia Æthiopis is a disease peculiar to negroes, affecting both mind and body, in a manner as well expressed by dysæsthesia, the name I have given it, as could be by a single term. There is both mind and sensibility, but both seem to be difficult to reach by impressions from without. There is partial insensibility of the skin, and so great a hebetude of the intellectual faculties as to be like a person half asleep, that is with difficulty aroused and kept awake. It differs from every other species of mental disease, as it is accompanied with physical signs or lesions of the body, discoverable to the medical observer, which are always present and sufficient to account for the symptoms. It is much more prevalent among free negroes living in clusters by themselves, than among slaves on our plantations, and attacks only such slaves as live like free negroes in regard to diet, drinks, exercise, etc. It is not my purpose to treat of the complaint as it prevails among free negroes, nearly all of whom are more or less afflicted with it, that have not got some white person to direct and to take case of them. To narrate its symptoms and effects among them would be to write a history of the ruins and dilapidation of Hayti and every spot of earth they have ever had uncontrolled possession over for any length of time. I propose only to describe its symptoms among slaves.

From the careless movements of the individuals affected with the complaint, they are apt to do much mischief, which appears as if intentional, but is mostly owing to the stupidness of mind and insensibility of the nerves induced by the disease. Thus, they break, waste and destroy everything they handle,—abuse horses and cattle,—tear, burn or rend their own clothing, and paying no attention to the rights of property, they steal other's to replace what they have destroyed. They wander about at night, and keep in a half-nodding sleep during the day. They slight their work,—cut up corn, cane, cotton or tobacco when hoeing it, as if for pure mischief. They raise disturbances with their overseers and fellow servants without cause or motive, and seem to be insensible to pain when subjected to punishment. The face of the existence of such a complaint, making man like an automaton or senseless machine, having the above or similar symptoms, can be clearly established by the most direct and positive testimony. That it should have escaped the attention of the medical profession, can only be accounted for because its attention has not been sufficiently directed to the maladies of the negro race. Otherwise, a complaint of so common occurrence on badly-governed plantations, and so universal among free negroes, or those who are not governed at all,—a disease radicated in physical lesions and having its peculiar and well-marked symptoms, and its curative indications, would not have escaped the notice of the profession. The northern physicians and people have noticed the symptoms, but not the disease from which they spring. They ignorantly attribute the symptoms to the debasing influence of slavery on the mind, without considering that those who have never been in slavery, or their fathers before them, are the most afflicted, and the latest from the slave-holding South the least. The disease is the natural offspring of negro liberty—the liberty to be idle, to wallow in filth, and to indulge in improper food and drinks.

In treating of the anatomy and physiology of the negro, I showed that his respiratory system was under the same physiological laws are that of an infant child of the white race; that a warm atmosphere, loaded with carbonic acid and aqueous vapor, was the most congenial to his lungs during sleep, as it is to the infant; that, to insure the respiration of such an atmosphere, he invariably, as if moved by instinct, shrouds his head and face in a blanket or some other covering, when disposing himself to sleep; that if sleeping by the fire in cold weather, he turns his head to it, instead of his feet, evidently to inhale warm air; that when not in active exercise, he always hovers over a fire in comparatively warm weather, as if he took a positive pleasure in inhaling hot air and smoke when his body is quiescent. The natural effect of this practice, it was shown, caused imperfect atmospherization or vitalization of the blood in the lungs, as occurs in infancy, and a hebetude or torpor of intellect—from blood not sufficiently vitalized being distributed to the brain; also, a slothfulness, torpor and disinclination to exercise, from the same cause—the want of blood sufficiently areated or vitalized in the circulating system. When left to himself, the negro indulges in his natural disposition to idleness and sloth, and does not take exercise enough to expand his lungs and to vitalize his blood, but dozes out a miserable existence in the midst of filth and uncleanliness, being too indolent and having too little energy of mind to provide for himself proper food and comfortable lodging and clothing. The consequence is, that the blood becomes so highly carbonized and deprived of oxygen, that it not only becomes unfit to stimulate the brain to energy, but unfit to stimulate the nerves of sensation distributed to the body. A torpor and insensibility pervades the system; the sentient nerves distributed to the skin lose their feeling to go great a degree, that be often burns his skin by the fire he hovers over, without knowing it, and frequently has large holes in his clothes, and the shoes on his feet burnt to a crisp, without having been conscious of when it was done. This is the disease called dysæsthesia—a Greek term expressing the dull or obtuse sensation that always attends the complaint. When aroused from his sloth by the stimulus of hunger, he takes anything he can lay his hands on, and tramples on the rights, as well as on the property of others, with perfect indifference as to consequences. When driven to labor by the compulsive power of the white man, he performs the task assigned him in a headlong, careless manner, treading down with his feet, or cutting with his hoe the plants he is put to cultivate—breaking the tools he works with, and spoiling everything he touches that can be injured by careless handling. Hence the overseers call it 'rascality,' supposing that the mischief is intentionally done. But there is no premeditated mischief in the case,—the mind is too torpid to meditate mischief, or even to be aroused by the angry passions to deeds of daring. Dysæsthesia, or hebetude of sensation of both mind and body, prevails to so great an extent, that when the unfortunate individual is subjected to punishment, he neither feels pain of any consequence, or shows any unusual resentment, more than by a stupid sulkiness. In some cases, anæsthesiæ would be a more suitable name for it, as there appears to be an almost total loss of feeling. The term 'rascality,' given to this disease by overseers, is founded on an erroneous hypothesis and leads to an incorrect empirical treatment, which seldom or ever cures it.

The complaint is easily curable, if treated on sound physiological principles. The skin is dry, thick and harsh to the touch, and the liver inactive. The liver, skin and kidneys should be stimulated to activity, and be made assist in decarbonising the blood. The best means to stimulate the skin is, first, to have the patient well washed with warm water and soap; then, to anoint it all over with oil, and to slap the oil in with a broad leather strap; then to put the patient to some hard kind of work in the open air and sunshine, that will compel him to expand his lungs, as chopping wood, splitting rails or sawing with the cross-cut or whip saw. Any kind of labor will do that will cause full and free respiration in its performance, as lifting or carrying heavy weights, or brisk walking; the object being to expand the lungs by full and deep inspirations and expirations, thereby to vitalize the impure circulating blood by introducing oxygen and expelling carbon. This treatment should not be continued too long at a time, because where the circulating fluids are so impure as in this complaint, patients cannot stand protracted exercise without resting frequently and drinking freely of cold water or some cooling beverage, as lemonade, or alternated with pepper tea sweetened with molasses. In bad cases, the blood has always the appearance of blood in scurvy, and commonly there is a scorbutic affection to be seen on the gums. After resting until the palpitation of the heart caused by the exercise is allayed, the patient should eat some good wholesome food, well seasoned with spices and mixed with vegetables, as turnip or mustard salad, with vinegar. After a moderate meal, he should resume his work again, resting at intervals, and taking refreshments and supporting the perspiration by partaking freely of liquids. At night he should be lodged in a warm room with a small fire in it, and should have a clean bed, with sufficient blanket covering, and be washed clean before going to bed: in the morning, oiled, slapped and put to work as before. Such treatment will, in a short time, effect a cure in all cases which are not complicated with chronic visceral derangements. The effect of this or a like course of treatment is often like enchantment. No sooner does the blood feel the vivifying influences derived from its full and perfect atmospherization by exercise in the open air and in the sun, than the negro seems to be awakened to a new existence, and to look grateful and thankful to the white man whose compulsory power, by making him inhale vital air, has restored his sensation and dispelled the mist that clouded his intellect. His intelligence restored and his sensations awakened, he is no longer the *bipedum nequissimus*, or arrant rascal, he was supposed to be, but a good negro that can hoe or plow, and handle things with as much care as his other fellow-servants.

Contrary to the received opinion, a northern climate is the most favorable to the intellectual development of negroes, those of Missouri, Kentucky, and the colder parts of Virginia and Maryland, having much more mental energy, more bold and ungovernable than in the Southern lowlands; a dense atmosphere causing a better vitalization of their blood.

Although idleness is the most prolific cause of dysæsthesia, yet there are other ways that the blood gets deteriorated. I said before that negroes are like children, requiring government in everything. If not governed in their diet, they are apt to eat too much salt meat and not enough bread and vegetables, which practice generates a scorbutic state of the fluids and leads to the affection under consideration. This form of the complaint always shows itself in the gums, which become spongy and dark, and leave the teeth. Uncleanliness of skin and torpid liver also tend to produce it. A scurvy set of negroes means the same thing, in the South, as a disorderly, worthless set. That the blood, when rendered impure and carbonaceous from any cause, as from idleness, filthy habits, unwholesome food or alcoholic drinks, affects the mind, is not only known to physicians, but was known to the Bard of Avon when he penned the lines—"We are not ourselves when Nature, being oppressed, commands the mind to suffer with the body."

According to unalterable physiological laws, negroes, as a general rule, to which there are but few exceptions, can only have their intellectual faculties awakened in a sufficient degree to receive moral culture, and to profit by religious or other instruction, when under the compulsatory authority of the white man; becuase, as a general rule, to which there are but few exceptions, they will not take sufficient exercise, when removed from the white man's authority, to vitalize and decarbonize their blood by the process of full and free respiration, that active exercise of some kind alone can effect. A northern climate remedies, in a considerable degree, their naturally indolent disposition; but the dense atmosphere of Boston or Canada can scarcely produce sufficient hematosis and vigor of mind to induce them to labor. From their natural indolence, unless under the stimulus of compulsion, they doze away their lives with the capacity of their lungs for atmospheric air only half expanded, from the want of exercise to superinduce full and deep respiration. The inevitable effect is, to prevent a sufficient atmospherization or vitalization of the blood, so essential to the expansion and the freedom of action of the intellectual faculties. The black blood distributed to the brain chains the mind to ignorance, superstition and barbarism, and bolts the door against civilization, moral culture and religious truth. The compulsory power of the white man, by making the slothful negro take active exercise, puts into active play the lungs, through whose agency the vitalized blood is sent to the brain, to give liberty to the mind, and to open the door to intellectual improvement. The very exercise, so beneficial to the negro, is expended in cultivating those burning fields in cotton, sugar, rice and tobacco, which, but for his labor, would, from the heat of the climate, go uncultivated, and their products lost to the world. Both parties are benefitted—the negro as well as his master—even more. But there is a third party benefitted—the world at large. The three millions of bales of cotton, made by negro labor, afford a cheap clothing for the civilized world. The laboring classes of all mankind, having less to pay for clothing, have more money to spend in educating their children, and in intellectual, moral and religious progress.

The wisdom, mercy and justice of the decree, that Canaan shall serve Japheth, is proved by the disease we have been considering, because it proves that his physical organization, and the laws of his nature, are in perfect unison with slavery, and in entire discordance with liberty—a discordance so great as to produce the loathsome disease that we have been considering, as one of its inevitable effects,—a disease that locks up the understanding, blunts the sensations and chains the mind to superstition, ignorance and barbarism. Slaves are not subject to this disease, unless they are permitted to live like free negroes, in idleness and filth—to eat improper food, or to indulge in spirituous liquors. It is not their masters' interest that they should do so; as they would not only by unprofitable, but as great a nuisance to the South, as the free negroes were found to be in London, whom the British government, more than half a century ago, colonized in Sierra Leone to get them out of the way. The mad fanaticism that British writers, lecturers and emissaries, and the East India Company, planted in our Northern States, after it was found by well-tried experiments, that free negroes in England, in Canada, in Sierra Leone and elsewhere, were a perfect nuisance, and would not work as free laborers, but would retrograde to barbarism, was not planted there in opposition to British policy. Whatever was the motive of Great Britain in sowing the whirlwind in our Northern States, it is now threatening the disruption of a mighty empire of the happiest, most progressive and Christian people, that ever inhabited the earth—and the only empire on the wide earth that England dreads as a rival, either in arts or in arms.

Our Declaration of Independence, which was drawn up at a time when negroes were scarcely considered as human begins, *"That all men are by nature free and equal,"* and only intended to apply to white men, is often quoted in support of the false dogma that all mankind possess the same mental, physiological and anatomical organization, and that the liberty, free institutions, and whatever else would be a blessing to one portion, would, under the same external circumstances, be to all, without regard to any original or internal differences, inherent in the organization. Although England preaches this doctrine, she practises in opposition to it every where. Instance, her treatment of the Gypsies in England, the Hindoos in India, the Hottentots at her Cape Colony, and the aboriginal inhabitants of New Holland. The dysæsthesia æthiopis adds another to the many ten thousand evidencies of the fallacy of the dogma that abolitionism is built on; for here, in a country where two races of men dwell together, both born on the same soil, breathing the same air, and surrounded by the same external agents—liberty, which is elevating the one race of people above all other nations, sinks the other into beastly sloth and torpidity; and the slavery, which the one would prefer death rather than endure, improves the other in body, mind and morals; thus proving the dogma false, and establishing the truth that there is a radical, internal, or physical difference between the two races, so great in kind, as to make what is wholesome and beneficial for the white man, as liberty, republican or free institutions, etc., not only unsuitable to the negro race, but actually poisonous to its happiness.

Notes

Samuel A. Cartwright, M.D., Chairman of the Committee appointed by the Medical Association of Louisiana to report on the above subject.

Read at the Annual Meeting of the Association, March 12th, 1851.

3.2   "The Plague Spot" (1878)

Mary P. Sawtelle

San Francisco located as it is under a generous sky, on a peninsula with a perfect natural drainage, and ventilated daily by sea breezes, ought to be and would be the healthiest city in the world, were its hygienic management in keeping with the progress of sanitary science. There are few men or women of ordinary intelligence in this enlightened city, who do not understand just what measures are necessary to prevent dyptheria, scarlet fever, small pox, typhoid fever and other contagions. The rules are very simple; personal cleanliness, cleanliness about private dwellings and yards, prompt removal and burning of garbage, clean streets, a correct sewer system, the removal of tanneries and other objectional manufactories from the limits of the city, ventilation of private and public buildings, and the enforcement of a rigid quarantine for all vessels from ports where there is the slightest suspicion of a prevailing epidemic.

The truth is that this city to-day is wholly at the mercy of any scourge that chooses to invade it. To say nothing of the prevalent horror of a bath tub among the masses, the little private pest holes about the dwellings, the outrageous condition of the public streets, the miserable system of sewerage, the horrible stenches that eminate from the scores of tanneries, the foul air generated in crowded tenements, and Christian white people, we are fostering in our midst an entire section, peopled with a race of semi savages, who have no idea of warding of disease by preventive measures. Chinatown is a standing invitation to cholera, small pox and other pests. There are comparatively few people in this city who have any conception of the real sanitary condition of that quarter of the city. The whole district inhabited by that strange race is reeking with corruption, the stench that issues from their basement dens to spread over the city is laden with the germs of death to another and better race. The only thing that saves San Francisco from a scourge more disastrous in its consequences than the one through which New Orleans and Memphis have been passing, is the pure winds of old ocean that sweep away the foul effluvia as fast as it arises.

But you ask what is the Board of Health doing all this time? Are there no steps being taken to purify that death engendering hole? Are our own people not being instructed what to do to preserve their health? Are not the quarantine laws strictly enforced? We fear not, we are not disposed to find fault with the health offlcers. They are all educated, honorable gentlemen and no doubt mean to do the best they can, and yet there is something wrong.

If we are to believe their own reports the Board is not quite satisfied with itself and surely the people of our fair city ought not to be, more especially the feminine

Mary P. Sawtelle, "The Plague Spot," *Medico-Literary Journal* 1, no. 4 (December 1878): 10–13.

element who can not tolerate filth any more than God can look upon sin with the least degree of allowance. Woman's nature abhors dirt so much so that if untoward circumstances surround her with "poverty, dirt and unwomanly rags" her shocked and wounded spirit will quit its tenement of clay and take up abiding place in purer climes. The masculine nature is wholy different in this respect.

Man can walk through reeking rottenness quite unconcerned, and this is proper enough, for if nature had endowed him with sensitive nerves that recoiled with horror at the unsightly things about him he would be wholly unfitted to do the rough work assigned him of cleaning away debris, making the earth a paradise where woman's purified spirit is willing to dwell. Her delicately attuned nature must hint, suggest, inspire his strong rough honest soul. Every man knows that all there is in him of the pure, the beautiful, or divine he caught through the inspiration of woman.

And we all know there is not an intelligent woman in San Francisco but can give directions, and they will be unhesitatingly obeyed, to have ever niche of ground over which her jurisdiction extends cleaned, purified and fit to be inhabited. Then why not extend her authority? give her a chance on the Board of Health? It is a reform much needed. Pacific Coast people are brave enough to establish customs of their own. Women may much better do this than sit broken hearted in their desolate homes, or plant flowers on little mounds in Lone Mountain and water them with their tears.

3.3 "Consumption Among the Indians" (1887)

Washington Matthews

The subject of consumption among our North American aborigines presents some interesting problems to the climatologist.

We have permanently established on our soil three of the most diverse varieties of the human species. Two of these have been introduced within a recent historic period; the third has dwelt in the land during a lapse of time which may be estimated only by the geologist; yet we find to-day among the autochthons a much higher death-rate than among the exotic races. From the census of 1880 we learn that the death-rate—i.e., the number of deaths during the year to one thousand of population—is for the three races as follows: Europeans, 17·74; Africans, 17·28; aboriginal Americans, 23·6.

Now the question arises, To what particular cause or causes is this high rate of mortality among the Indians especially due? On this point the Tenth Census seems to leave us not a moment in doubt. In Vol. XI—that on "Vital Statistics," by Dr. J. S. Billings," U.S.A.—we find a table (14) and a diagram (p. xxxvii) showing for whites, colored, and Indians, the proportions of deaths from specified diseases in one thousand deaths from known causes. The diagram is based on the table; but I will allude more to the former, since it gives at a glance the solution to our question. In this diagram there are twenty causes of death specified, and under each the three races are indicated by spaces differently shaded.

Under the heading of "Other Diseases of the Respiratory System" the mortality of the red and black races is about the same. Under eleven headings the black exceeds notably the red in mortality. Only under eight heads does the red notably exceed the black in its death-rate. Six of these are: accidents, diarrhœal diseases, measles, affections of pregnancy, scrofula and tabes, and venereal diseases. Of the latter Dr. Billings says (op. cit., p. xxxvi): "The high proportion of deaths among the Indians which is reported as due to venereal diseases is noteworthy, but probably a part of this is due to a greater readiness to name the true cause among these people than exists among the whites."

From my own experience of the ease with which Indian women travail, and the universal corrobative testimony of explorers and ethnographers, I marvel at the figures given under "affections of pregnancy," which, though not forming an important factor in the Indian death-rate, seem more fatal to the Indian than to the negro. In diarrhœal diseases the Indian rate is not greatly in excess of that of the other classes. Measles, although most fatal in the Indians, gives a mortality of only 61·78 in

Washington Matthews, "Consumption Among the Indians," *New York Medical Journal* 46 (January 1, 1887): 1–3.

a thousand. Notwithstanding the perils of a hunter's life, and of life under any circumstances on the frontier, we find that in deaths from injuries, although the rate for Indians is more than that for whites, it is less than for the colored race. But it is under the head of consumption that the Indian column is seen to rise conspicuously from 186 in the colored race to 286 in the Indian. A glance at the diagram shows that this is their specially fatal disease. Scrofula and tabes, being so closely allied to consumption, the numbers under this heading do little more than add to the testimony regarding the prevalence of the latter malady.

Comparing the Indian and white races, we find that from ten of the twenty causes the mortality of the latter is notably greater, under two headings it is nearly the same, and again we see under the title "Consumption" the Indian column rising far above the white, which is about 166 in a thousand—i.e., 20 less than the colored.

The probable inaccuracy of these Indian statistics is fully recognized in the Report, and it may be fairly urged in many cases, but with less justice, I imagine, with regard to consumption than with regard to many other causes of death. In its earlier stages consumption is a malady which often only the most skilled diagnostician can detect; yet in its later stages it is easily recognized. Above all, a death from consumption—using the term in the broad sense, in which it is necessarily employed here and in the vital statistics—is rarely assigned to another cause even by the layman.

We will next endeavor to determine if this disease always existed among the Indians to the same extent that it does now, or if it has increased of late years under the influence of the many complex causes which, not clearly analyzing, we are accustomed to epitomize in the expression "contact with civilization."

My own professional experience among our American aborigines includes a period of twenty-one years, and was gained among the Indians of a dozen different States and territories. Wherever I have sojourned I have always made it a point to give my professional services to Indians ungrudgingly and gratuitously, and for this reason I have had as good opportunities for observing their ailments as usually fall to the lot of the civilized physician. In no place where I have practiced among them have I failed to observe or learn of cases of consumption except in Owen's Valley, California, a locality which is fvaored with perhaps the most salubrious climate within our borders. It may have existed there, but it did not come to my knowledge during a residence of nearly one year in an Indian population of about eight hundred. Yet even here symptoms of scrofula were not entirely wanting.

My first experience with Indians as a physician was among some of the wildest tribes then existing on our continent, among those least influenced by civilization, prosperous, well nourished, dwelling in the heart of the buffalorange, and in what has proved to be—since the days of white occupation and the advent of the census-taker—a very healthful climate, the climate of the upper Missouri and Yellowstone Valleys. With certain preconceived notions of the healthfulness of the free out-door life and simple diet of the savage, and a conviction of the salubrity of the dry and elevated plains on which I found him, I was astonished to find that such a disease as consumption existed at all, and still more astonished to find it by no means infrequent.

As the years went by, and it fell to my lot to revisit, at long intervals of time, tribes which I had known in earlier days, I became impressed with the idea that this disease was on the increase among them. I well knew how easily I might be deceived in this matter. It was not in my power to collect complete data. I could only draw conclusions from the cases falling within my personal experience, and this experience was subject to limitations which had nothing to do with the prevalence of the disease. In former days the Indians had great confidence in their own shamans and little in white physicians; hence they consulted us less then than at present. In the old times they were wealthy and could afford to pay their extortionate medicine-men for their mummeries; in latter years their poverty compelled them to seek treatment which they could obtain for nothing. Furthermore, when they subsisted largely by hunting, they were much f their time abroad on the prairies and less under our observation.

Notwithstanding the possibility of my being led astray, it seemed evident to me that consumption increased among Indians under civilizing influences, and that its increase was not in a direct ratio to that of other diseases, but in a constantly augmenting ratio—again, that it varied greatly in different tribes.

I should have hesitated, however, to occupy your time with a recital of my convictions or impressions, based on personal experience, were it not that of late years some data have been collected which tend to strengthen them.

As the census reports for 1880 give the vital statistics for one year only, they can not afford any direct evidence as to the increase or decrease of any disease among Indians from year to year; but they give us some data from which we may draw reasonable inferences. They present us with two sets of tables for the Indians—one for those living on reservations, the other for those not on reservations, whom I will call Indians at large. Over two thirds of the latter class live in the States, less than one third in the territories. The Indians at large who reside in the States may broadly be said to represent those who have been brought most fully under the influence of civilization; those dwelling in the territories are for the most part residents of the most healthful sections within our borders (for instance, several thousand of the Pueblos of Arizona and New Mexico are included), and they must serve to reduce considerably the general death-rate and the consumption-rate of the class to which they belong. (By "consumption-rate" I mean the number of deaths from consumption in a thousand deaths from all known causes.) The reservation Indians, on the contrary, with some minor exceptions, are those who have been most recently subdued and brought under civilizing influences.

Let us compare the consumption-rates of these two classes. The rate for the reservation Indians is 184; that for the Indians at large is 373. In other words, the consumption-rate for the less civilized Indians is but 49 per cent. of that of the more civilized. But it may be urged that these figures are not so conclusive as they might, at the first glance, appear to be in determining the mere influence of civilization, since we have not taken into consideration the general consumption-rate of the different localities where the Indians in question are found, and it would be impossible to do so with any accuracy from the published data. I have, however, had access to some of the original reports, in manuscript, from which the statistics of the reservation Indians are drawn, and with these to aid me I have been able to compare the consumption-rates of different local groups of Indians with one another, and with the surrounding general consumption-rate. As far as I have instituted such comparisons, they seem to increase rather than diminish the force of the civilization factor.

The following is the consumption-rate among reservation Indians in thirteen different States and territories: Nevada, 45; California, 70; Arizona, 83; Colorado, 107; Nebraska, 150; Montana, 176; Dakota, 200; Oregon, 240; Idaho, 250; Washington, 302; Michigan, 333; Wisconsin, 361; New York, 625.

It is seen in the foregoing table that in States east of the Mississippi—the oldest States—where the Indians have been longest under civilizing influences, the consumption rate is the highest.

Now, the general consumption-rate of Dakota is 94, that of the rural districts of New York 152—much less than twice as great; while the Indian consumption-rate of New York is three times that of Dakota. Of the younger States and territories Washington has the highest Indian consumption-rate, yet it is only half the rate of New York, while the general consumption-rate of Washington (136) approximates that of New York closely. Of the States east of the Mississippi, Michigan has the lowest Indian consumption-rate, yet its rate is higher than that of Washington, while its general consumption-rate (137) is about the same. Again, take Wisconsin, in which the Indian consumption-rate is higher than in Washington, and the general consumption-rate (109) is lower.

In the office of the Indian Bureau in Washington, D.C., I have examined some of the reports of the agency physicians from the beginning of the fiscal year ending June 30, 1875, to the end of the fiscal year ending June 30, 1880. It becomes apparent in examining these reports that they are often very imperfect. In some it is evident that no deaths are recorded except those happening to occur in the practice of the physician—a practice often exceedingly limited. Again, there are often long hiatuses of many months in the series of reports, occasioned by the removal of a physician from office and tardiness in furnishing a successor. It is to be regretted, too, that it has not been always the policy of the Interior Department to furnish the Indians with regular medical graduates to treat their diseases and report on their sanitary condition. Still we may conclude that the reports are of some value for purposes of comparison. It is probable that they do not record proportionally more deaths from consumption than from other causes—in short, it is to be supposed that the consumption-rates are comparable.

Proceeding on this supposition, I have computed this rate for two sub-tribes of the great Dakota nation—subtribes of the same blood, not expatriated, living in climates not materially different from those which they have enjoyed for a century, and differing from one another only in degree of civilization. These are the people of Santee Agency, Nebraska, and those of Pine Ridge, Dakota. The former are much the more civilized. Many of them have taken lands in severalty, and are citizens of the United States. Nearly all the adults read and write, wear clothing like ours, and are professing

Christians. In the fiscal year of 1875 the consumption-rate of Santee was 631, that of Pine Ridge but 96; in the fiscal year of 1880 the rate of the former was 294, while in Pine Ridge no deaths were reported from consumption (but only 6 deaths from all causes in a population of over 7,000 are reported). In comparing, however, the statistics of Santee with those of Rosebud Agency, where there is another community of wild Dakotas, we find the showing for the civilized Indians not so bad. In the fiscal year of 1875 the wilder Indians had the lower consumption-rate (476 to 631), but in 1880 they had the higher rate (388 to 294).

As exhibiting progressive change in the consumption-rate in any one locality, the period of six years referred to above is to short. As far as I have worked out the consumption-rate for more than two years, I have found such great fluctuations that I do not hope for good data for generalization in so brief a period. I have, however, selected the reports from two Dakota agencies, Fort Berthold and Cheyenne River, hoping they might afford us some basis for an opinion. I have chosen these agencies for the simple reason that I have knowledge of the agency physicians, and feel confidence in their reports. The rate of Fort Berthold, computed from the records of the Indian Office, is as follows: For the fiscal years ending June 30, 1875, 41; 1876, 538; 1877, 500; 1878, 250; 1879, 133; 1880, 187.

Here we see that the rate of 1880 is considerably greater than that of 1875, but that during three of the intermediate years the rate is higher than in the last year.

For the rate of Cheyenne River I am indebted to the courtesy of my old and valued friend, Dr. C. E. McChesney, formerly agency physician and now Indian agent at Cheyenne River. The rate is as follows: For the calendar years 1878, 407; 1879, 550; 1880, 425; 1881, 417; 1882, not given; 1883, 561; 1884, 639; 1885, 649.

Here we find that, excepting a slight fall in 1880 and 1881, the increase in the consumption-rate is constant and marked.

In all the examples I have given I have endeavored to select tribes whose climatic environment has not been materially changed since the advent of the white man. The tribes of the Indian Territory, who are largely immigrants in that section, and other removed tribes, have been excluded.

If the evidence adduced is admitted to have value, it goes to show that consumption increases among Indians under the influence of civilization—i.e., under a compulsory endeavor to accustom themselves to the food and the habits of an alien and more advanced race—and that climate is no calculable factor of this increase.

It might be supposed that after many years of contact with this civilization, after several generations of "survival of the fittest" to cope with the new condition of things, and after no small dilution of the Indian blood by intermarriage with the exotic races, a state of tolerance to this disease would be produced; but the consumption-rate of the Indians of New York seems to indicate that a century is not time sufficient to establish such a tolerance.

Although I am not without some theories, based on personal observation, as to the special causes of this excessive tendency of the Indian race to consumption, I have not been able, in the brief hours I have devoted to this paper, to explain these theories fully, and show how they may be tested by the light of external evidence. Perhaps it is not necessary that I should do so before a body who are here to investigate chiefly the influence of climate. The term civilization is too broad, too inexact. What particular element of this civilization is the baneful one? is the question which will naturally be asked. Why does this civilization affect the Indian differently from the negro, who has as lately been introduced to its chastening influence, and is surrounded by conditions even more unsanitary? Recent investigations have demonstrated that the old notion of the red race being a dying race is incorrect. Ethnologically, it is a disappearing race; biologically, it is a living and increasing race. But, even if it were a dying race, why should consumption be its special enemy? Is it because of the meager rations of some poorly supplied agency? If so, why is it so prevalent in well-supplied agencies, and why most prevalent, or at least showing the highest rate, in New York, where the Indians are well-to-do, self-sustaining husbandmen? It is a general supposition on the frontier that it is change in diet which is the most potent remote cause of consumption among Indians. I have heard it said that hard bread killed more than hard bullets. It was a favorite expression of the late General Harney, the famous Indian fighter, that the cheapest way to settle the Indian question would be to take them all into New York and board them at the Fifth Avenue Hotel. His plan was excellent in more ways than one. I once knew a case of a previously healthy Indian camp of about two thousand people where, in one winter, when the buffalo left their country and they subsisted on flour and bacon furnished by the Government, the majority were attacked with scurvy, and about seventy died of the disease. Fine flour and bacon have, no doubt, had their share in the destructive work. But the consumption-rate, I find, is high at agencies where the supply of fresh beef is liberal—to judge from the annual reports of the Indian Commissioner—and it is high among the Indians of New York and Michigan, who have as varied a diet as their white neighbors. "Idiosyncrasy of race" and a score more of theories, trivial and profound, might be advanced and shaken at the first question.

Notes

Washington Matthews, M.D., Surgeon, United States Army.

Read before the American Climatological Association at its third annual meeting.

3.4   "Does Pulmonary Consumption Tend to Exterminate the American Indian?" (1887)

Thomas Mays

In an article published in "The New York Medical Journal" for January 1, 1887, entitled "Pulmonary Consumption among the Indians," Dr. Washington Matthews adduces evidence which "goes to show that consumption increases among the Indians under the influences of civilization," and that "where the Indians have been longest under civilizing influences the consumption rate is the highest." That the first proposition is quite in harmony with the operation of the law of adjustment between living bodies and their environment no one will, we think, call into question; but that the second proposition rests on an equally firm foundation neither follows from the truth of the first, nor is substantiated by facts as gleaned from the medical statistics reported by the commissioners of Indian affairs.

Dr. Matthews roughly divides the Indian population into two classes: (a) those living on reservations, and (b) those not living on reservations, or those at large. The latter constitute the class which has been most fully brought under the influences of civilization, two thirds of whom reside in the States. The former, or the reservation Indians, chiefly reside in the Territories, and have been most recently subdued and brought under civilizing influences. Following this the gives the consumption rate of 1880 among the Indian population in thirteen different States and Territories thus: "Nevada, 45; California, 70; Arizona, 83; Colorado, 107; Nebraska, 150; Montana, 176; Dakota, 200; Oregon, 240; Idaho, 250; Washington, 302; Michigan, 333; Wisconsin, 361; New York, 625." He concludes as follows: "It is seen in the foregoing table that in the States east of the Mississippi—the oldest States, where the Indians have been longest under civilizing influences—the consumption rate is the highest."

These figures reveal a startling condition of things, and, if true, would clearly show that the extermination of the Indian by natural means is only a matter of a comparatively short time, and they at once raise the question, Why should the fate of the Indian in respect to pulmonary consumption be harder than that of the white man? for we have[1] elsewhere given what we consider satisfactory proof that this disease is on the decrease among the white population in this country, owing to an adjustment of internal to external relations.

Before proceeding any further it is important to consider the methods which Dr. Matthews employed in getting the above-mentioned results; and this will serve to explain the variations in the calculations which each of us obtains. His "consumption rate is the number of deaths from consumption in a thousand deaths from all known causes." This obviously may become a very unreliable standard for comparison,

Thomas Mays, "Does Pulmonary Consumption Tend to Exterminate the American Indian?" *New York Medical Journal* 45 (May 7, 1887): 508–510.

especially when our estimates are to cover the statistics of a number of years. If the number of deaths from all causes were unvarying, or very nearly so, from year to year, or would necessarily bear a proportionate relation to the number of deaths from consumption, the plan would answer admirably. This not being so, results vary in accordance with the prevalence or absence of other diseases than consumption; hence more reliable results can be obtained when the number of deaths from any given disease is compared with the whole population or class among which it prevails. This latter method is the one which we adopted here.

In looking up the data for this paper we encountered a great many obstacles. In the first place, we found that up to 1882 the reports of the Indian Commissioners gave no statistics in regard to consumption among the Indians, for prior to that time consumption and scrofula were classed together under the heading of tubercular diseases; and at no time do these reports furnish the number of deaths from consumption—only giving the number of those suffering from this disease in each agency. Imperfect as the work therefore is, and brief as the period is over which it extends, we think sufficient information may be gathered to show that consumption pursues the same general course among the Indians as it does among the white race— viz., first contact with the influences of civilization increases its death-rate, and prolonged contact diminishes it.

For the sake of clearness and comparison we add the above table, in which are given the population, the proportionate average number of deaths from all causes, and the manner in which the Indians were brought into civilization, in each of twenty Indian agencies. These agencies are divided into three groups: (1) those which existed prior to 1863, (2) those which were established from 1863 to 1880, and (3) those which have been established since 1880. This division is made for the purpose of showing the different degree of effects produced by civilization on the Indian race. This is very natural, for, if civilization has any deteriorating tendency in this respect, it must be granted that a few years are necessary for its development: hence the third group should manifest no or very little deterioration; the second group more; and the first group, if prolonged contact with civilization increases deterioration, should show most of all; while, on the other hand, if there is any tendency of adjustment between the constitution of the Indian and the causes which generate consumption, the first group should be comparatively free, or at least more so than the second group.

From this tabular arrangement it will be perceived that the Indian follows the same law of adjustment concerning consumption as that which is followed by his white neighbor. The Indians of the first group may be divided into two classes—those belonging to the Mission, Navajo, and Pueblo agencies, and those belonging to the rest of the agencies. Those of the first division are socially of a higher type than those of the second division. They bear a strong resemblance to the Mexican Indians, from whom they acquired many arts, and they are principally engaged in civilized pursuits. The Mission Indians are said to be the longest-lived people in the world—one per cent of them are reported to be centenarians. As a rule, they live now as they have lived during the last three centuries. The Navajos are like the Pueblo and Zuni Indians. They pursue

Table 3.4.1

Names of agencies	Population	Proportionate average no. of cases of consumption to population from 1882 to 1886	Proportionate no. of deaths f'm all causes to population in 1886	How and when they were brought under the influences of civilization
First group				
1. Mission, Cal.	2,958	1 to 1,494	1 to 73	By treaty with Hidalgo.
2. Mackinac, Mich.	4,000	1 to 896	1 to 93	By executive order in 1855.
3. White Earth, Minn.	5,885	1 to 840	1 to 49	By treaty in 1855.
4. Nevada, Nev.	3,757	1 to 535	1 to 79	By executive order in 1859.
5. Navajo, New Mex.	23,000	1 to 1,200	1 to 82	By treaty in 1863.
6. Pueblo, New Mex.	7,762	1 to 1,500	1 to 46	Received under old Spanish grant in 1848.
7. New York, N.Y.	4,960	1 to 681	1 to 34	By treaty in 1797.
8. Umatilla, Or.	902	1 to 301	1 to 54	By treaty in 1855.
9. Green Bay, Wis.	3,036	1 to 303	1 to 41	By treaty in 1848.
Second group				
10. Colorado River, Ariz.	1,012	1 to 253	1 to 23	By congressional act in 1865.
11. Round Valley, Cal.	602	1 to 120	1 to 33	By congressional acts in 1864 and 1873.
12. Cheyenne River, Dak.	3,288	1 to 125	1 to 20	By treaty in 1868.
13. Pine Ridge, Dak.	7,000(?)	1 to 100	1 to 30	By treaty in 1868.
14. Fort Hall, Idabo.	1,432	1 to 238	1 to 45	By treaty in 1868.
15. Osage, Ind. Ter.	1,582	1 to 258	1 to 20	By congressional act in 1872.
16. Colville, Wash. Ter.	3,568	1 to 298	1 to 46	By executive order in 1872.
17. Shoshone, Wyom.	1,800	1 to 267	1 to 50	By treaty in 1868.
Third group				
18. Pima and Maricopa, Ariz.	5,050	1 to 2,500	None.	By congressional act in 1883.
19. Papago, Ariz.	7,300	None.	None.	By executive order in 1882.
20. Sac and Fox, Iowa.	380	One case reported.	1 to 48	By purchase deeds, 1876, 1882.

agriculture, spin wool, and weave cotton, and are famous for the fine blankets which they manufacture. On account of the higher state of their civilization, the Indians of this division never underwent that marked transition which those of the second division encountered when confronted by the higher plane of civilization.

The Indians of the second division of the first group more definitely represent that type of the savage with which we are familiar at the present day, and are the descendants of those with whom our Indian wars were carried on in earlier times, and they are analogous in nature to the Indians of the second group. An examination of the table shows quite a uniformity in the death-rate of nearly all these agencies. Thus, among the New York Indians, which have been longest under the jurisdiction of the Government, the consumption rate is exceedingly low (1 to 681). So is the consumption rate of the Mackinac (Michigan), White Earth (Minnesota), and Nevada (Nevada) Indians, while that of the Umatilla and Green Bay Indians is higher, but still makes a remarkably favorable showing. The conclusion, then, which can be drawn from these statistics, unless they are entirely unreliable, is that the influence of civilization on the American Indian in the long run is not detrimental to his well-being, so far as pulmonary consumption is concerned.

The agencies of the second group represent those Indians who have been brought under civilizing influences in more recent times—from 1863 to 1880—and, in contradistinction to the first group, the reports show that their consumption rate is high.

Group third represents those Indians who have been subjugated most recently. They are still leading a very primitive life, in many respects they bear a strong resemblance to the second division of the first group of Indians, and their consumption rate, as shown by the table, is almost nil. The Pimas are agriculturists and vegetarians, and live in adobe houses. The Papagos are Catholics, industrious and friendly, and their form of government is much like that of the Mexicans and Pueblos. The Sac and Fox tribe of Iowa are said to be physically as fine a class of men and women as it is possible to find. They live in the rude huts of their ancestors, cook their food on the ground floor, and leave the smoke to escape through the roof, thus securing good ventilation.

While it is much to be regretted that the reports of the Indian Commissioners contain no medical statistics concerning the Cherokee, Choctaw, Chickasaw, Creek, and Seminole Indians, who probably represent the highest grade of civilization yet attained among this race, these statistics show that the Indian in his primitive condition is almost free from pulmonary consumption; that his first contact with civilization vastly increases his liability to the disease, and that a prolonged contact diminishes this liability. And we see, therefore, that the Indian follows the same law of adaptation as that which is followed by the white and colored races, and does not occupy that exceptional position in this respect which is ascribed to him by Dr. Matthews.

Probably one of the chief causes of disintegration among the Indians when first coming in contact with civilization consists in an entire reversal of their previous habits and customs. The life of physical open-air activity, which invigorates the Indian's respiratory organs as well as his whole body, is now exchanged for a reservation life, where his nomadic instincts are curbed and his wants are fully satisfied, and in consequence he sinks into a state of lethargy and idleness from which the soon merges into pulmonary disease. After having endured the first shock of the conflict, a reaction begins to show itself. He gradually becomes accustomed to his new relation, assumes an industrious and peaceful life, and so elevates himself out of his physical and moral degradation.

It is not true, as is often stated, that the Indian only acquires the vices and not the virtues of the white man. It is no doubt true that the acquires his vices first, and consequently sinks early into disease and crime; but his history shows, too, that after the is adjusted to his new condition he also becomes capable of leading a highly moral and virtuous life—a life which compares very favorably with that led by his white neighbor.

Another important factor which tends to bring the Indian into harmony with his surroundings is a blood mixture with the white race. Mixture with white blood, which is already adapted to a higher plane of civilization, will certainly improve the Indian, and serve to increase his resistance to disease; and there is sufficient proof to show that this process of conservation, or blood adjustment, is going on at a rapid rate among those Indians who have been longest in contact with the white race, like those of New York and Michigan, who are largely composed of mixed bloods. There can be no doubt that his influence contributes largely to the greater immunity of these Indians from pulmonary consumption. This observation is full of meaning when it is linked with the opinion of one who has had a wide experience among the Indians, that "the half-bloods resist disease and death from pulmonary troubles longer than full-bloods."[2]

Notes

Thomas J. Mays, M.D., Philadelphia, PA.

1. "Study of Pulmonary Consumption in the City of Philadelphia," "Trans. of the College of Physicians," Nov. 3, 1886.

2. Captain Pratt's report of the Indian school at Carlisle, Pa. See "Report of Indian Commissioner for 1886," p. 22.

3.5 "The Study of Consumption Among the Indians" (1887)

Washington Matthews

A Reply to Dr. Thomas J. Mays, of Philadelphia

In the "New York Medical Journal" for May 7, 1887, appears an article entitled "Does Pulmonary Consumption tend to exterminate the American Indian?" by Thomas J. Mays, M. D., of Philadelphia, which purports to be a refutation of a paper of mine entitled "Consumption among the Indians," published in this journal in the issue of January 1, 1887.

It would occupy to much valuable space to review all the items in Dr. Mays's paper which invite criticism. It will only be necessary to examine a few of the more important in order to show the nature of the work and allow the reader to estimate its value.

Dr. Mays objects to the system I pursue in endeavoring to estimate the prevalence of consumption among the Indians, and announces as a substitute for it a method of his own in the following words: "Hence more reliable results can be obtained when the number of deaths from any given disease is compared with the whole population or class among which it prevails; this latter method is the one adopted here." This method would undoubtedly be the best if it were possible to obtain data to sustain it. It is a method which in the beginning of the investigation occurred to the present writer, but which he soon felt constrained to abandon for want of material to work with. In order to utilize this method it is necessary that all deaths occurring on a reservation during a year or some other specified time, with their causes, should be correctly reported; otherwise the ratio between deaths and population can not be computed. In the majority of Indian reservations, however, this is practically impossible. In some of the largest reservations, where the Indians are scattered over wild tracts larger than many of our States, it is impossible that the agent or physician, no matter how zealous or active he may be, can become cognizant of one death in ten that occurs; and the census-takers during the census year were not able to do much better. The agency physicians are supposed to report only cases that come under their observation or treatment, and it is from their reports that the statistics which Dr. Mays has consulted are prepared in the office of the Commissioner of Indian Affairs. If, in some of the smaller reservations, the physicians can approximate more closely to the actual number of deaths in making their, reports, it only renders comparison more uncertain.

Washington Matthews, "The Study of Consumption Among the Indians," *New York Medical Journal* 45 (July 30, 1887): 127–128.

But I must do Dr. Mays the justice to say that he seems to be well aware of the practical worthlessness of his "method," for he tells us of the only documents he professes to have consulted—"The Reports of the Indian Commissioners"—that "at no time do these reports furnish the number of deaths from consumption"; nowhere else in his paper does he make any allusion to his "adopted" method, nowhere does he base any argument on it, and nowhere does he advance data to sustain it. A "table" which he gives might seem at a first glance to be the desired data, but a little examination shows that it is not. The only column headings in this table are the following: "Names of Agencies," "Population," "Proportionate Number of *Cases* of Consumption to Population from 1882 to 1886," "Proportionate Number of *Deaths* from *all Causes* to Population in 1886," and "How and When they were brought under the Influences of Civilization" (Italics mine). A column of "deaths from consumption," which would be an essential element in the table to render the method of Dr. Mays valid, is not to be found.

There is no part of the paper under consideration on which more labor seems to have been expended than on this "table," yet there is no part on which labor has been expended more vainly. Its value should depend chiefly on the chronological order of the column headed "How and When they were brought under the Influences of Civilization," for upon this order are based three "groups" from the disease- and death-rates of which we are asked to draw important inferences. Let us take for consideration the first item on the list—the Mission Indians of California ("Mission, Cal."). These, we are told, "were brought under the influences of civilization" "by treaty with Hidalgo." Although this statement is worded so as to make it appear that the treaty was with some individual of the name of Hidalgo, and although no date is given, it may be presumed that reference is made to the treaty between the United States and Mexico, which was concluded at the town of Guadalupe de Hidalgo, Republic of Mexico, in 1848, and which is often called by American historians the treaty of Guadalupe-Hidalgo. By this treaty the present State of California (with its Mission Indians) was formally ceded to our republic. But why has Dr. Mays seized upon this episode in history as showing when these Indians were brought under the influences of civilization? The Round Valley Indians of California, the Colorado River Indians of Arizona, the Nevada Indians of Nevada, and the Navajo Indians of New Mexico, all passed into our jurisdiction under this same treaty, yet Dr. Mays assigns to the four peoples just mentioned dates for their dawn of civilization quite different—so different in two cases as to throw the individuals concerned into this second group, while the Mission Indians come in the first group. The Pueblo Indians of New Mexico, too, fell to our lot by the same treaty, and, while the date in their case is given correctly (1848), the irrelevant and misleading remark, "Received under old Spanish grant," is added.

But this is not all that our author has to say about the Mission Indians. He further informs us concerning them: "As a rule, they live now as they have lived during the last three centuries." The facts in the case are that few unexpatriated peoples on the face of the earth have, within the last three centuries, suffered greater vicissitudes or experienced greater changes in their mode of living than these Mission Indians.

They are the people in whose cause Mrs. Jackson wrote her immortal work of "Ramona," and in which she has attempted to portray but a few of their more recent calamities. When, in 1769, the Franciscans began to establish missions in Alta California, the natives were in a low state of savagery, having no knowledge of agriculture, living on the spontaneous productions of the land and water, going naked, or dressed only in skins and rushes (*tule*). At the close of the eighteenth century the converted Indians of the country numbered thirteen thousand well-fed and industrious people, with habits of life totally changed, tending vast herds of cattle, practicing weaving, agriculture, and other civilized arts, and conforming to the rites of the Catholic Church. They continued to increase rapidly in numbers until the fall of the Spanish power in 1822, since which time, and particularly since they have come under our jurisdiction, they have been robbed of their lands, reduced to poverty, and vastly diminished in numbers.

We might very properly assign as an initial date to the civilization of the Mission Indians that of the first permanent establishment of these missions among them, for the missions wrought marked and rapid changes. But, if we do so with the Californians, we must do the same in the case of the Papago Indians of Arizona. Among this tribe similar missions were established long before the Californian missions came into existence—as early, undoubtedly, as 1720, and perhaps as early as 1694. Their fine old church of San Xavier del Bac, which took fourteen years to build, was finished in 1797. Dr. Mays properly describes them as "industrious and friendly," and as having a form of government "much like that of the Mexicans and Pueblos"; but such as they are now they have been for over a century and a half. Nevertheless, this long-civilized tribe comes next to the last in Dr. Mays's third group (as civilized since 1880), while the Mission Indians head his first group.

But time fails for the further pursuit of this criticism. The samples given fairly represent the character of the whole compilation, which the author assures us he has added "for the sake of clearness and comparison." How it can serve either purpose it is left to the reader to imagine, and also to determine for himself what value the deductions can have which are based on such material.

Note

Washington Matthews, M.D., U.S. Army.

4 EVOLUTION AND DEGENERATION

Introduction

Evelynn M. Hammonds and Rebecca M. Herzig

Well before Darwin published his landmark *Origin of Species* in 1859, theorists of race posited causal relationships between ecological conditions and human variation. (Recall Samuel Cartwright's assertion that "Negroes" possession of an "anatomical peculiarity" of the eye provided protection from solar rays.) At the heart of many these discussions, as with so many scientific studies of race, was the contentious matter of change: whether or not particular characteristics or capacities were amenable to alteration. The question of the adaptability of races to various "climatic pressures" was exacerbated not only by the harsh realities of the trans-Atlantic slave trade (which led profit-minded slave holders to study the relation between race and place) but also the staggeringly high mortality rates of European colonial troops: one study reported that scarcely 7,000 of the original 32,000 French soldiers invading Santo Domingo survived their first two months in the West Indies. With matters of evident economic and political import riding on the answers, investigators scrutinized relative abilities in "acclimation" to new environs. In the first selection below, a review essay from an edition of the *Medical and Surgical Reporter* appearing just two years after *Origin of Species*, notice how the writer discusses the individual and collective effects of migration across latitude. What forms of "deterioration" appear to concern the writer? And to what extent might racial improvement (or individual perfectibility) appear possible through deliberate activity?

While the relations between nature and human change occupied American writers since the earliest days of the union, these relations found new force with the dissemination of Darwin's landmark *The Descent of Man and Selection in Relation to Sex* (1871). This two-volume tome outlined views on subjects treated only vaguely in the 1859 *Origin of Species*—namely, man's advent and subsequent development. As Darwin delved more deeply into the topic of human variation, he necessarily entered heated debates about the nature of racial difference. Were modern races altogether separate species with distinct ancestors, or did they derive from common progenitors? How did such races develop and what evolutionary purpose, if any, did racial differences serve? Given the pressures of natural selection, what would be the fate of so-called "civilized" and "savage" races?

As the book's title suggests, Darwin's answers to such questions lay primarily in sexual selection. Darwin touched on the idea of sexual selection briefly in the earlier *Origin*, suggesting that characteristics which were otherwise inexplicable in terms of survival could be explained in terms of the advantage conferred in the struggle for mates: the result of sexual selection, unlike natural selection, is "not death to the unsuccessful competitor, but few or no offspring." A decade later, Darwin had given sexual selection an absolutely central role in the larger process of natural selection (departing from fellow evolutionist Alfred Russel Wallace in doing so). In the general conclusion to *Descent*, he attributed a vast array of aesthetic, physical, and moral qualities to sexual selection: "Courage, pugnacity, perseverance, strength, and size of body, weapons of all kinds, musical organs, both vocal and instrumental, bright colours and ornamental appendages, have all been indirectly gained by the one sex or the other through the exertion of choice, the influence of love and jealousy, and the appreciation of the beautiful in sound, colour, or form." As historian Evelleen Richards has shown, throughout his discussions of sexual selection Darwin stressed the primacy of male choice; female preference, such as it appears, is generally depicted as resulting from the efficacy of a male's successful display.

Despite the radical challenge that theories of natural and sexual selection posed to creationist views of man's unique status, Darwin hardly dislodged the typological and hierarchical understandings of race and sex held by most nineteenth-century anthropologists, natural historians, and physicians. Arguing that savages differed little from animals in their mental and moral capabilities, Darwin's assertion of the evolutionary continuity between man and "lower" animals reinvigorated more traditional racial schemas, in which white Europeans occupied the highest rung of physical, intellectual, and moral development, and black Africans the bottom. Darwin also continued the tradition of considering the future of racial types, speculating on the role that the pre-existing "grade of civilisation" played in deciding sexual choice and racial competition. As he wrote in an 1864 letter to Wallace, "I suspect that a sort of sexual selection has been the most powerful means of changing the races of man. I can show that the different races have a widely different standard of beauty. Among savages the most powerful men will have the pick of the women, and they will generally leave the most descendents."

Darwin's ideas about civilization and evolution arrived to an America in tumult. In the twelve years between the publication of the *Origin* and *The Descent of Man*, the United States had experienced its bloodiest and most divisive conflict to date, the Civil War (1861–5); seen the emancipation and enfranchisement of 4 million former slaves (1864–70); and begun the transformation from a set of mostly agrarian, locally governed communities to an increasingly urban, industrial and centralized society. Questions of natural difference remained central to these changes and their aftermath, and the meaning of differences between races remained a key problem for American intellectuals, politicians, and scientists. By the 1870s and 1880s, countless American writers embraced and extended Darwinian concepts of variation, struggle and survival in their discussions of races and their futures.

At the iconic center of such discussions stood the figure of the "Negro." Even before the end of the Civil War, American statisticians, anthropologists, and physicians heralded the Negro's gradual extinction as an "unerring certainty," an inevitability that would be further compounded by emancipation and social integration. According to these commentators, the reasons for the Negro's imminent disappearance were to be found primarily in "his" (most commentators held males as representative) physical structure. With black Emancipation, the future of the Negro became an even more heated question in scientific circles. American anthropologists and anatomists elaborated study after study of the facial features, hair texture, and limb length of the figural Negro, in order to distinguish his "simian" character from the white man, whose orthognathous jaw, 90° facial angle, and large calf muscles allegedly bespoke a more advanced stage of evolutionary development.

As further evidence of the Negro's "degenerate" character, some scientists also asserted a lower degree of sexual dimorphism in Negroes as compared to whites. The naturalist Edward Drinker Cope argued in an 1881 lecture that the female members of the "most civilized" races displayed the highest degree of "feminine" traits such as smooth skin and high-pitched voices. According to Cope and most other late nineteenth-century Anglo-American naturalists, physicians and anthropologists, the more pronounced the differences between the sexes, the more evolutionarily advanced the race. Such commentators often emphasized the significance of such sexual differences in order to argue against middle-class white women's access to higher education.

To post-bellum evolutionary scientists, the Negro did not simply display degenerate physiological characteristics; his "inferior" constitution predisposed him to disease and infant mortality. A growing reliance on numbers as the sign of "objective" knowledge contributed to these assessments of degeneration. Physicians began pointing to newly aggregated vital statistics as evidence for the rapid decline of the Negro population in the post-Reconstruction era. In seemingly unassailable numbers, these commentators saw proof of the Negro's inability to adapt once removed from his "natural" environments (the tropics and Southern plantations) and the superior stock of white Anglo-Saxons. As the German-born statistician Frederick Hoffman concluded in this 1892 article in the Boston magazine *The Arena*, "No one can foretell the probable future of the colored population of the United States with any degree of *absolute* accuracy, but the facts presented in this article indicate tendencies which warrant us to believe that the time will come, if it has not already come, when the negro, like the Indian, will be a vanishing race." Hoffman's essay was a precursor to his most infamous work, *Race Traits and Tendencies of the American Negro*, a massive compendium of comparative morbidity, mortality, and birth-rate data published by the American Economic Association in 1896. Both works remind us that questions about evolution and degeneration were never limited solely to white-black racial comparisons. Amerindians, Australians, Polynesians, and other so-called "savage races" were also generally regarded by American scientists and physicians as evolutionarily inferior groups, destined to vanish from the earth.

Despite their ubiquity, the perspectives on racial evolution and degeneration discussed above did not go unchallenged. Other natural and social scientists, physicians, and intellectuals, among them several prominent black Americans, disputed the conclusion reached by Hoffman and his colleagues that the Negro race was evolutionarily inferior and bound for extinction. In 1897, the black sociologist and mathematician Kelly Miller published a substantial review of Hoffman's work, in which he challenged the figures on black birth and mortality rates, highlighting Hoffman's errors of calculation and raising questions about numbers as neutral arbiters of fact. Miller also questioned Hoffman's claim that high black morbidity and mortality statistics resulted from innate racial deficiencies. Instead, Miller pointed to the harsh social conditions in which most blacks lived during the late nineteenth and early twentieth centuries. As late as the 1890s, the vast majority of black Americans lived in the South, where Jim Crow laws instituted after the failure of Reconstruction kept most blacks socially segregated, politically disenfranchised, and economically impoverished. According to Miller, lack of access to adequate nutrition, housing, employment, and medical care, rather than anatomical inferiority, contributed to the elevated morbidity and mortality rates Hoffman observed among blacks. Similar arguments were made by other black intellectuals, most notably W. E. B. Du Bois, one of the era's most influential social scientists. In his 1906 publication, *The Health and Physique of the Negro American*, Du Bois argued that tuberculosis, long viewed as *the* disease of racial degeneration, was in fact "not a racial disease but a social disease," springing from the poor conditions in which the Negro resided, not from a fixed racial characteristic.

Du Bois' claims were echoed by contemporary black physicians such as C. W. Birnie, who, in his 1910 article on "The Influence of Environment and Race on Diseases," identified social inequality, not racial degeneration, as the real factor producing higher rates of tuberculosis, syphilis and pneumonia among African Americans. Birnie's article appeared in the *Journal of the National Medical Association*, the mouthpiece of the only national Negro physicians' organization (the AMA, which deferred to the discriminatory policies of local medical societies, excluded most black physicians until the late 1960s). The *Journal of the National Medical Association* was one of the few journals in which black physicians could publish dissenting views on race, evolution, and disease prior to successes of the post-WWII civil rights movement.

Ideas of racial "improvement" acquired material form with the emergence of a widespread, popular eugenics movement over the first three decades of the twentieth century. Eugenics, from the Greek word for "well born," implied the improvement of heritable traits through selective breeding—either the prohibition of reproduction by the "unfit" or the encouragement of breeding of the "fit." While eugenics harkened back to the work of Francis Galton and other nineteenth-century thinkers, by the 1910s it had acquired the status of a full-fledged public obsession. High schools, colleges, and universities offered courses on eugenic family planning; county and state fairs held eugenic competitions; hundreds of "eugenic field workers" (mostly women) fanned out across the country conducted home interviews to collective pedigree information from individuals and families.

While the popularity of eugenics in the United States began falling off even before the atrocities of the Nazi regime came to light, at its height the eugenics movement appealed to myriad segments of American society—black as well as white, poor as well as affluent. Part of the movement's appeal lay in the resonance between the bold idea of selective human reproduction and received practices of animal husbandry. A post-Darwinian fascination with racial degeneration or improvement converged with readily apprehensible principles of breeding, as evident in the selection by Alexander Graham Bell. Bell, a Scottish born inventor, is now primarily remembered for his 1876 patent on the telephone (the actual invention of the device has since been credited to the Italian-American Antonio Meucci). But his interest in deafness also led him to a study of heredity. In this 1914 article from the monthly journal of the American Breeder's Association, Bell spells out his vision for increasing the mass of "normals" in society. In what ways does he appear to take up, expand, and/or reject Darwinian concepts? How are racial and sexual differences linked in his essay? What kinds of relations does he present between humans and animals, men and women, normals and desirables? Finally, as always, we might consider the shifting meanings of the term *race* itself: how does Bell use the word here? To what does it appear to refer?

Acknowledgment

Abigail Bass prepared an earlier draft of this introduction.

Bibliography

Anderson, E. D. "Black Responses to Darwinism, 1859–1915." In R. L. Numbers and J. Stenhouse, eds., *Disseminating Darwinism: The Role of Place, Race, Religion, and Gender*. Cambridge: Cambridge University Press, 1999, 247–266.

Athey, Stephanie. "Eugenic Feminisms in Late Nineteenth-Century America: Reading Race in Victoria Woodhull, Francis Willard, Anna Julia Cooper, and Ida B. Wells." *Genders* 31 (2000).

Bederman, Gail. *Manliness & Civilization: A Cultural History of Gender and Race in the United States, 1880–1917*. Chicago: University of Chicago Press, 1995.

Bix, Amy Sue. "Experiences and Voices of Eugenics Field-Workers: 'Women's Work' in Biology." *Social Studies of Science* 27, no. 4 (August 1997): 625–668.

Boudin, M. "Etudes de Pathologie Comparée." *Annales d'Hygiene et de Médecine Legale* (July 1849), reviewed in "Man's Power of Adaptation to Different Climates." In *Pennsylvania Journal of Prison Discipline and Philanthropy* 5, no. 1 (January 1850): 5–9.

Byrd, W. Michael, and Linda A. Clayton, *An American Health Dilemma: Race, Medicine, and Health Care in the United States, 1900–2000*, vol. 2. New York: Routledge, 2002.

Clarke, Edward H. *Sex in Education; Or, A Fair Chance for the Girls*. Boston: J. R. Osgood, 1873.

Cope, Edward D. "The Developmental Significance of Human Physiognomy." *The Origin of the Fittest: Essays on Evolution*. New York: D. Appleton & Co., 1887.

Darwin, Charles. *The Descent of Man and Selection in Relation to Sex*. Princeton: Princeton University Press, 1981 [1871].

Degler, Carl N. *In Search of Human Nature: The Decline and Revival of Darwinism in American Social Thought*. New York: Oxford University Press, 1991.

Fredrickson, George M. *The Black Image in the White Mind*. Hanover, N.H.: Wesleyan University Press, 1987.

Gaines, Kevin K. *Uplifting the Race: Black Leadership, Politics, and Culture in the Twentieth Century*. Chapel Hill: University of North Carolina Press, 1996.

Gallagher, Nancy L. *Breeding Better Vermonters: The Eugenics Project in the Green Mountain State*. Hanover, N.H.: University Press of New England, 1999.

Haller, John S., Jr. *Outcasts from Evolution: Scientific Attitudes of Racial Inferiority, 1859–1900*. Chicago: University of Illinois Press, 1971.

Haller, Mark. *Eugenics: Hereditarian Attitudes in American Thought*. New Brunswick, N.J.: Rutgers University Press, 1963.

Hasian, Marouf A. *The Rhetoric of Eugenics in Anglo-American Thought*. Athens: University of Georgia Press, 1996.

Hoffman, Frederick. *Race Traits and Tendencies of the American Negro*. New York: American Economic Association, 1896.

Holt, Thomas C. "W. E. B. DuBois's Archeology of Race: Re-Reading 'The Conservation of Races." In Michael B. Katz and Thomas J. Sugrue, eds., *W. E. B. DuBois, Race, and the City: The Philadelphia Negro and Its Legacy*. Philadelphia: University of Pennsylvania Press, 1998.

Hunt, James. *The Negro's Place in Nature*. New York: Van Evrie, Horton, and Co., 1864.

Kline, Wendy. *Building a Better Race: Gender, Sexuality, and Eugenics from the Turn of the Century to the Baby Boom*. Berkeley: University of California Press, 2002.

Larson, Edward J. *Sex, Race, and Science: Eugenics in the Deep South*. Baltimore: Johns Hopkins University Press, 1995.

Love, Spencie. *One Blood: The Death and Resurrection of Charles R. Drew*. Chapel Hill: University of North Carolina Press, 1996.

Miller, Kelly. "A Review of Hoffman's *Race Traits and Tendencies of the American Negro*." *The American Negro Academy, Occasional Papers, No. 1*. Washington, D.C., 1897, 3–36.

Mitchell, Michele. *Righteous Propagation: African Americans and the Politics of Racial Destiny after Reconstruction*. Chapel Hill: University of North Carolina Press, 2004.

Newman, Louise Michele, ed. *Men's Ideas/Women's Realities*. New York: Pergamon Press, 1985.

Porter, Theodore. *Trust in Numbers: The Pursuit of Objectivity in Science and Public Life*. Princeton: Princeton University Press, 1995.

Richards, Evelleen. "Darwin and the Descent of Woman." In D. Oldroyd and I. Langham, eds., *The Wider Domain of Evolutionary Thought*. Dordrecht: Reidel, 1983.

Rowold, Katharina, ed. *Gender and Science: Late Nineteenth-Century Debates on the Female Mind and Body*. Bristol: Thoemmes Press, 1996.

Russett, Cynthia. *Sexual Science: The Victorian Construction of Womanhood*. Cambridge: Harvard University Press, 1989.

Selden, Steven. *Inheriting Shame: The Story of Eugenics and Racism in America*. New York: Teachers' College Press, 1999.

Stepan, Nancy. "Biological Degeneration: Races and Proper Places. In J. E. Chamberlin and S. L. Gilman, eds. *Degeneration: The Dark Side of Progress*. New York: Columbia University Press, 1985, 97–120.

Stepan, Nancy. "Race and Gender: The Role of Analogy in Science." In Evelyn Fox Keller and Helen E. Longino, eds., *Feminism and Science*. New York: Oxford University Press, 1996, 121–136.

Stepan, Nancy, and Sander L. Gilman, "Appropriating the Idioms of Science: The Rejection of Scientific Racism." In Sandra Harding, ed., *The "Racial" Economy of Science*. Bloomington: Indiana University Press, 1993, 183–184.

Stocking, George. *Race, Culture, and Evolution: Essays in the History of Anthropology*. Chicago: University of Chicago Press, 1982, [1968].

4.1   "The Impossibility of Acclimatizing Races" (1861)

In a recent review of some modern narratives of arctic life and explorations, in the *British and Foreign Medico-Chirurgical Review*, the writer presents arguments against the possibility of actually and permanently acclimatizing any race which is foreign to the soil. Taking, for illustration, regions in which the opposite intensities of solar light and heat exist, as in the tropics and the arctic regions, and their inhabitants, as the Esquimaux and the Negro, it is admitted that a transfer of either from the opposite extreme of climate would soon show their inadaptation to the change, and the vicissitudes to which they would be subjected would soon prove fatal to both. Yet each seems, by nature, to be fitted to withstand the deleterious influences which surround him in his native region. The Negro luxuriates in the parching rays of an equatorial sun, can subsist in health on a light diet of tropical fruits, and is invulnerable to the pestilential miasms of the jungle. The Esquimaux, perhaps, even enjoys life in regions where darkness reigns during a quarter of the year, without any vegetable food and with but a snow-hut for his dwelling.

It is evident that races are born with constitutions peculiarly adapted to resist the surrounding vicissitudes of light and heat, and with ability to subsist on such material for nutriment as their respective localities supply. While thus individual races, which inhabit the extremes of latitude, cannot bear transplantation, the inhabitants of the temperate zone possess, temporarily at least, an adaptability to almost all latitudes, and may, provided the persons so situated adopt the habits of life of the autochthones, continue in an ordinary condition of health. But a thorough transplantation and acclimatization of a race from north to south, or from south to north, has never taken place; for, although individuals may, as has been said, exist under certain circumstances in health, yet physiological incompatibility will, ere a generation expires, be evident, and failure in propagation of the race will soon bring about its extinction.

The circumstances which prevent any race from becoming actually cosmopolite, are evidently those connected with latitude—with the extremes of insolation. That a perfect transplanting and acclimatization in the direction of longitudinal measurement cannot be accomplished, is not so evident, but the article to which we have alluded, insists on the same impossibility of adapting nations to other localities, east or west on the globe, to that in which they have habitually resided. The Anglo-Saxon, he asserts, can only exist in health, and the race be propagated with vigor, on its hereditary soil in Europe. He asks, "Is the Celt or Saxon thriving in North America?" "Is the Red man fitted for a large portion of the Eastern hemisphere?" "Is the standard of

"The Impossibility of Acclimatizing Races," *Medical and Surgical Reporter* 5, no. 23 (March 9, 1861): 623–625.

health as high among the natives of the Union as it is among their progenitors in the British Isles?" That the standard of health is as high among the *natives* of the Union, cannot be denied. If the general longevity in the United States is not so great as in Northern Europe, the failing may be accounted for by circumstances independent of the difference in longitude, and the average of human life is much lessened by immigration from our ancestral soil, flooding the country with a cachectic population. But the reviewer thinks that immigration, by continually replenishing the vigor by engrafting from the old stock, is the only means of saving the American people from dying out in a generation or two. He says:—"We believe it would fail and gradually die out, and that the time would come, however distant, when the Saxon would no longer be found in Australia, in Kentucky, in Tennessee."

Whether the Red man is "fitted for a large portion of the Eastern hemisphere," cannot be from practical observation answered, but that the White man "waxes strong amid the forests of the Far West," is well known to every one who has seen the hardy "backwoodsmen" who form the van-guard in the westward march of civilization. The best physically developed class of men we have ever seen, were not only natives of the very States which are instanced as fatal to the Saxon, but were descendants from an ancestry of pioneers whose axes broke a way for the sunlight to reach, for the first time, the soil of those forest states.

In perfection of physical development, America has produced instances which are presented to the world as unequalled. Some which have lately appeared, far exceed any precedent. The strongest men known in modern times, are Dr. Windship, of Boston, and Mr. Thompson, of Chicago—the former about twenty-eight years of age, and weighing less than one hundred and fifty pounds. We have seen him pile up ten kegs of nails, weighing eleven hundred pounds, tie a rope around them to keep them together, and then, seizing the rope with his hands alone, swing them in the air. Each of these gentlemen has lifted, with the aid of straps on his shoulders, about a ton, and a number of other persons have accomplished feats that approximate these efforts which throw into the shade those European Samsons, Thomas Topham and the Belgian Giant. In a late pugilistic contest, an American, whose wonderful physical powers were developed even on the very western verge of this continent, showed himself, on English soil, superior to the previous fistic champion of the world.

While admitting that differences in latitudes, with their climatic accompaniments, rendergreat changes of residence, from north to south, or from south to north, intolerable, yet there is no evidence that a migration, east or west, on the same parallel, will at all impair health or longevity, or produce a procreative degeneration of a race.

Such removal, with its attendant changes of habit and diet, certainly does produce some physiological changes; but these do not necessarily involve a sacrifice of health or longevity. It has been remarked that Americans whose ancestry was the primitive settlers of the country, and have descended unmixed with foreign stock, have acquired a resemblance to the real autochthones of the continent—the American Indian. Of such descent are numerous families in the New England and Western States, and the tall, bony, athletic figure, high-cheek bones, and long, straight hair, so frequently seen in those regions, seem rather in favor of the observation. Indeed, the typical "Yankee" is always depicted in this outline.

We may assert, from observation, that such physiological and physiognomical change is not accompanied with a deterioration in health or abbreviation of longevity. We do not think that the model "John Bull," with his dumpy figure, round, plump face, capacious abdomen, and short limbs, is, notwithstanding his advantage of residence in the locality which, according to the reviewer, nature intended for our race, any nearer than the preceding portrait to the Apollonian pattern.

4.2 "Vital Statistics of the Negro" (1892)

Frederick L. Hoffman

A prominent writer has defined science as a knowledge and classification of *facts*, and it is in the sense of this definition that I propose to present the so-called race problem.

Statistics relating to the colored population are, however, difficult, and in many instances impossible to be obtained; and in consequence this attempt to present the race problem from the standpoint of vital statistics will necessarily be wanting in completeness.

The census office reports for the past year a total population of the United States of 62,622,250, of which 7,500,000 are stated to be persons of color. At the beginning of the present century, the colored element in the United States numbered about 1,000,000, and has consequently increased at an average decennial rate of a little over 700,000. The actual increase since 1880 has been in round numbers 1,000,000, and for the past 30 years this increase has been equal to nearly 3,100,000.

Seven eighths of the total colored element are to be found in the Southern States, and of these only the six Gulf States contain colored majorities or large colored minorities. In 1800 the Southern States contained 54,258 colored persons to every 100,000 of white population, while the count just completed states the ratio for the present decade to be 41,475 colored to every 100,000 white. The table below [table 4.2.1] will clearly show the distribution of the two races in the ten States containing eight tenths of the total colored population. The District of Columbia is added merely for comparison. The ten States cover nearly 500,000 square miles, being about one sixth of the total area of the United States exclusive of Alaska.

The total increase of the white element exceeds in round numbers the colored increase by 700,000, and shows a numerical relation of 58 white persons to almost 42 colored population.

Thus the predicted phenomenal increase of colored population has actually proven to be a decrease from 46,000 in 1880 to 41,500 to every 100,000 white in 1890, much to the disappointment of that class of writers who for the past ten years have been frightening the Southern people with the prospect of an early negro supremacy. To show the misconception and utter absurdity of some of these predictions as to the future increase of negro population, I will quote the calculations of Mr. Darby, placing the same side by side with the census figures, and also the estimate of Mr. De Bow, superintendent of the seventh census. The first column of figures [table 4.2.2] shows the *actual* condition; the second, guesswork, and the third, a scientific estimate, almost equal to absolute accuracy.

Hoffman, Frederick L., "Vital Statistics of the Negro," *Arena* 29 (April 1892): 529–542.

Table 4.2.1
White and Colored Population of Ten Southern States and District of Columbia, Census 1890

	White				Colored			
	Population 1890	% of total	Increase since 1880	% of increase	Population 1890	% of total	Increase since 1880	% of increase
Alabama	830,706	54.91	168,611	25.46	681,431	45.04	81,328	13.55
Arkansas	816,517	72.37	224,986	38.03	311,227	27.59	100,561	47.73
Dist. of Columbia	154,352	66.99	36,346	30.80	75,927	32.96	16,331	27.40
Florida	224,461	57.35	81,856	57.40	166,678	42.58	39,988	31.50
Georgia	973,462	52.98	156,556	19.16	863,716	47.01	138,583	19.11
Louisiana	554,712	49.59	99,758	21.93	562,893	50.32	79,238	16.38
Mississippi	539,703	41.85	60,305	12.58	747,720	57.98	97,429	14.98
North Carolina	1,049,191	64.85	181,949	20.98	567,170	35.05	35,893	6.76
South Carolina	458,454	39.82	67,349	17.22	692,503	60.16	38,171	14.59
Tennessee	1,332,971	75.42	194,140	17.05	434,300	24.57	31,149	7.73
Virginia	1,014,680	61.27	133,822	15.19	640,867	38.70	9,251	1.46
Total and average	7,949,299	57.95	1,405,678	25.07	5,744,432	41.99	717,922	18.29

Table 4.2.2
Colored Population of the United States

United States Census		Darby's Estimate	De Bow's Estimate
1860	4,441,800	7,860,000	4,319,000
1870	4,880,000	10,600,000	5,296,000
1880	6,580,000	14,000,000	6,494,000
1890	7,500,000	19,000,000	7,962,000

Mr. Darby over-estimated the probable increase of the colored race, but almost correctly calculated the increase of the white element. Arguments like those of Mr. Tourgee in his "Appeal to Cæsar" and of the author of an "Appeal to Pharaoh" are only guesswork, and are proven such by an appeal to facts and to history.

Professor Gillian some years ago, as quoted by Mr. Tourgee, estimated the colored element in the United States for 1980 at 192,000,000, and the total population at 528,000,000; whereas the highest *reliable* estimate places the total population for 1980 at 296,000,000, or just about 130,000,000 less. A writer in the *American Statistician* for 1891, a San Francisco publication, calculates the probable colored element of the total population for 1920, only thirty years hence, at 50,000,000; when it could, in all human possibility, hardly exceed 15,000,000. The *American Statistician* goes even higher in its estimate than Mr. Darby, whose estimate for 1920 is 47,000,000.

The principal factors in the miscalculation of the probable future colored population have been the *over-estimate* of the birth rate and the *under-estimate* of the death rate. Again, most writers on this subject have ignored the important fact that the colored population of the United States is an isolated body of people, receiving no addition in numbers by immigration, and in consequence present conditions essentially different from those of other races and nationalities that have settled on American soil. The Indian is on the verge of extinction, many tribes having entirely disappeared; and the African will surely follow him, for every race has suffered extinction wherever the Anglo-Saxon has permanently settled.

Up to the year 1830, the negro increased at a greater rate than the white race of the South; but since then the white race has been slowly gaining on the colored element, and this gain has been due to the *natural* increase of population, and not, as may be argued, to Northern settlers or European immigration. But for the enormous losses sustained by the Southern people during the late war, the result for the past thirty years would have been still more astounding.

For some generations the colored element may continue to make decennial gains, but it is very probable that the next thirty years will be the last to show total gains, and then the decrease will be slow but sure until final disappearance.

Vital statistics of the colored race are, perhaps, the most difficult body of facts to collect in the United States. It would be a comparatively easy matter to collect a body of figures and facts relating to horses or mules, and to show the prevalence of the most fatal diseases among them, for there is not a Southern State without a bureau of agriculture; but on the other hand, there is but *one* Southern State, Alabama, in possession of a State Bureau of Registration of Vital Statistics; and it is to the cities we shall have to turn for the material necessary to gain an understanding of the conditions as they exist to-day. Such figures as may be introduced in the following tables, have been principally obtained from the registration reports of the Southern cities. The State of Alabama is the *only* Southern State in which a fairly successful attempt is being made to collect vital statistics; and in course of time the present State Board of Health, under the superintendence of its efficient health officer, Dr. Cochrane, of Montgomery, will undoubtedly succeed in accomplishing exceptionally valuable results.

In Florida and North Carolina attempts are being made to secure registrations of births and deaths, and the respective States deserve much credit; but what can be said of a State like Tennessee, where a Bureau of Vital Statistics in successful operation was abolished by its own Legislature? In all discussions of the race problem from the standpoint of population statistics, the birth rate of the negro is usually held out as the most conclusive proof of a probable future numerical negro supremacy; yet, notwithstanding the most earnest effort, I have failed to secure reliable data from a *single* State or city from which to arrive at anything like a birth rate of the colored population.

The fact is, in most instances no record of births is required; and when such registration is attempted, the returns seldom exceed more than half of the actual births for both races. To show how misleading such statistics may be, I will quote the white and the colored birth rates of the State of Alabama, which for the year 1889 is stated to be 26.47 per 1,000 for the white race, and 21.53 per 1,000 of the population for the colored race, showing an actual excess of white births,—a result no one familiar with the actual conditions would for an instant consider a possibility. Thus a consideration of a colored birth rate is out of place; but we can arrive at proper conclusions in regard to the future of the colored race by a *thorough* examination of the mortuary reports of the Southern cities, and in this respect we have access to considerable statistical material of great value. The great prolificness of a race means absolutely nothing, unless it be counterbalanced by a correspondingly low death rate; for it is not so much the number of children that are born, as the number that maintain individual life, on which depends the future of a race. It matters not how many are born if most of them die. What is the result? There is an old saying that "it is not what a man earns, but what he saves, that makes him rich."

An examination of the mortuary statistics of the South for both races will convince the most superficial reader that the death rate of the negro is out of all proportion to the rate of mortality of the white race. The table below [table 4.2.3] exhibits the comparative mortality of whites and blacks in eight different Southern municipalities.

The eight cities from whose annual registration reports I have compiled this table, are representative centres of the colored population of the regions embraced

Table 4.2.3
Annual Rate of Mortality per 1,000 of Living Population, by Race

	White (rate per 1,000)	Colored (rate per 1,000)
Birmingham, Ala.	14.85	26.64
Washington, D.C.	17.25	32.87
Atlanta, Ga.	15.71	36.28
New Orleans, La.	21.27	30.93
Wilmington, N.C.	13.90	28.50
Charleston, S.C.	19.05	43.66
Memphis, Tenn.	19.33	26.15
Richmond, Va.	19.53	27.81
Average	17.61	31.60

under the table of population statistics at the beginning of this article. From Arkansas, Florida, and Mississippi, I found it impossible to obtain returns.

It will be seen by the above table that in not a single instance does the negro mortality come anywhere near the white race, but almost without exception exceeds it by from 30 per cent to 100 per cent. In Washington, D.C., the colored mortality exceeds that of the white by 15.6 per 1,000 of living population, and in Charleston, S.C., by over 130 per cent. The average rate for the eight cities is 17.61 for the white and 31.60 for the colored race, an excess of about 79 per cent on the part of the latter. The colored race is certainly in need of a high degree of prolificness, to make good a loss of nearly *two* deaths to every *one* of the white race.

It cannot be argued that these conditions do not apply to the country population, for then the white race would certainly enjoy the same favorable circumstances as the blacks, and thus the relations would be substantially the same. Again, it must not be forgotten that the drift of the colored population is into the cities; and the increase in urban population of the South, during the past ten years, has been principally due to an influx of colored country population. Already colored farm labor is becoming scarce in certain sections of the country, and the loss of the farmer or planter will be the gain of the undertaker, for the drift of the negro into the cities is usually a drift into an early grave. If we seek for the causes of this frightful rate of mortality, the data are not wanting.

The colored population is placed at many disadvantages it cannot very well remove. The unsanitary condition of their dwellings, their ignorance of laws of health, and general poverty are the principal causes of their high mortality; but there are a few specific causes which a careful analysis of death returns clearly demonstrates, and it is with these we shall principally concern ourselves. There are *two* main causes of mortality among the adult negro which cannot but be of the most momentous influence upon the future generations; these are venereal diseases and deaths from con-

Table 4.2.4

Mortality of Whites and Blacks by Consumption

	White			Colored		
	Total deaths	Deaths from consumption	% of total	Total deaths	Deaths from consumption	% of total
Washington, D.C.	2,713	305	0.11	2,439	392	0.16
Atlanta, Ga.	708	80	0.11	907	136	0.15
New Orleans, La.	3,925	463	0.12	2,150	369	0.17
Charleston, S.C.	516	43	0.08	1,431	213	0.15
Memphis, Tenn.	638	76	0.12	706	126	0.18
Richmond, Va.	1,094	113	0.10	1,224	133	0.11
Average	—	—	0.10^7	—	—	0.15^2

sumption. Any physician who has practiced among colored people will bear me out in my statement that at least three fourths of the colored population are cursed with one kind or another of the many diseases classified as venereal. The gross immorality, early and excessive intercourse of the sexes, premature maternity, and general intemperance in eating and drinking of the colored people are the chief causes of their susceptibility to venereal diseases; and the want of proper medical attendance makes him, in most cases, a victim for life. They have a common habit of drugging, and usually take the reverse of what they ought to have taken to effect a cure.

The following table [table 4.2.4] will show the relative percentage of deaths due to consumption, out of the total mortality. Consumption is becoming more and more a constitutional disease of the negro; and if the deaths from pneumonia and other lung troubles were added, it would be found to constitute the major portion of fatal diseases.

The average percentage of white deaths due to consumption in the six cities furnishing available returns, will be found to equal 10.7 per cent of deaths from *all* causes; the colored percentage of the *same* causes will be found to reach 15.1 per cent of total deaths, being nearly 50 per cent in excess of the white ratio. The reports from the six cities show a singular similarity of returns; and they seem to prove, more conclusively than any other argument, the fearful predominance, of this deadly malady among the colored population. The figures show also indisputable facts as to the inferior constitution and vitality of the colored race. Something must be radically wrong in a constitution thus subject to decay. Aside from consumption, nearly all of the other fatal diseases will be found to be more fatal to the negro than to the white man. Thus to malarial and other fevers from which the colored race is commonly supposed to be exempt, the negro of to-day is as much, if not more, subject than the whites. Under the

Table 4.2.5
Mortality of Whites and Blacks from Zymotic Diseases

	White			Colored		
	Total deaths	Deaths from zymotic diseases	% of total	Total deaths	Deaths from zymotic diseases	% of total
Richmond, Va.	1,094	438	0.40	1,224	326	0.27
Washington, D.C.	2,934	678	0.23	2,630	670	0.25
Charleston, S.C.	492	100	0.20	1,375	230	0.17
New Orleans, La.	3,925	601	0.15	2,150	300	0.14
Knoxville, Tenn.	356	115	0.32	213	50	0.23
Nashville, Tenn.	581	123	0.21	640	98	0.15

heading of "Zymotic Diseases," the figures given below [in table 4.2.5] for six cities will show an almost equal ratio with the white population.

If it is argued that, granted the same conditions and the same opportunities as the white race, the colored race would prove itself of a more enduring vitality, the proof can be furnished that even if he be placed on equal grounds he still will exhibit what an eminent writer calls "his race proclivity to disease and death."

The experience of the army during the war and its twenty years' experience of peace and normal condition since 1870, will furnish the proof that the colored race, even under the most advantageous conditions, will fail to hold its own against the white race.

I have compiled from the "Medical and Surgical History of the War" the figures given below, to show the fatality of four principal diseases for each race. The figures in the first column state the total number of cases of each disease; the second column contains the total number of deaths resulting; the third column shows the percentage of deaths to cases, affording an easy means of comparison between the two races.

It will be seen from this table [table 4.2.6] that the liability of the negro to death is almost double that of the white race. Even if under the same treatment, under the care of the same physician, and in the same hospital, negroes die almost two to one of the white element of the army. In his liability to death when attacked with consumption, the negro will rarely be found to escape at all, and in not a single instance does he exhibit anywhere a vitality or power of resistance equal to the white race.

The records of the United States Army for the past twenty years tell the same story, exhibit the same relative proportion of deaths to disease, and increased mortality over the white portion of the army. The surgeon-general, in his annual report for 1889, refers to the matter in the following language: "The death rate of the people of African descent is always higher than that of the whites living in the same settlement. This is

Table 4.2.6
Disease and Mortality in the United States Army during the War

	White			Colored		
	Total cases	Deaths	%	Total cases	Deaths	%
Consumption	13,499	5,286	0.39	1,331	1,211	0.91
Typhoid fever	75,368	27,056	0.36	4,094	2,280	0.56
Chronic diarrhœa	170,488	27,558	0.16	12,098	3,278	0.27
Inflammation of lungs	61,202	14,738	0.24	16,136	5,233	0.32
Average	—	—	0.29	—	—	0.51[5]

ascribed, for the most part, to the comparative poverty of the colored people, which crowds them into dwellings in the less desirable parts of the locality. It would seem, however, from the records of the army, that there is a race proclivity to disease and death; for although the colored troops are in all respects subject to the same influence as the white troops at the same station, the cases of sickness, and notably the death rates, are greater among them than among the whites."

The average rate of mortality, for the past twenty years, of white and colored troops, according to statistics furnished me by the War Department, was 14.36 per 1,000 of mean strength for colored troops and 11.50 per 1,000 of mean strength for the white troops. The rate of mortality due to consumption was 0.66 per 1,000 for the white and 1.19 per 1,000 for the colored element of the army. This agreement of facts and figures shows whither the colored race is drifting. We have seen that the average total mortality and the ratio of death due to consumption exceed that of the white race by a very high percentage. We have also seen that, even under the same conditions, the negro exhibits the same excessive tendencies to disease and death.

Thus we reach the conclusion that the colored race is showing every sign of an undermined constitution, a diseased manhood and womanhood; in short, all the indications of a race on the road to extinction. Additional proofs, more convincing still, are furnished by separating the death returns of the two races according to age and sex.

If in the beginning it was shown that the average total *mortality* of the colored race exceeded that of the white race, the following table will show that the *average age* of the colored element is much below that of the white. In Washington, during a period of eleven years, the white population maintained an average age of thirty-two years and nine months, exceeding the colored average of twenty-one years and eleven months by nearly ten years.

Since the average ratio of deaths by age is so much alike in the several cities embraced under Table 4.2.7, we may well assume the difference in the average duration of life to be the same for the two races in other portions of the South. Thus the percentage of deaths under twenty years for Savannah and Charleston is almost equal to the 56 per cent of Washington.

Table 4.2.7
Mortality of Whites and Colored, According to Age

	White							Colored						
	Total deaths	Still-births	% of total	Deaths under 5 years	% of total	Deaths under 20 years	% of total	Total deaths	Still-births	% of total	Deaths under 5 years	% of total	Deaths under 20 years	% of total
Washington, D.C., 1890	2,934	183	0.06	895	0.31	1,113	0.38	2,630	288	0.11	1,172	0.45	1,474	0.56
Memphis, Tenn., 1890	638	32	0.05	157	0.25	219	0.34	706	49	0.07	244	0.35	341	0.48
New Orleans, La., 1889	4,122	285	0.07	1,401	0.34	1,709	0.41	2,179	223	0.10	723	0.33	958	0.44
Savannah, Ga., 1889	479	34	0.07	145	0.30	182	0.38	870	116	0.13	372	0.43	451	0.52
Charleston, S.C., 1889	516	40	0.08	158	0.31	191	0.37	1,431	153	0.11	592	0.42	756	0.53
Richmond, Va., 1890	1,094	72	0.07	332	0.31	430	0.39	1,224	142	0.12	496	0.41	631	0.52
Average	—	—	0.07	—	0.30	—	0.38	—	—	0.11	—	0.40	—	0.51

Table 4.2.8

	Male deaths	Female deaths
Massachusetts, 1889	29,017	28,042
Brooklyn, N.Y., 1889	9,605	8,875
Philadelphia, Penn., 1888	10,566	9,806
St. Louis, Mo., 1890	4,611	3,798
Toronto, Can., 1889	1,609	1,323
Hamilton, Can., 1889	365	309
Total	55,773	52,153

A close examination of Table 4.2.7 will show a similarity of conditions aston-ishing to one unacquainted with comparative mortuary statistics. We find but slight variation in the colored returns; and with the exception of the low rate of early deaths for Savannah, we have a corresponding agreement of returns for the white race.

According to these tabulated returns of six Southern cities, but one third of the white deaths occur under twenty, against a colored ratio of more than one half. In not a single instance does the white race reach even the lowest ratio of the colored; and even the high ratio of New Orleans, of 41 per cent, is 3 per cent less than the lowest colored mortality, under twenty, reported from the same city. The infantile mortality, or deaths under five years, for the two races presents the same condition of an exces-sive ratio on the part of the colored element.

The rate of ante-natal to the total mortality of the two races, one of the most important features in population statistics, places the colored element at still more fear-ful odds to the white, being everywhere in excess by from 30 to 200 per cent.

It now only remains for me to present the mortality of the two races according to sex; and it is here that one of the strongest points, as to the future of the colored race, will find its basis.

The white population, as will be seen by the figures under Table 4.2.8, shows a male ratio of mortality higher than that of the female portion of the population. Tak-ing figures at random from States and cities in America and Canada, as the following table will show, these relations are commonly prevailing.

No additional argument is needed to assure the reader of the correctness of the assertion that the *white* female mortality never and nowhere, under normal conditions, exceeds that of the other sex, but, on the contrary, usually falls considerably below. In Massachusetts this excess of male deaths is on an average 1,000 per annum, and has been the same, according to registration returns, for the past twenty years. We can therefore safely lay it down as an axiom that the white race depends on the mainte-nance of this favorable ratio for its natural increase, and we may assume the same to hold good for the colored race, whose future will therefore depend on its ability to maintain existence under the condition that its female mortality be less than its male.

Table 4.2.9

Mortality by Color and Sex

	White		Colored	
	Male	Female	Male	Female
Charleston, S.C.	275	251	526	676
New Orleans, La.	2,225	1,700	1,057	1,093
Atlanta, Ga.	423	370	522	520
Memphis, Tenn.	386	252	380	326
Nashville, Tenn.	304	277	325	315
Washington, D.C.	1,631	1,303	1,292	1,338
Richmond, Va.	557	537	604	620
Baltimore, Md.	3,249	3,105	1,047	1,125
	9,050	7,795	5,753	6,013
		F. 0.46%		F. 0.51%

But proofs are not wanting to show that just the reverse is its present and inevitable future condition.

As will be seen from the data compiled under Table 4.2.9, the colored female mortality is in excess, in many instances, against a male excess of the white. In not a single instance does the white male mortality exceed that of the white female, but in *nearly every* instance does the male negro mortality fall below that of its female.

This comparative exhibit of mortality by race and sex demonstrates the inferiority of the constitution of the colored female; and this being true, the whole body politic of the colored race is undermined and finally doomed. Additional evidence of the deteriorated physique of the colored female is gained by an investigation of the comparative ratio of still-births prevailing among the two races. The high rate of female mortality, together with the high rate of still-births, is a convincing proof of an inferior womanhood. The enormous losses sustained by the colored population from these causes may be better understood when we add together the losses for a number of years.

During the period 1880 and 1890 the colored still-births numbered in Richmond, Va., 1,265, and in Washington, D.C., for the same period, nearly 3,000. In the latter city during the last decade, the rate of illegitimates to total births was equal to 21.34 per cent, against a white rate of only 3 per cent.

If a high rate of still-births is a proof of a weakened female constitution, a high rate of illegitimate births proves the cause of this growing debility and frailty of the colored female. The laws of morality can no more be violated than the physical laws of nature; and the whole life of the negro is a constant violation of both. The penalty paid by the unfortunate and ignorant is premature death.

Table 4.2.10

Mortality Statistics, by Sex and Color, for 1850, U.S. Seventh Census

	White		Slaves	
	Male	Female	Male	Female
Baltimore, Md.	1,567	1,376	299	244*
Charleston, S.C.	189	131	54	34
Louisville, Ky.	494	337	77	54
Memphis, Tenn.	205	112	50	40
Mobile, Ala.	310	163	77	73
New Orleans, La.	2,666	992	243	171
Norfolk, Va.	102	98	54	38
Richmond, Va.	175	139	61	37
Wilmington, N.C.	48	33	35	27
Total	5,756	3,381	950	718

*Free colored population.

The female deaths form 46 per cent of the total white mortality, whereas the colored rate of female deaths is 51 per cent. The white female, embraced under Table 4.2.10, gained in total numbers 1,257 on the white males, whereas the colored female shows a net loss of 260. The three cities, New Orleans, Baltimore, and Washington, report, for a combined total experience of twenty-three years, a white female gain of nearly 5,000, against a loss on the part of the colored female of nearly 1,300, a difference in favor of the white race of 6,300 females in these three cities alone.

What else but final extinction can be the future of the negro, thus presenting all the evidences of a vanishing race? What fears need we entertain when we take into consideration the fact that for the white race the indications point in the opposite direction?

Such are the conditions of to-day, and this statistical review of the race question will prove that no one need fear a possible negro supremacy, impossible under the prevailing conditions.

In conclusion I will present some data comparing the conditions of the past with those of the present, illustrating the changed conditions affecting the life and well-being of the colored race. The following statistical exhibit is taken from the mortality report of the Seventh United States Census, showing the relative mortality of males and females of nine Southern cities.

It is fortunate that we possess at least some statistics relative to the negro antedating the war. A comparison of the past with the present affords us the most valuable clue as to the future tendency of the race. According to De Bow, superintendent of the seventh census, the mortality of Charleston, S.C., for the period 1830–45, was, on an average for the white race, one death to every forty-three living, and one to every fifty living on the part of the colored. For the past ten years this ratio has been among the white population one death to every forty-seven living, and among the colored, *one to every twenty-two living*, showing a decrease of the white mortality and an increase of the colored mortality of over 100 per cent. It seems almost as if the period could be calculated when the death rate will be *one* death to every *one* living. If it has increased from one to every fifty living to one to every twenty-two or twenty-eight living, in forty-five years, may we not conjecture what it will be forty-five years hence?

No one can foretell the probable future of the colored population of the United States with any degree of *absolute* accuracy, but the facts presented in this article indicate tendencies which warrant us to believe that the time will come, if it has not already come, when the negro, like the Indian, will be a vanishing race.

4.3   "The Influence of Environment and Race on Diseases" (1910)

C. W. Birnie

I desire most earnestly and sincerely to discuss a question that is of vast and vital importance to us because of the influence and impressions that are made on the minds of people generally; and not only to us as a race is the free, full and frank discussion of this question necessary, but to each and every thinking citizen, as the matter concerns all.

The subject is, "The Influence of Environment and Race on Diseases." There must be no delicacy in the study and investigation of such a subject; the false must be eliminated and the true stand out in its real character. Charges that are made must be refuted, if false; if true every effort must be made to produce a wholesome and effective change. Causes must first be investigated in order that effects may follow that are consistent therewith. We should probe to its greatest depth every accusation that is made and use correctives as far as possible. Unfortunately for us, when we have studied the question in all its bearings; laid bare every weak point or supported every strong one, yet even then we lack the means of disseminating the facts to the country. We have not the same opportunities of reaching the public mind and eye as those have who make statements that are illogical and sometimes wanting a foundation in fact. We do not desire to cover up a weak spot, and wherever one is made clear to us we rather offer thanks especially if a practical remedy is suggested.

Almost daily, we find newspapers, and magazines teeming with articles endeavoring to prove that the Negro race stands as a menace to the white men socially, morally and physically. We are held up in scorn before the world as lepers. The results of such criticisms are extremely pernicious and damaging. Public sentiment is being educated against us. The unthinking take the argument without the power or capacity to investigate and accept it as a truth.

Now, we owe it to ourselves, to our race, to our profession, that we should come before the public and say if these things are true, and if they are, frankly and honestly admit them; and bend every possible effort; use every knowledge that we possess to remedy the evil. If they are not true; point out by the strictest of reasonings, the presentation of strongest of facts to counteract any charge or part of a charge that cannot bear the light of scrutiny.

Let us ask ourselves, Is there a degeneracy of the Negro race going on? Is he physically, morally, and socially losing his place in the race of life? Will the Negro race be eliminated, and his place taken by the white man as a survival of the fittest? Surely these are questions of vital importance to us.

Birnie, C. W., "The Influence of Environment and Race on Diseases." *Journal of the National Medical Association* 2, no. 4 (1910): 243–251.

To prove the position, our only means would be to institute comparison; to take hold of the statistical tables prepared by those who have specially investigated the subject. Even statistics, we find in some cases absolutely worthless, so colored by race predjudice as to be unreliable. When we find a disposition to be true, to be accurate, to be governed by facts, we gladly accept them, even when against us.

The Board of Health in the following cities furnish a large part of the information desired and the statistics used are deduced from them. Memphis, New Orleans, Augusta, Baltimore, Washington, D.C., and the U.S. Army. In all of these cities there are large colored populations, giving abundant opportunity to study the question.

. . .

From these figures certain conclusions can be deduced. It is apparent that the proportion of deaths from tuberculosis, pneumonia and kindred diseases among the Negro race is appalling. Now, how shall we account for this terrible death-rate? Is it a deteriorating physical condition? I cannot bring myself to the point of wholly assuming it to be this cause. It is well known to us as physicians, and a large number of laymen are beginning to realize the fact, that tuberculosis is largely a disease of poverty. I do not mean that persons of affluence are exempt, but it is more prevalent among the poor. Take the combination of ignorance and poverty and we have a fertile soil; poor food, poor housing, poor water, and a total lack of knowledge or disregard for the fundamental laws of hygiene will make a high death-rate among any people. Dr. Guiteras lecturing before a class at the University of Pennsylvania gave as his opinion, that all things being equal the death-rate among Negroes and whites would be about the same. That the large death-rate among Negroes was almost wholly a matter of environment.

Often you will see the statement, and 'tis probably true, that prior to the Civil War tuberculosis was almost unknown among Negroes. The cause is not hard to find. The owners insisted on hygienic manner of living as a purely business matter. The slave was so much property that had to be hedged by every possible protection. From a financial point of view it was necessary that he be kept in the best marketable, physical condition. It was purely a matter of dollars and cents. But take the same uneducated, inexperienced, ignorant people, throw them, without preparation, on their own resources, and it is the expected that has happened. But while the former slave owner or his descendants are making all the charges, they are not willing to assume their responsibility for a condition that came most naturally; primarily the burden is upon them and ultimately they feel the result. Bear in mind that I am now speaking of places that have a large former slave population and where they are yet employed largely as servants.

You will find that the tenements and other places of residence of the colored people are owned largely by the white man. With him it is only a question of getting his rent. The condition of the place matters not at all. The poor by reason of their poverty must rent from him and he knows it. Tell him of rendering a house comfortable or of disinfecting a home or a room that had been occupied by tubercular patients and he will dismiss at once the request. That same man will employ as his cook, nurse or washer an occupant of such a home; and he forms a medium of infection to him and his family. Surely if a man is fettered he cannot use his strength as one that is not encumbered. The Negro is largely fettered by his environments. Far be it from me to charge the whole Southern people with a purpose to do the Negro injustice or to be inhuman, because I recognize the fact that among the white people of the southland, we have some of the kindest hearts, the warmest friends that can be found in the world. But there are also those who never stop to consider the poor and lowly.

The large death-rate among the Negro from pneumonia is another lamentable evidence of neglect to furnish comfortable homes by those who rent them. The contributing causes are so similar to tuberculosis that they can hardly be separated; poor houses; underfed people, inability to provide by reason of the smallest of wages, comfortable clothing, etc., these make the high mark of mortality.

We have now come to two diseases: syphilis and gonorrhea which of all in the list have been used most to the Negro's detriment by both medical and lay writers; newspapers and magazines have been made to circulate the idea that its prevalence should be laid at the door of the Negro race. The extreme delicacy of the case, the possible offence that a free statement controverting some of the baseless accusations made has caused me to almost conclude to withhold any statement. No other diseases have given the same opportunity to make of the Negro a scape goat as the ones mentioned. If these diseases are freely and fairly discussed it would not require much argument to prove that to turn aside blame, both medical and lay writers have set at defiance every known principle of right. Comparisons made from statictics of Northern hospitals and physicians, put the blame on other shoulders than the Negro. I am afraid that too many physicians of both races have been prone to pronounce patients syphilitic when if the case had been patiently and properly investigated a very different conclusion would have been reached. Every skin disease appearing on a Negro is at once charged to the province of syphilis.

The death-rate among Negroes is increased alarmingly by the number of children under 2 years of age. Such is due to neglect and privations that parents generally are powerless to prevent. It does not stand to reason that the Negro children are less healthy than the whites—by no means. Take the condition under which the average Negro child is born and reared, and the wonder is indeed that the death-rate is not much higher. Their mothers are compelled by necessity to go to work within a month or two after confinement, leaving the young one in charge of an older brother or sister, who is little beyond babyhood, or possibly some feeble old woman, superannuated, almost unable to take care of herself. Thus at a time of the child's life when it needs the greatest care and attention, it receives none. Fed almost anything during the day, having its natural food from its mother after a hard day's work and early in the morning, the only surprise is that more of them do not die. And even in cases when conditions are a little better, ignorance, want of proper food, lack of knowledge that 'tis better to employ a skilled physician rather than to trust to untrained and uneducated nurses. Such things as these swell the death-rate of Negro children.

In studying and investigating this subject we find some facts and figures that we cannot account for, and some notions and beliefs handed down from generations shattered.

It has long been thought and published that the Negro is less prone to malaria than the Caucasian. The idea obtained that there was some hereditary immunity obtained from his African ancestors, that by repeated infection some anti-toxin has been developed that in a degree rendered him less susceptible. But neither reports from private practice nor statements furnished by boards of health bear out this supposition. The fact is that comparative tables would lead me to suppose that if there is any immunity, the white man possesses it. My own opinion is that neither race is less susceptible. As is well known to all present it has been conclusively shown that the mosquito is the agent that carries malaria. And that it breeds in damp low places, especially when there is stagnant water, and those are the conditions under which the bulk of the colored population live in our larger cities. They are the last to get sewers, water or any conveniences or necessities conducive to health.

Why diabetes should be so extremely rare among the Negroes, is impossible to say, except as is suggested by Dr. J. A. Robinson, it bears out the nervous theory of the origin of this disease. And a reference to any table will show that nervous diseases form a very small part of the death-rate. But almost any physician either in private or hospital practice can count on the end of his fingers the number of cases of diabetes he has noted within the Negro race. On the other hand nephritic troubles are much more common though there does not seem to be any marked disproportion between the races in these cases. Among the Caucasian we will find that cancer is much more common. Both cancer and ulcer of the stomach being extremely rare among Negroes; which can probably be accounted for by the simple diet and manner of living. We will find diphtheria, scarlet fever, appendicitis, and cerebral congestion all more frequent among Caucasians.

The question that concerns us most now is, Is there a remedy for these things, and what is it? One of the remedies I would suggest is the establishment of day nurseries where the working mother may leave her little ones, another that I would suggest is that the Negro children in every school be thoroughly educated in the fundamental principles of physiology and hygiene. To this end every teacher should be required to study these subjects and to teach them not in a perfunctory way, but to impress upon the minds an understanding of the pupils in order that a radical change might be brought to the lives of the children, and through and by them carried into the homes. The physician's duty does not end with a diagnosis of the case, he too should be a teacher to lead his patients up to a better living, a clear observance of the laws of health.

Discussion

The discussion of the paper was opened by Dr. S. S. Thompson, of Washington, D.C., who said, "We are compelled to take into consideration, in the selecting of statistics, the method employed in obtaining them, and secondly, the inclination of the individual taking the statistics. Dr. Birnie has ably set forth that it is refuted that the Negro is a menace to the city. This has been written in medical journals of the opposite race and in newspapers. The statement that the Negro is a menace to the city is false; that the Negro is a carrier of disease. If the Negro was a carrier of disease and a menace to the city, there would be a complete anihilation of the whites in the South, because it is in the South and among Southern whites that the largest proportion of our race lives; consequently the Negro is eliminated from that accusation. In Alabama the ratio of disease between the races is 10 to 64, that would be in numbers about 19 white and 31 colored. In the District of Columbia, the ratio is 1 to 65, or 16.9 white and 27.8 colored; nearly 50 per cent. Taking Atlantic City, New Jersey, the ratio of disease is 16.1 to 16.5, only a difference of .4. I want to say this in defence of the Negro—they claim the Negro is fast dying out. From statistics today it shows that the Negro is not dying out, and fading before this civilization as the American Indian is. Take tuberculosis. Statistics show that the alien is dying twice as fast as the Negro. Some of this is true, but there are some statistics which ought to be compared to see whether we are getting a square deal. The statistics of the District of Columbia show that the colored population die faster than the white population, which is due to the large number housed in alleys. So there must be some way to improve these alley conditions."

Dr. M. O. Dumas, of Washington, D.C., said:

"Mr. President: There is no inherent quality in the Negro that makes him die faster than the other races. He was born into this world with the same amount of vitality and the same amount of resisting power that other races are endowed with. The high death-rate of the Negro resolves itself into a very large measure into the matter of environment, which has been emphasized by Dr. Thompson. When you take into account the poor housing facilities that we are obliged to put up with, the nature of the Negro's occupation, which takes him out into the most foreboding weather; when we note the poorly clad man on the coal cart; I am persuaded to believe that these are some of the elements which enter into the high death-rate of the Negro. Now, Mr. Chairman, another thing which accounts for the high death-rate is the Negro's inclination in many instances to the vicious habit of drinking—excessive use of alcohol will certainly lower his vitality and make him a rich prey for any disease that comes along. In proportion as we are helping to abate the alcoholic traffic about us, which is lowering the vitality of the race, just in that proportion will we be able to ward off the inroads which disease is making upon our people. We ought not forget that while the death-rate is high on account of tuberculosis, that it is not due to any inherent quality of the race. I hope that those of us who are interested in this matter will take up the matter with our people and show them the proper way to live and teach them what the inevitable result will be if they do not heed the advice we give them."

Dr. John B. Hall, of Boston, Mass., said: "Perhaps we are more concerned with every-day life in a sense. We do not spend enough time giving consideration to the manner of living about us. I think it devolves upon the physician more than any other set of professional men. We have got to defend ourselves. We have heard read the statistics of death-rate in different cities. The Northern cities seem to be better. I had occasion to look up the death-rate in Boston. In one section colored people lived almost entirely; in another section known as South Boston, Irish people lived. The death-rate from tuberculosis among the Irish people and among the colored people differed but very little, which made me think environment had a great deal to do with it. I had occasion to look up the death-rate of children under one and under five years old. The death-rate of the white was 16 per thousand; among the black 18 per thousand. We ought to give more consideration to the manner of living of our people. Look after them when they are well."

Dr. S. Leroy Morris of Atlantic City, New Jersey said:

"The condition and environment with which our people are compelled to put up, are principally the cause of the high death-rate. I do not take much stock in the death-rate stated by some white men. If you will notice when you go back to your homes the conditions surrounding our people, especially the lower classes, you can readily see why these conditions exist and why the death-rate is so high. I have seen robust and healthy children come into this world and die soon after, not because they did not have the vitality, but because of improper environment, which makes them more susceptible to disease. I lived among colored people and we had no street and our back yards were not cleaned out. I took it upon myself and went to the Board of Health and made complaint and the result was that all that square has asphalt pavement and the back yards have been cleaned out. The thing is, we must go to the front when we want anything. We must impress upon the Board of Health that they are maintaining a nuisance. Gentlemen, if you want a healthy city, you must show an interest in it by helping to improve its surroundings."

On motion by Dr. Sterrs, the discussion was closed.

On motion the meeting adjourned until 2:30 p.m.

Note

C. W. Birnie, M.D., Sumter, S. C.

4.4 "How to Improve the Race" (1914)

Alexander Graham Bell

Success Possible, but not by Processes Employed with Lower Animals—Little Gain from Preventing Marriage of Undesirables—Important Point Is Formation of a Prepotent, Desirable Stock by Marriages of Desirable People with Each Other—This Prepotent Stock Will Then Raise the Level of the Great Bulk of Normals

Living organisms have proven so plastic in the hands of scientific breeders that we have learned to improve our breeds of plants and animals by suitable selection controlled by man.

Human beings, also, are undoubtedly capable of modification by selection; but it is manifestly impossible to apply to them the processes employed with the lower animals.

The difficulties of the problem may perhaps be appreciated if we consider for a moment how far it would be possible to improve our breeds of domestic animals under the conditions which prevail among human beings.

Given, for example, a flock of sheep to be improved, but under human conditions.

First, we must not butcher any of the animals. Ovine life is to be considered as sacred as human life. We must not mutilate the animals; nor do anything to them that is inconsistent with the humanitarian spirit of the age.

The weaklings are to be preserved and given special care. In fact, all of the animals, including the poor little deformed lambs, are to be kept alive as long as possible. They are to be treated with kindness and consideration until they die of old age, or from other causes beyond our control.

To these conditions we may add the following: polygamous unions must not be permitted; nor unions between individuals related in various ways.

A man, for example, may not marry his grandmother; nor his mother; nor his sister; nor his daughter; and if we apply all the human restrictions to sheep, we shall have our hands full indeed in merely examining the ancestry of the flock, and the relationships of the individuals to one another, so as to avoid the prohibited unions.

While we are forbidden to allow certain classes of unions, we are not permitted to select the individuals that should be mated together to improve the stock. Each individual of the flock, under the restrictions referred to, must be free to choose its own mate; and the pairing shall be for life.

Alexander Graham Bell, "How to Improve the Race," *Journal of Heredity* 5, no. 1 (January 1914): 1–7.

We may confidently assert that under such conditions no scientific breeder would undertake to improve the flock,—it would not be possible.

The Human Problem

But these are the conditions we must face in attempting to improve our own race; and we may as well recognize, first as last, that we have no power to compel improvement.

A gleam of hope, however, appears in this connection when we realize that there is one great and fundamental difference between a community of human beings, and a flock or herd of animals: The individuals of the human community possess intelligence.

The individuals have power to improve the race, but not the knowledge of what to do. We students of genetics possess the knowledge but not the power; and the great hope lies in the dissemination of our knowledge among the people at large.

Another important difference between human beings and the lower animals, arising from intelligence, is that human beings give some thought to their unborn progeny. All desire that their offspring may be of the best; and no one wishes to have degenerate or defective children.

The attitude of the public mind is therefore favorable to voluntary compliance with plans which appeal to the intelligence of the community as reasonable and right; and favorable to the formation of a public opinion which will compel compliance.

These are such hopeful conditions that they will bear recapitulation.

The members of a human community, both individually and collectively, desire that their descendants may, if possible, be better than themselves.

They possess intelligence to understand the laws of heredity as applicable to man; and a willingness to adopt any reasonable and practicable measures that may be formulated for the benefit of future generations.

All recognize the fact that the laws of heredity which apply to animals also apply to man; and that therefore the breeder of animals is fitted to guide public opinion on questions relating to human heredity. Without power to control, he has power to advise; and the public generally will accept his statements as sound, because based upon special knowledge and experience in the breeding of animals.

What an opportunity for the members of the American Genetic Association to benefit the human race! Most of the disputed questions of human heredity can be settled by them, and their verdict will be acquiesced in by the general public.

Statistics relating to the effect of inbreeding among animals, for example, could surely be made to guide public opinion rightly on the subject of consanguineous marriages among human beings.

So, too, statistics relating to the effect upon the offspring of maturity and immaturity in the parents of animals, would seem to have a bearing upon the question of early *versus* late marriages among human beings.

The first thing for us to do, is to make known to the public the processes that are needed to improve the race; and then to show how, by intelligent cooperation among the members of the community, these processes may be applied.

Improving Racial Stature

In considering the question of improvement, it may be well to begin by taking some specific quality of an inheritable nature and examining its distribution among the population at large.

Take stature as an example. We have pigmy races of men, and it is quite conceivable that some such race might deem it desirable to increase the general height of the population.

The members of the race all possess the desirable characteristic (height), but in varying degrees; and upon this variability depends the possibility of improvement. The difference between the extremes shows the amplitude of the variation; and if we sort out the population in accordance with the degree in which they possess the quality, we shall find a continuous series from the lowest to the highest. Some intermediate point represents the average degree in which the quality is possessed by the race.

The people who are markedly above the average height will, in this case, constitute the desirable class; and those who are markedly under the average would be the undesirable.

We are accustomed to focus our attention so exclusively upon the desirable and undesirable classes that we are apt to forget that there is an intermediate class, the normal, which is many times greater than both of the others put together, constituting, indeed, the bulk of the population.

The accompanying diagram may perhaps be of assistance in realizing the relative proportions of these classes. Let the large square represent an enclosure completely filled with the people under consideration. The square then represents by its area the whole population to be improved in height.

Now if we look the people over, we shall find here and there exceptional individuals who stand well above the general level. Collect them together into a pen in one corner of the enclosure. These constitute the desirable class represented by the small shaded square at the top of the diagram, which expresses, by its area, the number of tall people found.

In a similar manner, collect the markedly undersized individuals, and place them in a separate pen represented by the shaded square in the lower corner of the diagram. These constitute the undesirable class.

The rest of the population, occupying the unshaded portion of the large square, are normal people of somewhat about the average height.

Stature is convenient as a typical illustration because in this case the desirable quality, height, is capable of measurement.

Application Universal

If, however, any other inheritable quality be taken as an illustration, the people can, in a similar manner, be sorted out into the three classes shown in the diagram:

1. The great normal class possessing the quality in a normal or average degree.
2. The desirable class, possessing the quality in a markedly greater degree than the average.
3. The undesirable class, possessing it in a markedly less degree than the average.

On the scale shown in the diagram [figure 4.4.1] the desirable and undesirable classes each constitute 1% of the population and the normal 98%. Whatever may be the actual relative proportions, the diagram expresses the undoubted fact that the normal class constitutes the bulk of the population; and that the desirable and undesirable classes are very small as compared with the normal.

In the case considered the people generally have been of small stature as far back as their history extends. There has been no substantial change in the average

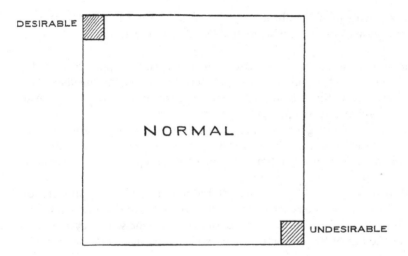

Figure 4.4.1
The make-up of the human race. In discussions of eugenics, the relatively small part of the population made up by the superior and the inferior is sometimes allowed to occupy so prominent a place that we forget that the great bulk of the race is made up of normal people, as the diagram clearly shows. Mr. Bell points out that because of the overwhelming numerical preponderance of the normal people, the easiest, quickest, and most natural method of raising the level of the whole race is to raise the level of this huge mass of normals, instead of devoting all our attention to reducing the relatively insignificant number of inferiors.

height of the race within historical times. From this we may conclude that the ancestors of the present generation were mainly of the present normal height; and that only a few of them were much taller or shorter than this.

The above diagram, then, represents substantially the relative proportions of the three classes at each successive generation of the population. It typifies the future distribution, as well as the past, unless some means can be found to change it.

The desirable and undesirable classes, like the normal, are sprung mainly from normal parents; so that it is obvious that no process of interference with the marriages of these classes could much affect the relative proportions of the three classes in the next generation of the community. If, for example, the desirables and undesirables should all decide to lead celibate lives so as to leave no descendants, we would have just about as large a proportion of desirables and undesirables in the next generation of the community, born from the normal class.

Selection in Marriage

The individuals belonging to the desirable and undesirable classes are not only few in number, but are scattered throughout the community. They appear only here and there as exceptional cases, and are not segregated from the others in their actual distribution in the population. If, then, they decide to marry, it is obvious that most of the desirable and undesirable individuals will marry normal persons, because normal people constitute the bulk of the community with whom they come in contact; and the offspring will tend to revert to the normal type of the race. From this it follows that, on the whole, the offspring of the desirables will be less desirable than themselves; and the offspring of the undesirables more desirable; most of the offspring will be of the normal type.

Given a large normal class, and two small classes, the desirable and undesirable, the problem is, how to increase the proportion of desirable children born from the normal population.

This can be accomplished by marriage with members of the desirable class.

In the typical case considered, this would mean that persons of normal height would increase their liability to have tall children by marrying tall people.

Where normals marry normals a small proportion (say 1%) of the offspring will belong to the desirable class.
Where normals marry desirables the percentage of desirable offspring will be increased (say to 10%).

Of course, it is only possible for a small proportion of the normal population to marry persons belonging to the desirable class, on account of limited numbers. The range of choice, however, may be extended by marriages with brothers or sisters or close blood relatives of desirable persons. That is upon the assumption that we are here dealing with an inherited characteristic.

Proof of Inheritance

The late Professor W. K. Brooks, of Johns Hopkins University, said.[1]

"An inherited characteristic may, or may not, have been manifested by the parents or other ancestors.... If it is more common either among the ancestors or the brothers and sisters and cousins of the organism than it is in the race at large, this fact is scientific proof that it is an inherited characteristic."

Where a peculiarity manifests itself in only one member of a family we are dealing with a sporadic case; and the peculiarity may, or may not, be transmitted to the descendants: But where a number of people in the same family exhibit the same congenital peculiarity we have good reason to believe that it is an inherited characteristic, and therefore liable to be handed down to some of the descendants by any member of the family, even by those members who do not exhibit the peculiarity in their own persons.

These considerations lead us to the conclusion that persons of normal height will increase their liability to have tall offspring by marrying into families containing a number of tall persons. In such a family the tendency to produce tall people is evidently an inherited characteristic; and the normal and undersized members of the family, as well as the taller members, will have a tendency to transmit the characteristic to some of their descendants. The tendency is in the blood, and the whole family possesses it.

Influence of Ancestry

In this connection the influence of ancestry is most marked:

1. Where Normals (whose ancestors on the whole were normal people) marry Desirables (whose ancestors were chiefly normal), the normal partners will prove prepotent over the desirable partners in affecting the offspring. The bulk of the offspring will be normal; and only a small proportion (say 10%) desirable.
2. But where Normals (with normal ancestry) marry Desirables (whose parents were desirable) the proportion of desirable offspring will be increased (say 20% desirable). The potency of the desirable partner to have desirable children is practically doubled, because of inheritance from both parents; and the prepotency of the normal partner is correspondingly reduced.
3. Where Normals (with normal ancestors) marry Desirables (whose grandparents as well as parents were desirable) the proportion of desirable offspring will be still further increased (say to 45%); and the prepotency of the normal partners will be reduced almost to zero.
4. Where Normals (with normal ancestry) marry Desirables (whose ancestors for several generations back were all desirable) it is the desirable partner who becomes prepotent over the normal partner in affecting the offspring. The vast majority (say 90%) will be desirable.

The influence of the normal partner will be less and less according to the number of desirable ancestors possessed by the desirable partner. With a sufficient number of generations of desirable ancestors, the desirable partner is what is known as "thoroughbred" in respect to the inherited characteristic. The normal partners, on the other hand, are not usually thoroughbreds for the normal condition; for if we examine their ancestors for the same number of generations back, we shall find, as a rule, that some of the ancestors were not normal: Some few were desirable, and some few even undesirable.

Thoroughbreds Prepotent

We breeders are familiar with the prepotency of the thoroughbred animal over the normal animal; and we can therefore confidently assert that the human throughbred will be equally prepotent over the normal human being in affecting the offspring.

In spite of the smallness of the desirable class, its influence in improving the normal population would be considerable should it contain a number of highly prepotent individuals: For their marriages with normal people would result in the production of offspring almost exclusively of the desirable type (say 90 per cent desirable). We should also remember that the brothers and sisters of the prepotent individuals will also have a strong tendency to produce desirable offspring where they marry normals with normal ancestry.

The liability of normal persons to have desirable offspring will be increased by marriages with persons of the desirable class; and diminished by marriages with undesirables.

In the latter case the proportion of *undesirable* children will be increased and if the undesirable partner is descended from undesirable ancestors the proportion of undesirable children will be still further increased.

The establishment of even a small body of prepotent individuals within the undesirable class would exert a considerable harmful influence, because their marriages with normal persons would result chiefly in the production of undesirable children.

Upsetting the Equilibrium

In considering the influence of marriage in affecting the distribution of the three classes in the next generation of the community, we may notice that in general the marriages of normals tend to keep things just as they are. That is, they tend to produce a large class of normals and two small classes of desirables and undesirables, in just about the same relative proportions as in the preceding generation.

The marriages of the desirables, on the whole, tend to raise the average of desirability in the community; and the marriages of the undesirables tend to lower it. The desirables pull the average upwards, the undesirables pull it downwards; and when these upward and downward tendencies are just equal they neutralize one another and a static condition prevails: The race as a whole neither advances nor recedes.

If the potency of the desirable class to produce desirable children is greater than the potency of the undesirable class to produce undesirable children, then we have a case of unstable equilibrium, and the whole race begins to move upwards.

Conversely, if the potency of the undesirable element is greater than that of the desirable element, then the race begins to move downwards.

Thus, it is the *difference* between the potencies of the desirable and undesirable classes that controls and determines the amount and direction of the racial movement.

If, then, we wish to improve the race the aim should be *to increase the potency of the desirable class to produce desirable children*; and this can be accomplished by promoting the marriages of the desirable with one another.

The moment we have a body of desirable persons *whose parents were also desirable*, improvement of the race begins through the marriage of such persons with the normal population: for the proportion of desirable offspring born from the normal partners will be greater than in cases where the desirable partner had no ancestors belonging to the desirable class.

The improvement will be still greater when we have a body of desirable persons who had grandparents as well as parents desirable; and still greater with each increase in the number of desirable ancestors.

Thus, the simple process of promoting the marriages of the desirable with the desirable will, through the mixture of the descendants with the rest of the population, inaugurate an improvement of the whole race; and the movement will advance with accelerated velocity as we have more and more potent individuals of the desirable class. This process continued through a number of successive generations would ultimately result in the establishment of a prepotent stock within the desirable class, and then the improvement would be very marked indeed.

Chief Object of Eugenics

Here it is to be noted that the elevating tendency is due to the desirable class alone; and that improvement depends upon *increasing the number and proportion of desirables born in successive generations of the population*. Hence, this should be the chief object of eugenics; and it is to be regretted that the efforts of eugenists have been mainly directed to the diminution of the undesirable class.

So much has this been the case that the very word "eugenics" is suggestive to most minds of hereditary diseases and objectionable abnormalities; and of an attempt to interfere, by compulsory means, with the marriages of the defective and undesirable. This relates to cacogenics ("badly born") rather than to eugenics ("well born").

The utmost that could be even hoped for from such a process would be to lessen the tendency to retrogression and degeneration; and even this result would not be attained, at least in any great degree, for the simple reason that the undesirables, as a rule, are descended from normal parents.

Prohibition of marriage would not, therefore, have much effect upon the continued production of an undesirable class. We would have just about as many undesirable people appear in the next generation, *born from the normal population*.

Then again, the tendency to reversion to the normal type of the race is so strong that the children of undesirables are mainly of the normal type; so that prohibition of marriage would prevent the production of very many more normal children than undesirable children.

Whatever processes may be employed to improve the race, we shall always have the undesirable with us, because they are sprung mainly from the normal class; and *it is more practicable to improve the undesirable strains than to eradicate them.*

If undesirables marry normal or desirable partners they will not only have fewer undesirable children than if they married one another, but the potency of the offspring to produce undesirable grandchildren will be reduced. The undesirable blood is diluted, so to speak, by admixture with normal blood; and most of the offspring will be of the normal type.

Conclusion

A public sentiment already exists that persons possessing inherited characteristics of a desirable kind should marry and have large families. This sentiment undoubtedly is favorable to the improvement of the race; but it does not go far enough.

We should impress upon the public the point that one certain means of increasing the prevalence of any hereditary characteristic in a community is to induce the individuals who possess it *to marry one another*; and thus produce a more potent stock in the next generation.

It is neither practicable nor advisable that the individuals referred to should marry exclusively among themselves, but only to a much greater extent than now prevails; and the public policy should be: Promote the marriages of the desirable with one another.

Note

1. See appendix to Rep. Royal Comm. on Blind, Deaf and Dumb. etc., London, 1889, ii, 322; also Education of Deaf Children, Pt. II, p. 104, Volta Bureau, Washington, D.C., 1892.

5 TECHNIQUES OF MEASUREMENT

Introduction

Introduction

Evelynn M. Hammonds and Rebecca M. Herzig

Underlying most scientific investigations of racial difference are very practical problems of *method*: when, how many, and which measurements are to be taken, by whom, in what conditions. How might one go about tabulating the number of hairs on a human head? Or weighing 800 slippery, perishable, human brains? Or assessing the differences in color between brown and blue eyes? Each of these technical questions necessarily raises larger dilemmas about the nature of difference. As the selections included here indicate, although measurements are often undertaken in order to resolve political and social debate, techniques of measurements themselves reflect and reproduce received controversies about race.

And such measurements carry definite material consequences. In the antebellum South, where legal distinctions between bodies meant the difference between enslavement and liberty, judges were often summoned to stabilize racial categories for a ruling class that required a world comprised of black and white. To settle disputes over property, inheritance, divorce, and self-determination, courts routinely sought reliable standards of classification, a tool that might irrefutably determine disputed racial identity. Tools for measuring difference have also been critical in calculating life insurance premiums, determining medical treatment, gauging fitness for military service, and deciding Indian tribal membership.

Understandably, then, tools and methods have long been matters of contention in sciences of difference. Studies often belabor the minute details of their tools (cephalometers, craniometers, craniophores, craniostats, parietal goniometers) and techniques (discriminant functional analysis, calibration, electrophoresis). They also must endeavor to explain the relationship between individuals and groups—or, in contemporary parlance, the selection of a "representative sample." In the first selection, from a statistical survey of Civil War soldiers completed on behalf of the U.S. Sanitary Commission, author Benjamin Gould explains the relationship between statistical averages and the discernment of "types" of man. Gould (1824–1896), who studied with the Harvard mathematician and philosopher Benjamin Peirce, is remembered by historians of science for his massive compendia of star positions. That pre-war work for the U.S. Coast Survey appears to have informed his actuarial studies for the U.S. Sanitary Commission. In this selection, to whom does he give credit for the development of his statistical methods? Where do statistically generated ideal racial types exist, according to Gould?

In an important early article, the influential anthropologist Franz Boas similarly discusses the practical dilemmas of assessing human difference. Where Gould worked with living subjects, Boas focuses on skeletons, describing the difficulties of obtaining anatomical materials (the "soft parts" of bodies). What role does Boas allot to measurement within physical anthropology? How does he distinguish measurement from description, and what limitations does he place on the utility of measurement? What kind of emphasis is placed on the idea of "frequency"? Boas's later essays, particularly his 1911 article "Instability of Human Types" and his 1922 "Changes in the Bodily Form of Descendents of Immigrants," are often credited with producing a radical break in theories of race in western science, by introducing distinctions between race, language, and culture. Do you detect any such distinctions here?

The emphasis on measurement and quantification evident in the work of Gould and Boas, part of a much broader late nineteenth-century interest in anthropometry, continued among psychologists, criminologists, physical anthropologists and others over the opening decades of the twentieth century. Some researchers attempted to develop new techniques to quantify abstract qualities such as self-control, inhibition, or volition. Others instead sought to measure more readily identifiable portions of human anatomy such as sweat glands or pelvises. Still others struggled to come up with newly precise tools for measuring characteristics with a range of infinitesimal variations, such as pilosity (hairiness) or skin color.

With the 1898 annexation of Hawai'i and the 1898 takeover of the Philippines, Puerto Rico, Guam, and Cuba, researchers devoted increasing attention to racial comparisons across the growing American empire. The University of Chicago anatomists Elbert Clark and Ruskin H. Lhamon exemplified this trend with their 1917 essay in the journal *Anatomical Record*. Notice here the recurrent question of relations between specific physiognomic characteristics and adaptability to particular climatic conditions. A subsequent essay by H. A. Bowman of the Anatomical Laboratory at Western Reserve University in Cleveland instead addresses the technical problems associated with gauging skin pigmentation, including the difficulties of printing true colors on chip samples, the inaccuracies presented by varying light sources, and the personal errors introduced by individual researchers. Publishing in a journal edited by Aleš Hrdlička, the pre-eminent anthropometrist of the day, Bowman joined others in promoting a specialized "color top," in part because it de-emphasized the implicit European or white standard previously used in skin pigmentation studies.

The final selection touches on a subject of ongoing controversy: the relation between innate physiological characteristics and athletic ability. Here Montague Cobb (1903–1990), one of the only professional black anatomists of his era, explores whether "Negroes" have "special physical endowments fitting them for sprinting and jumping." What forms of evidence does he summon when addressing this question? What techniques of measurement does he describe? How have questions concerning the relations between training, innate endowment, athletic success, and race persisted into the present? Finally, some scholars have argued that each act of measurement not only assesses difference, but also serves to establish the difference it purports to measure. Would you agree or disagree? In what senses do these tools *generate* difference in the act of measuring? Do the instruments such as the color top or the statistical table themselves contain racial politics?

Bibliography

Cole, Simon A. *Suspect Identities: A History of Fingerprinting and Criminal Identification.* Cambridge: Harvard University Press, 2001.

Crane, Albert Loyal. "Race Differences in Inhibition." *Archives of Psychology* 9, no. 63 (1922–1923): 9–84.

Danforth, C. H., and Mildred Trotter, "The Distribution of Body Hair in White Subjects." *American Journal of Physical Anthropology* 5, no. 3 (1922): 259–265.

Gere, Cathy. "Bones That Matter: Sex Determination in Paleodemography, 1948–1995." *Studies in History and Philosophy of Biological and Biomedical Sciences* 30C, no. 4 (December 1999): 455–471.

Gilman, Sander L. *Making the Body Beautiful: A Cultural History of Aesthetic Surgery.* Princeton: Princeton University Press, 1999.

Goodman, Alan H. "Bred in the Bone?" *The Sciences* (March/April 1997), 20–25.

Gould, Stephen Jay. "American Polygeny and Craniometry before Darwin: Blacks and Indians as Separate, Inferior Species." In Sandra Harding, ed., *The "Racial" Economy of Science.* Bloomington: Indiana University Press, 1993, 84–115.

Guthrie, Robert V. *Even the Rat was White: A Historical View of Psychology.* Boston: Allyn and Bacon, 1998.

Hammonds, Evelynn M. "New Technologies of Race." In Jennifer Terry and Melodie Calvert, eds., *Processed Lives: Gender and Technology in Everyday Life.* New York, 1997.

Harrison, Ira E., and Faye V. Harrison, eds. *African-American Pioneers in Anthropology.* Urbana: University of Illinois Press, 1999.

Rafter, Nicole Hahn. *Creating Born Criminals.* Urbana: University of Illinois Press, 1997.

Sheldon, William H. *The Varieties of Human Physique: An Introduction to Constitutional Psychology.* New York: Harper & Brothers, 1940.

Spady, J. G. "Dr. W. Montague Cobb: Anatomist, Physician, Physical Anthropologist, Editor Emeritus of the Journal of the National Medical Association, and First Black President of NAACP." *Journal of the National Medical Association* 76, no. 7 (July 1984): 739–744.

5.1 Investigations in the Military and Anthropological Statistics of American Soldiers (1869)

Benjamin Gould

The value of the results of these measurements will depend chiefly upon the degree of approximation with which their mean represents the normal dimensions of the classes of men under consideration. These normal dimensions would, for any one class of persons, be afforded by the arithmetical mean, or average value, of the corresponding dimensions of all men of the same class, provided an indefinitely large number could be obtained; and it becomes an important problem to ascertain the limits within which our finally adopted determinations would probably be varied by an indefinite increase in the number of men measured,—or, in other words, to obtain some numerical expression of the degree of reliance which should be placed on the mean values derived from our respective measurements, as indicating the normal dimensions.

It seems, therefore, not amiss to offer here a few words concerning the true significance of averages, and the nature of typical forms. The subject has been so thoroughly elaborated, both in its mathematical and its philosophical bearings, that few, if any, remarks on its elementary principles may claim the credit of originality. Even the mode of presenting the ideas involved in a popular form offers little unoccupied ground, since the elegant and learned treatises by Quetelet, De Morgan, and others. And the only endeavor in this place will be to present such considerations as are requisite for proper criticism of our materials.

. . .

Now we may regard the laws of Nature, to which the Supreme Being has assigned the duty of carrying out his creative mandates, as occupying, in the almost infinitely varied circumstances under which they find application, a position analogous to that of marksmen aiming at a target. There exists, for plant and beast and man, a type,—not necessarily clothed with a material body, yet none the less a real entity. And as, among hundreds of thousands of shots, no single one may centrally strike the target, while their grouping may indicate its center, with a precision greater than our senses permit us to appreciate; so, by a sufficient number of measurements, under circumstances sufficiently varied, upon a sufficient number of subjects, we may arrive at a knowledge of the form and dimensions of the ideal, typical plant, or animal, or man,—to which all individuals are approximations, although no one of them may ever have attained, or hoped to attain, its accurate impersonation. Varieties and individual dissimilarities here occupy positions relatively analogous to the constant and variable

Gould, Benjamin, *Investigations in the Military and Anthropological Statistics of American Soldiers* (New York: Hurd and Houghton, 1869), 240–241, 244–247, 249.

errors of aim on the part of the marksman; and possibly in the exalted scheme of Nature, even species and genera, to go no higher, may in their turn occupy the same relative stations, when our field of view is adequately magnified.

Applying these principles to the present investigation, we see that there is a human type to be sought, though attainable only by the combination of results from many races; a type of race, attainable through the study of many nationalities; a type of nationality, and a type of each class within its bounds. Our measurements pertain almost exclusively to American soldiers, and these not of the same age, nor all of them of mature growth; yet they are from wide-spread regions of the continent, and many of them belonged by birth to other nations. Our aim has been to deduce the types for as many as may be of these various classes of men, and to test the trustworthiness of the results by the accordance between the series of observed and theoretical deviations of the several measurements from their mean.

The existence of types for man, and for the races and classes of men, was first demonstrated by Quetelet, who has done more than any one else to study and discuss the average man, in his various relations, physical, social, and moral. He has illustrated the relation of the theoretical laws of chance to investigations like the present so happily, that, even at the risk of prolixity, it seems well to reproduce the illustration here. It must first be premised that, by the mean or average result of measurement, two distinct kinds of inference may be denoted. The mean result may be the mean of many measurements of a single object,—and thus afford the closest attainable representation of a material thing,—or it may be a mean of the measurements of many different, although similar objects, and thus represent no particular thing. In the first instance, the individual measures, and in the second, the measures of individuals, group themselves about the mean in conformity with the law of error; but there is this wide distinction, that while in the former case the several values are closely connected, varying only by the errors of the measurer, they are in the latter case devoid of all mutual connection of a material kind; and the existence of any mutual connection must be determined by the degree and nature of the accordance of the measures. When such connection exists, the accordance or discordance of the several measures follows precisely the same laws in the two instances; and the adoption of the idea of a type, in approximate conformity to which all individuals of a class are fashioned, abolishes the practical distinction between the two sorts of means.

To borrow Quetelet's illustration, let us suppose that it is desired to obtain by measurement the dimensions of a statue. Measuring any portion ten or twelve times successively, with all possible care, it is improbable that any two of the results would be identical; and in a thousand repetitions of the process we should obtain a series of numerical values, the mean of which would differ very little from the true one, while the amount of discordance in individual cases would be inversely proportional to the precision of the measures. And assorting the results by order of discordance from the mean, we should find their distribution to follow the law of probability, since the only deviations would be those due to want of skill, or care, or to imperfection of the senses.

If, instead of a statue, a living person be taken as the subject of measurement, the chances of error are much more numerous, and the magnitude of the errors would be increased by the absence of rigidity of the flesh, and by the real fluctuations of the dimensions in consequence of respiration and other involuntary motions, and unconscious changes of attitude by the subject. Yet the mean of a thousand measurements of each dimension would afford an approximation to the true average dimensions of the living person, nearly as close as to those of the statue in the former instance, and the variations of the several results would follow a similar law.

Modifying the supposition, imagine a thousand sculptors employed to copy the statue or the person, with all possible precision, and their copies measured in the place of the original. Then, to the original sources and chances of error would be added the inaccuracies of the copyists; still from the mean of all we should derive essentially the same value, and the discordances would be similarly grouped about this mean.

Finally, suppose that while the number of the copyists is adequately increased, many of them are hampered by the prejudices or prepossessions of their several schools of art; that their material varies in character, both for the different copies and for the different portions of the same copy; that many are supplied with improper tools; that some are partially blind, others crippled in their hands and arms; and that their degrees of skill are very diverse; still the mean of all the results would enable the archetype to be reproduced with much accuracy, and the agreement, in number and amount, of the variations with those prescribed by the law of error, would establish the fact that such a common model had actually existed.

Thus it is that we may hope to discover the type of humanity, as well as the types of the several classes and races of man. In the present research we are dealing only with some of his external physical manifestations, but we aim at the deduction of the numerical expressions of these as a step toward constructing the typical or average man, who, though probably never clad in flesh, is yet a reality, not merely existing in the Divine mind, but capable of perception and recognition by human sense. Indeed the external form of this average man may legitimately be adopted as a standard of beauty and a model for art. The eminent scientist already named has shown that we may discover not merely the outward semblance of this abstract being, but his needs, capacities, intellect, judgement, and tendencies; and Quetelet may thus be regarded as the founder of statistical anthropology, indeed of social science, in the true significance of the word, according to which science depends upon the investigation of laws, not upon the consideration of isolated facts, nor the dissemination of correct principles.

It is only when statistical research conducts to the discovery of types, or when the inferences drawn from it may be tested, and confirmed by detection of some systematic subordination to law in their variations, that statistics afford a safe guidance. The discredit in which this mode of investigation is held by many able men, and the errors in which it has frequently involved candid inquirers, may thus be accounted for. To hold any means of research in disrepute is unphilosophical; to regard any process as responsible for the results of its misapplication is absurd. Many moral, social, political, and physical laws seem only deducible, and are certainly only demonstrable, by statistical investigation, although no methods in the whole range of science require more caution and skill in their employment, or can more easily delude the unwary.

"The average man," says Quetelet, "is for a nation what the center of gravity is for a body; to the consideration of this are referred all the phenomena of equilibrium." The full discussion of many of the data collected in these examinations, and preserved in the archives of the Sanitary Commission, would doubtless bring many important facts clearly to light. But various considerations, especially that of financial means, restrict the present discussion to some of the more important physical characteristics.

. . .

The numerical values of some of the quantities here described are given, with some of the mean results of measures of the several dimensions, in order to aid the student in estimating the degree of reliance to which the results are entitled. But he must remember that the average discordances, being deduced from the variations of individual measures from their mean, show the numerical values, not of the tendency to error in the measurements, but of the tendency of single members of a class to vary from the mean or type corresponding to that class. So, too, the quantity which we call the Probable Error of the Mean denotes the value of this probable error, as deduced from intrinsic evidence alone, this same degree of variation in individual results furnishing the basis. Whether the value obtained is a typical value or not, must be inferred from the degree of accordance between the system of computed and the system of observed variations. This degree of accordance between the two systems is itself capable of expression in a concise numerical form, by deducing its modulus from the series of differences between the theoretical and actual values, after each difference has been affected with its proper weight; but such computation is somewhat laborious, and it has appeared unadvisable to undertake it here.

5.2 "Some Recent Criticisms of Physical Anthropology" (1899)

Franz Boas

During recent years a number of severe attacks against the methods of physical anthropology have been made, which are directed mainly against two points—(1) the possibility of classifying mankind according to anatomical characteristics, and (2) the practicability of description of types by means of measurements.

Before we attempt to reply to these criticisms, it may be well to make a few brief remarks on the development of the methods of physical anthropology. The living representatives of the various races of man were originally described according to their general appearance—the color of the skin, the form and color of the hair, the form of the face, etc. Later this general description was supplemented by the study of the skeletons of various races, and a number of apparently characteristic differences were noted. One of the principal reasons that led to a more detailed study of the skeleton and to a tendency to lay the greatest stress upon characteristics of the skeleton, was the ease with which material of this kind could be obtained. Visitors to distant countries are likely to bring home skeletons and parts of skeletons, while not much opportunity is given for a thorough examination of a considerable number of individuals of foreign races. The difficulty of obtaining material relating to the anatomy of the soft parts of the body has had the effect that this portion of the description of the anatomy of man has received very slight attention. In comparatively few cases have we had opportunity to make a thorough study of the characteristics of the soft parts of the body of individuals belonging to foreign races. The desire to find good specific characters in the skeleton has also been stimulated by the necessity of studying extinct races. The conditions in these cases are the same as those found in paleontological studies, where the osseous remains alone of extinct species are available. Researches into the earliest history of man must be based on studies of the skeleton.

Studies of the human skeleton had not been carried very far when it was found to be not quite easy to determine racial characteristics with sufficient accuracy by mere verbal description. This led to the introduction of measurements as a substitute for verbal description. With the increase of the material, the necessity of accurate description became more and more apparent, because intermediate links between existing forms were found with increasing frequency. These conditions have led to a most extensive application of the metric method in the study of the human skeleton and also in the study of the external form of the living.

Boas, Franz, "Some Recent Criticisms of Physical Anthropology," *American Anthropologist* 1 (January 1899), reprinted in Franz Boas, *Race, Language, and Culture* (Chicago: University of Chicago Press, 1982), 165–171.

The results of the minute studies that have been carried on in this manner appear discouraging to many students, because we have not been able to find any criterion by which an individual skeleton of any one race can be distinguished with certainty from a skeleton belonging to another race, except in a very general way. A typical full-blood Negro may be distinguished from a White man, and an Indian of Florida from an Eskimo; but it would be difficult to distinguish the skeleton of a China-man from that of certain North American Indians.

This lack of definite individual descriptive features has led many investigators to conclude that the method is at fault, and that the skeleton cannot be used as a satisfactory basis for a classification of mankind. This view has been strengthened by the belief, frequently expressed, that the characteristic features of each race are not stable, but that they are influenced to a great extent by environment, geographical as well as social.

It seems to me that these views are not borne out by the observations that are available. The first objection, which is based on the lack of typical characteristics in the individual, does not take into consideration the fact that anthropological study is not a study of individuals, but of local or social varieties. While it may be impossible to classify any one individual satisfactorily, any local group existing at a certain given period can clearly be characterized by the distribution of forms occurring in that group. I do not hesitate to say that, provided we had satisfactory statistics of the distribution of human forms over the whole globe, an exhaustive description of the physical characteristics of any group of individuals belonging to one locality would enable us to identify the same without any difficulty. This clearly emphasizes the fact that anthropological classification must be considered as a statistical study of local or social varieties. But it will be asked, How does this help in classifying individual forms? The problem must be considered in the following way:

Each social unit consists of a series of individuals whose bodily forms depend on their ancestry and on their environment. If the opinion of the critics of physical anthropology regarding the predominant effect of environment is correct, then we cannot hope to make any discoveries as to ancestry of local or social groups by means of anatomical investigations. If, on the other hand, it can be shown that heredity is the predominant factor, then the prospects of important discoveries bearing on the early history of mankind are very bright indeed. It seems to the writer that a biological consideration makes it very probable that the influence of heredity should prevail, and thus far he has failed to find conclusive proof to the contrary.

The critics of the method of physical anthropology will of course concede that a Negro child must be a Negro, and that an Indian child must be an Indian. Their criticism is directed against the permanence of types within the race; for instance, against the permanence of short or tall statures, or against the permanence of forms of the head. It must be conceded that muscular development may exert an important influence on the form of bones, but it does not seem likely that it can bring about an entire change of form. The insufficiency of the influence of environment appears in cases where populations of quite distinct types inhabit the same area and live under identical conditions. Such is the case on the North Pacific coast of our continent; such was the case in successive populations of southern California and of Utah.

While this may be considered good evidence in favor of the theory of predominance of the effect of heredity, the actual proof must be looked for in comparisons between parent and offspring. If it can be shown that there is a strong tendency on the part of the offspring to resemble the parent, we must assume that the effect of heredity is stronger than that of environment. The method of this investigation has been developed by Francis Galton and Karl Pearson, who have given us the means of measuring the degree of similarity between parent and child. Wherever this method has been applied, it has been shown that the effect of heredity is the strongest factor in determining the form of the descendant. It is true that thus far this method has not been applied to series of generations, and under conditions in which a considerable change of environment has taken place, and we look forward to a definite solution of the problem of the effect of heredity and of environment through the application of this method. In the study of past generations we cannot, on the whole, compare directly parent and offspring, but we have to confine ourselves to a comparison between the occurrence of types during successive periods. The best available evidence on this subject is found in the populations of Europe. It does not seem likely that the present distribution of types in Europe can be explained in any other way than by the assumption that heredity had a predominant influence. Much has been made of the apparent change of type that takes place in the cities of Europe in order to show that natural selection may have played an important part in making certain types of man predominant in one region or another. Ammon has shown that the city population of southwestern Germany is more short-headed than the country population, and concludes that this is due to natural selection. All the phenomena of this character that have been described can be explained satisfactorily by the assumption that the city population is more mixed than the country population. This point has been brought out most clearly by Livi's investigations in Italy. He has proved that in regions where long-headed forms prevail in the country, in the city the population is more short-headed; while in regions in the country in which short-headed forms prevail, in the city the population is more long-headed.

Under present conditions, it seems best not to start the study of the anatomical characteristics of man from far-reaching assumptions in regard to the question of the effect of heredity and environment, but first of all to ascertain the distribution of types of man. This is a definite problem that requires treatment and investigation just as much as the study of languages or the study of the customs of various tribes. At the present time we are far from being familiar with the distribution of types on the various continents. No matter what the ultimate explanation of the distribution of types may be, we cannot evade the task of investigating their present distribution and of seeking for the explanation of the reasons for such distribution.

Before entering into this subject more fully, it may be well to take up the second criticism of the method of physical anthropology, which has been made with increasing frequency of late years. A number of investigators object to the metric method of anthropology, and desire to bring about a substitution of description for measurements. This proposition is based on a misunderstanding of the function of measurements. The necessity of making measurements developed when it was found that the local varieties of mankind were very much alike—so much so that a verbal description failed to make their characteristics sufficiently clear. The process by means of which measurements have been selected has been a purely empirical one. It has been found that certain measurements differ considerably in various races, and are for this reason good classificatory criteria. The function of measurements is therefore solely that of giving greater accuracy to the vague verbal description. It is true that in the course of time a tendency has developed of considering as the sole available criteria of race the measurements which by experience have been found to be useful. This is true particularly of the so-called cephalic index; that is, the ratio of width to length of head. There are anthropologists who have subordinated everything else to the study of the cephalic index, leaving out of consideration altogether the forms of the skull and of the skeleton as expressed by their metric relations or as expressed by means of drawings or diagrams. It has frequently been pointed out that the same cephalic index may belong to forms that anatomically cannot be considered as equivalent. We find, for instance, that the same cephalic index belongs to the Eskimo, to the prehistoric inhabitant of southern California, and to the Negro. Still these three types must be considered as fundamentally different. Anthropologists who limit their work to the mechanical application of measurements, particularly of single measurements, and who try to trace the relationships of races by such means, do not apply the metric method in a correct way. It must be borne in mind that measurements serve the purpose only of sharper definition of certain peculiarities, and that a selection of measurements must be adapted to the purpose in view. I believe the tendency of developing a cast-iron system of measurements, to be applied to all problems of physical anthropology, is a movement in the wrong direction. Measurements must be selected in accordance with the problem that we are trying to investigate. The ratio of length and breadth of head may be a very desirable measurement in one case, while in another case it may be of no value whatever. Measurements should always have a biological significance. As soon as they lose their significance they lose also their descriptive value.

The great value of the measurement lies in the fact that it gives us the means of a comprehensive description of the varieties contained in a geographic or social group. A table that informs us of the frequency of various forms as expressed by measurements that occur in a group gives us a comprehensive view of the variability of the group that we are studying. We can then investigate the distribution of forms according to statistical methods; we can determine the prevalent type and the character of its variation. The application of rigid statistical methods gives us an excellent means of determining the homogeneity and the permanence of the type that is being studied. If a group of individuals who present a homogeneous type is not subject to changes, we must expect to find the types arranged according to the law of probabilities; that is to say, the average type will be the most frequent one, and positive and negative variations will be of equal frequency. If, on the other hand, the homogeneous type is undergoing changes, the symmetry of arrangement will be disturbed, and if the type is heterogeneous we must expect irregularities in the whole distribution. Investigations of this character require the measurement of very extensive series of individuals in order to establish the results in a satisfactory manner. But the character of the distributions that may thus be obtained will furnish material for deciding a number of the most fundamental questions of physical anthropology.

I may now revert to the question previously under discussion. I have tried to show that the metric method may furnish us material proving the homogeneity or heterogeneity of groups of certain individuals. This test has been applied to a number of cases. I have examined from this point of view the North American half-bloods, that is, individuals of mixed Indian and White descent. I have shown that the transverse development of the face, which is the most distinctive difference between Indian and White, shows a tendency in the mixed race to revert to either of the parental races, and that there is no tendency toward the development of an intermediate form. Bertillon has shown similar irregularities to exist in France. On the other hand, extensive series of measurements of enlisted soldiers of Italy show in many parts of the kingdom a comparatively homogeneous series. Hand in hand with this phenomenon go remarkable differences of variability. In places where we have reason to believe that distinct types have intermingled, we find a great increase in variability, while in regions occupied by homogeneous populations the variability seems to decrease. These facts are strong arguments for the assumption of a great permanence of human types. It is necessary that the analysis of distributions of measurements be carried much further than it has proceeded up to the present time; this done, I believe we shall obtain a means of determining with considerable accuracy the blood-relationships of the geographical varieties of man.

I wish to say a word here in regard to the question of the relationship between the earliest prehistoric races and the present races. In so far as the reconstruction of the characteristics of prehistoric races can be based on extensive material, there will be a certain justification for a reconstruction of the soft parts, if a detailed comparison of the osteological remains of prehistoric types and of present types proves them to be conformable. Where, however, the similarity is based on a few isolated specimens, no such reconstruction is admissible, because the attempt presupposes the identity of the prehistoric race with the present. Since remains of the earliest man are very few in number, it is hardly possible to gain an adequate idea of what the characteristics of the soft parts of his body may have been except in so far as the forms of muscular attachments allow us to infer the size and form of muscles.

When we base our conclusions on the considerations presented in this paper, we must believe that the problem of physical anthropology is as definite as that of other branches of anthropology. It is the determination and explanation of the occurrence of different types of man in different countries. The fact that individuals cannot be classified as belonging to a certain type shows that physical anthropology cannot possibly lead to a classification of mankind as detailed as does the classification based on language. The statistical study of types will, however, lead to an understanding of the blood-relationship between different types. It will consequently be a means of reconstructing the history of the mixture of human types. It is probable that it will lead also to the establishment of a number of good types which have remained permanent through long periods. It will be seen that that part of human history which manifests itself in the phenomena that are the subject of physical anthropology is by no means identical with that part of history which manifests itself in the phenomena of ethnology and of language. Therefore we must not expect that classifications obtained by means of these three methods will be in any way identical. Neither is it a proof of the incorrectness of the physical method if the limits of its types overlap the limits of linguistic groups. The three branches of anthropology must proceed each according to its own method; but all equally contribute to the solution of the problem of the early history of mankind.

5.3   "Observations on the Sweat Glands of Tropical and Northern Races" (1917)

Elbert Clark and Ruskin H. Lhamon

The present preliminary report on the sweat glands was begun as a part of a joint investigation suggested by the late Paul C. Freer, Director of the Bureau of Science of Manila. The general problem was the supposed untoward effect of a tropical residence on the white man and the supposed constitutional adaptability of the dark races to a tropical climate. The investigations were undertaken in a coöperative way by chemists and physicists of the Bureau of Science, the departments of Anatomy, Pharmacology, Physiology and Physics of the University of the Philippines and the United States Army Medical Board for the Study of Tropical Diseases.

Certain claims of Daubler ('00) and Aron ('11) seem to indicate at least one definite adaptation of the dark races to the tropics and lead us to investigate the sweat glands. Daubler states that the size of the sweat glands of the native of tropical Africa is much greater than that of the European. Aron finds that the sweating apparatus of the aborigines of the Philippines, the Negritos, is much superior to that of the white man. This superiority he says is shown by the difference in the manner of sweating rather than in the amount of sweat produced. According to Aron, the Negrito secretes small beads of sweat over the entire body, which soon forms a thin film. As the whole surface of the body is covered by this water film, the maximum cooling effect from evaporation is obtained. In the case of the white man, on the other hand, the sweating is practically limited to certain areas of the body surface. In these areas the sweating may be quite profuse, but, as most of it drops off, comparatively little cooling effect from evaporation is produced. He suggests that the Negrito has a greater number of sweat glands which are more equally distributed over the entire body.

We have attempted to compare the sweat glands of the tropical races with those of the northern races. Our observations at present scarcely extend beyond a comparison of the number of sweat glands in certain definite skin areas of various races. Similar comparisons of other areas will be made as material is collected. A comparison as to size is an almost endless task as is shown by the work of Huber and Adamson ('03). These investigators have found from measurements of reconstruction a great variation in the size of the sweat glands in the Caucasian. The length of the tubule in the coiled portion of the gland from the plantar region of the foot of an adult was 4.25 mm. A gland from the hairy portion of the pubic region of a woman was found to measure 10.4 mm. They have further shown, and we have confirmed their findings, that it is almost impossible to determine with any degree of accuracy the size of sweat glands without making reconstructions of them. It is even

Clark, Elbert, and Ruskin H. Lhamon, "Observations on the Sweat Glands of Tropical and Northern Races," *Anatomical Record* 12, no. 1 (1917): 139–147.

now and then exceedingly difficult in a series of sections to trace with certainty a single gland, especially toward the beginning and end of a series of sections of any one gland. Loops of neighboring glands are often in close proximity and are apparently surrounded by a common layer of somewhat denser connective tissue so that a separation of tubules belonging to two, or now and then even three, contiguous glands can be made with certainty only by reconstruction.

In maceration preparations of skin from the palmar region and from the chest in both American and Filipino we have noted a great variation in the size of the glands in each piece of skin taken. Frequently a given gland was fully twice as large as its neighbor. It is thus apparent that a racial comparison of sweat glands as to size to be of any pretence to accuracy must be based upon measurements of a vast number of glands in different portions of the skin of the several races.

Technique

In our study of the number and distribution of the sweat glands we find that there is less variation in those occurring on the plantar surface of the foot and the palmar surface of the hand than in other regions of the body. The sweat glands first make their appearance in the embryo in these regions. These areas also lend themselves readily to an extensive enumeration of the sweat glands. Our observations so far have been made principally on these regions. In a warm climate prints of the fingers, palms, toes and plantar surfaces of the feet can be made which in many cases will give a negative impression of every gland duct in the area printed. We have employed the method which is in use in the recruiting office of the United States Army. A very thin layer of the best printer's ink is rolled out with a small hand roller upon a glass plate. The subject's finger (hand or toe, etc.) is carefully and lightly pressed first upon the ink, and next upon a special type of glazed paper. The impressions, when made under suitable conditions, are remarkably clear and the duct of every sweat gland can usually be accounted for. Skin with much cornified epithelium will not make a satisfactory print unless pains are taken to macerate with a dilute caustic and scrape off the surface tissue.

The gland ducts may be counted directly upon the hands and feet with the aid of a good hand lens of about 12 diameters magnification. This is facilitated by first rubbing a little powdered graphite upon the surface of the skin to be examined. This method is far more tedious and has been used only as a check upon the print method. The print method is obviously not adapted to those areas of skin where hair and wrinkles occur. As the ducts of sweat glands frequently open into hair follicles the glands can not be counted by direct inspection. Here we have resorted to maceration methods and to stained sections of skin cut parallel to the surface. The latter method was used by Krause in his study of the number of sweat glands in the skin of the various regions of the body.

Through the courtesy of the surgeon-general's office United States Army and of the office of the chief surgeon, Philippien Division of the United States Army we have had the opportunity of examining finger prints of 300 American white soldiers, 200 American negro soldiers, 150 Filipino (Christian) soldiers and 100 Moro scouts. In addition we have ourselves made many prints of fingers, toes, palms and plantar surfaces of the feet of Americans, Filipinos, Igarotes, Negritos and Hindoos.

In counting sweat glands a glass slide with a graduated square (0.5 cm each way) was placed over the print, graduated surface down, and the specimen magnified 10 diameters. In good prints, as stated above, every sweat duct can be counted (fig. 5.3.1). The graduated square was always placed over the print of the distal phalanx in counting the glands of the fingers, and over that area where the cristae cutis form a whorl or delta.

Racial Variation in Sweat Glands

We have counted 248,998 sweat glands in $\frac{1}{4}$ square centimeter areas of 1,572 fingers and 38,736 in the palms, toes and plantar surfaces of the feet. Table 5.3.1 shows the average counts per square centimeter of skin area for the various races. Rather uniform variations have been observed in the distribution of sweat glands in the fingers. The number of glands varies in different areas of the volar surfaces of the fingers. The number is greatest near the distal ends and smallest in the immediate region of the flexion groove at the joints. As stated above, the number is most nearly constant in the region of the whorl or delta than in other regions. Comparatively few prints of the thumb have been examined. In all of these, however, the number of sweat glands has been distinctly lower than in any of the other fingers. Of the other fingers the second or index finger has shown the lowest average number of glands per unit of skin area, while the fourth or ring finger has shown by far the greatest number of glands. A more detailed comparison of the number of sweat glands per square centimeter of skin for the different fingers in the several races is given below in table 5.3.1. The average number of glands per square centimeter of skin area for the finger for all the races examined is 624.4.[1] The greatest number was found in prints of the fingers of Negrito children. It was found that the number of glands per unit of skin area for the hands, feet and toes bears a rather close racial relation to the number on the fingers, and in the different individuals varies directly with the number on the fingers. As regards the different races our results show a greater number of sweat glands in all the tropical than in the northern races.

Taking the American white soldier as the standard, the number of sweat glands per unit of skin area was found to be 6.83 per cent greater in the American negro soldier, 16.61 per cent greater in the Filipino soldier, 22.34 per cent greater in the Moro soldier, 26.81 per cent greater in the adult Negritos, 31.72 per cent greater in the Hindu and 69.82 per cent greater in the Negrito youths and children. Additional details of these counts will be found in table 5.3.1. The greater number of sweat glands per unit area with the Negrito youth and child is no doubt due to a corresponding difference in size of the individuals. As all the sweat glands are fully formed at birth[2] it is

Figure 5.3.1
Photograph of a finger print of the distal phalanx of the fourth finger (left) of an American white soldier. ×6.5.

Table 5.3.1

Average Number of Sweat Glands per Square Centimeter of Skin Area in the Various Races As Shown by Fingers

Nationality	Fingers of left hand				Fingers of right hand				Individual average
	5th	4th	3rd	Index	Index	3rd	4th	5th	
American (white)	569.6	600.4	564.4	511.2	519.2	546.4	590.4	552.4	558.2
American (negro)	605.2	634.4	586.0	576.6	561.6	590.4	631.2	594.8	597.2
Filipino	652.0	691.2	653.6	618.4	610.4	630.4	692.8	637.2	653.6
Moro	674.4	755.2	682.8	650.8	651.2	665.2	725.6	640.8	684.4
Negrito (adult)	670.4	752.0	718.8	682.4	666.0	720.0	736.8	701.6	709.2
Hindu	722.0	813.6	744.0	657.6	650.0	737.6	782.0	738.4	738.2
Negrito (youth)	1010.0	996.0	942.8	942.0	865.2	961.2	982.8	881.2	950.0
Average per finger	707.67	749.9	696.3	661.4	646.2	693.0	734.5	678.0	
General average									698.5

merely a question of the increase in skin area during growth bringing about a dispersion of the glands. The ratio of the number of sweat glands in the American white soldier (100 per cent) to that of the Filipino soldier (116.93 per cent) and the Negrito adult (124.05 per cent) shows a wider variation than the differences in size[3] of the individuals of these respective groups. The American negro soldier is of the same approximate size as the American white soldier, and there are 6.83 per cent more sweat glands per unit of skin area in the former. The Hindus examined were of a larger average size[4] than the American soldiers and showed the highest sweat gland count for adults (131.76 per cent). Thus racial variation in size does not account for the difference in ratio of the sweat glands per unit of skin area.

The number of sweat glands was determined in a similar manner from prints of the palmar surface of the hand and the plantar surface of the feet of Americans (white) and Filipinos. Successful prints were made from these areas of 6 American university men and 6 Filipino students; 325 separate areas were counted, giving a total of 38,736 glands. The average number of sweat glands on the palmar surface was 438.0 per square centimeter in the American, and 473.6 in the Filipino. On the plantar surface of the feet there were 436.4 glands per square centimeter in the American and 498.4 in the Filipino. On the plantar surfaces of the toes there was an average of 527 and 525.5 sweat glands per square centimeter in the American and Filipino respectively. Thus the number of sweat glands in the Filipino was in this series 8.1 per cent greater on the palm and 14.1 per cent greater on the plantar surface of the feet than in corresponding areas in the American. In all these counts there was very little individual variation. Different individuals of the same group gave almost the same count.

We are not able to confirm Aron's observation that the tropical aborigines secrete only small beads of sweat over the entire body. On two tramping expeditions in

Figure 5.3.2
Drawing of a finger print to show orifices of sweat gland ducts. ×10.

the mountains of the Philippines which we were fortunate enough to arrange with a number of Negritos we observed streams of sweat running down the back, and copious sweating on scalp, forehead and face and sweat dripping from the chin. When making finger prints in camp it was necessary repeatedly to dry off droplets of sweat from the fingers of the Negritos.

From the few maceration preparations mentioned above we are not able to discern any difference in the size of the sweat glands of the American and the Filipino.

As regards number, all our observations show a higher count in all the tropical races. These counts, furthermore, were made on those areas in which the number of sweat glands is the most nearly constant.

Notes

1. This average is much lower than the estimation of earlier authors, thus—"Über die Menge der Knäueldrüsen haben wir ältere Angaben von Krause senior denen zufolge ihre

Zahl zwischen 400–600 (Rücken, Wange, erste zwei Abschnitte der unteren Extremitaten) und 2600–2736 auf I □'' Haut schwankt und die grösseren Zahlen an der hanffläche und Fussohle sich finden. Neure Zählungen von Hörschelmann ergeben viel näher stehende Grenzzahlen von 641 Fussrucken, und IIII (vola manus) auf I □ cm und viel mehr drüsen."—Koelliker.

2. At the fourth month according to Wilder ('16).

3. Captain Davis of the recruiting office of the United States Army in Manila tells us the average weight of the American white soldier is approximately 150 pounds and of the Filipino scout approximately 130 pounds. The Negritos are smaller and can be estimated at about 120 pounds.

4. All these Hindoos were tall, large and portly and averaged 160–165 pounds in weight.

Bibliography

Aron, Hans 1911 Investigation on the action of the tropical sun on man and animals. Phil. Jour. Sc., Sec. B, vol. 6, p. 101.

Däubler 1900 Die Grundzüge der Tropenhygiene, Berlin, 1900. Cited by Aron.

Huber, G. Carl and Adamson, Edward William 1903 A contribution on the morphology of suderiparous and allied glands. Contrib. Med. Research dedicated to Victor Clarence Vaughan, 1903.

Koelliker, A. v Gewebelehre des Menschen.

Rubner, Max 1900 Vergleichenden Untersuchungen der Hautthätigkeit des Europäers und Negers. Archiv. f. Hygiene, Bd. 38, s. 148.

Wilder, Harris H. 1916 Palm and sole studies. Biol. Bull., vol. 30, p. 135.

Diem, F. 1907 Schweissdrüsen an der behaarten Haut der Saugertiere. Anat. Hefte, Bd. 34, s. 187.

5.4 "The Color-Top Method of Estimating Skin Pigmentation" (1930)

H. A. Bowman

Introduction

In spite of the doubt expressed, by almost all authors who have used the color top in estimating skin pigmentation, concerning the theoretical value of the estimations used, the method continues to be employed. Whatever the theoretical objections, the practical advantages are clear. It is a method readily applied in the field and in the laboratory. Whether it give an accurate estimate or not, it does at least give some relative quantitative precision to measurements of skin pigmentation. It is a far easier problem to match skin color by the color top than by von Luschan's scale. And most practical of all, it raises no resentment in the subject, for it never brings home to him, as the von Luschan scale invariably does, that his color is being matched against a European or white standard. For these reasons it has seemed worth while to spend a little time and effort on improving the technique and on investigating the practical reliability of the method.

There are now so many estimates registered upon the preserved samples of skin in this laboratory that the records provide an inexhaustible mine of material upon which to draw for such a purpose. At Doctor Todd's request, therefore, I have undertaken to make an investigation into the theoretical and practical reliability of the test in order to give the method its validation. A list of the more important contributions to the color-top method is printed at the end of this paper. It is to Doctors Davenport, Herskovits, and Todd, together with the various investigators who have worked in their respective laboratories, that we owe our knowledge of the method. To no one of these has the principle involved appealed, yet all acknowledge that, for the present, the method is the most practical available. A great many observations find no place in our list of literature, since the investigators, while using the method, have concentrated their attention upon the practical application of the method rather than upon its theory. Since there is already an excellent introduction by Davenport(2) to the principle and technique of the method, it is unnecessary to do more by way of introduction than refer the reader to this article.

[While this work was in progress there came to my attention two articles presenting alternative methods of quantitatively estimating pigmentation(8, 9). No reference to these is made in the body of this article, since neither method lends itself as yet to the purpose in hand.

Bowman, H. A., "The Color-Top Method of Estimating Skin Pigmentation," *American Journal of Physical Anthropology* 14, no. 1 (1930): 59–71.

There is, of course, no doubt that the color top is, in principle, an inacceptable method. Indeed I believe that it may be replaced by the technique of Shaxby and Bonnell after the photometric method has been more extensively explored and critically validated. Doctor Shaxby writes to me an excellent criticism which should be quoted word for word.

The photometric method is quick and seems a little nearer representing the colour in spectral terms than the colour top. It avoids the use of "Black" by merely lessening the intensity of the comparison screen to equality of brightness with the skin reflection. My chief quarrel with the colour-top is in the employment of "yellow" which (spectrum yellow though it is called by Milton Bradley) is of course, like all yellow pigment colours, chiefly "red + green"; hence its use disguises the true total red reflection (as does the "white disc", including all colours) so that an equation as $9W + 8Y + 37R + 46N$ gives little direct information as to the reflection in the different parts of the spectrum. Apart from the correction of 37R to about 40 per cent of 37, the 9W and the 8Y each has a contribution which it owes to R, and the actual true spectrum Y is not in the least represented by 8Y.

All this is true, but it is not so damaging to the color-top method of ranking the N value, since we are not concerned with the interplay of $R + Y + W$, but only with the relation of $R + Y + W$ to N. Technically it is not practicable to gauge N on the color top without these other discs. And indeed Doctor Shaxby continues: "As a consistent mode of description I have very little to say against the colour-top method."

Dr. Lee R. Dice has very kindly explained to me the Ives tint photometer which he uses in the University of Michigan. The principle is satisfactory, but the technical application, like that of Shaxby's method, is not yet appropriate for human skin-pigmentation estimation, especially in the field. Hence, in acknowledging the definite contributions of these investigators to the subject of skin pigmentation we are not yet able to put their methods to direct use.

After the work was finished, Miss Barnes' very important memoir(10) was published. Since Miss Barnes' observations touch only occasionally upon the subject matter of the present article, Mr. Bowman has been unable to take advantage of the significant data therein set forth. I should however like to draw attention to the uniformity of conclusion by Doctor Herskovits and Miss Barnes (her p. 333) on the one hand and by Mr. Bowman and myself on the other, regarding an observational error of ± 3 per cent in the N value. I should also like to suggest, on the basis of collateral work here on hybrid skulls and skeletons, that when the negro inheritance is sufficiently diluted it breaks down into great variability (see Miss Barnes' table XIII, Sf). This seems to be in accordance with the quantum theory of physics. T. W. Todd.]

The Problem of Printing Reds

Professor Davenport(2) has pointed out that even in the carefully printed monograph by Ridgway on color standards(4) the standard colors change during the process of repeated exposures to light, and in the different printings of the same standard the color

Table 5.4.1

Corrected Values for R Readings, Assuming the Red Disc to Contain 41 Per Cent Red and 59 Per Cent Black (H. Beecher)

Sector reading	41 per cent of sector reading	Sector reading	41 per cent of sector reading	Sector reading	41 per cent of sector reading	Sector reading	41 per cent of sector reading
1.0	2.1	3.3	6.8	5.6	11.5	7.9	16.2
1.1	2.3	3.4	7.0	5.7	11.7	8.0	16.4
1.2	2.5	3.5	7.2	5.8	11.9	8.1	16.6
1.3	2.7	3.6	7.4	5.9	12.1	8.2	16.8
1.4	2.9	3.7	7.6	6.0	12.3	8.3	17.0
1.5	3.1	3.8	7.8	6.1	12.5	8.4	17.2
1.6	3.3	3.9	8.0	6.2	12.7	8.5	17.4
1.7	3.5	4.0	8.2	6.3	12.9	8.6	17.6
1.8	3.7	4.1	8.4	6.4	13.1	8.7	17.8
1.9	3.9	4.2	8.6	6.5	13.3	8.8	18.0
2.0	4.1	4.3	8.8	6.6	13.5	8.9	18.2
2.1	4.3	4.4	9.0	6.7	13.7	9.0	18.5
2.2	4.5	4.5	9.2	6.8	13.9	9.1	18.7
2.3	4.7	4.6	9.4	6.9	14.1	9.2	18.9
2.4	4.9	4.7	9.6	7.0	14.4	9.3	19.1
2.5	5.1	4.8	9.8	7.1	14.6	9.4	19.3
2.6	5.3	4.9	10.0	7.2	14.8	9.5	19.5
2.7	5.5	5.0	10.3	7.3	15.0	9.6	19.7
2.8	5.7	5.1	10.5	7.4	15.2	9.7	19.9
2.9	5.9	5.2	10.7	7.5	15.4	9.8	20.1
3.0	6.2	5.3	10.9	7.6	15.6	9.9	20.3
3.1	6.4	5.4	11.1	7.7	15.8	10.0	20.5
3.2	6.6	5.5	11.3	7.8	16.0		

is not always the same. This latter objection refers to reds more than to any other color. In consequence of his observation of this defect, Doctor Todd referred the problem to Mr. James Redfern, a printer of very considerable experience. Mr. Redfern tells us that while a printer will undertake, in poster or color work, to match yellows, blues, or greens, he will give no guarantee that reds will be accurately matched nor will he accept responsibility for uniformity of the red colors even in a single batch of posters. Apparently the art of printing reds is the most elusive problem of color printing to-day. We cannot, therefore, hope for uniformity in the red disc of the color top and the surprising thing is that the Bradley firm have succeeded so well with their "ox-blood" disc. Harris(3) has investigated the slight differences which the Bradley firm have found unsurmountable and Todd(6) has warned the worker that each batch of tops, at least,

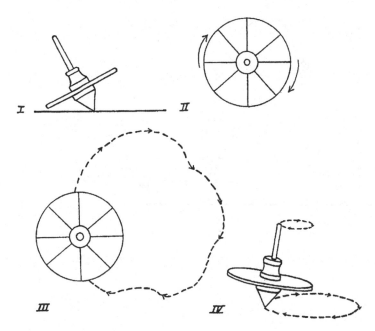

Figure 5.4.1
The several possible movements of a spinning color top. II always occurs. III is usually present. I and IV generally occur also.

should have its red standard checked against Ridgway's scale. It is also important, as Todd(6) has pointed out, to have uniformity of background color in the cardboard base of the top on which the color sectors are laid. Todd himself prefers to use always the grey rather than the brown background, although this is largely due to the fact that all the work in this laboratory has been done on grey backgrounds which the Bradley Company has courteously supplied upon request.

Davenport and his associates still prefer to give color values in direct readings of the color-top sectors, whereas all other workers have followed Todd's lead in correcting the N (nigrum) value for black by adding to the N reading the approximately 59 per cent of black which is present in the red (ox-blood) disc.

Since this corrected value is of some significance in evaluating the degree of pigmentation, I present in table 5.4.1, worked out by Beecher, the true red values for each reading, by sectors, of the red disc. The difference between the reading and the corrected R value is to be added to the N estimate.

The Stabilization of Technique

The very simplicity of spinning the color top enforces a realization of the errors to which the readings are liable. It is a sure scientific acumen which distrusts the results

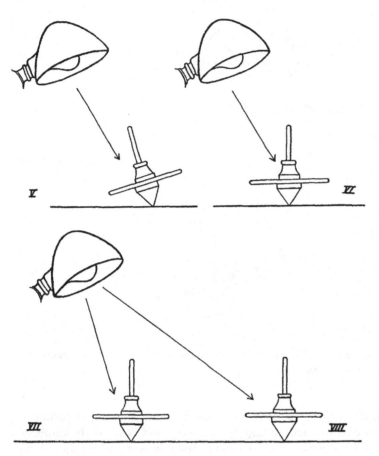

Figure 5.4.2
Effect of incidence of light. Tops V and VII register a lighter shade than tops VI and VIII since the latter are more obliquely illuminated.

obtained by so simple a technique. One cannot have both a simple technique and an easy analysis, though it is also true that elaborate method does not necessarily result in ease of interpretation.

Through the artistic ability of Miss Helen Williams I have attempted to demonstrate graphically some of the most obvious errors to which the reading of a simply spun color top is liable. Figure 5.4.1 illustrates the various possible movements in spinning. Movement II always occurs. The top usually executes movement III and generally also I and IV. In addition the speed varies greatly and the reading is darker as the top slows down. Color matching is therefore complicated by the movement of the top itself. But in addition the problem of incidence of light affects the reading. This is illustrated in figure 5.4.2. Top no. V tends to give a lighter reading than top no. VI

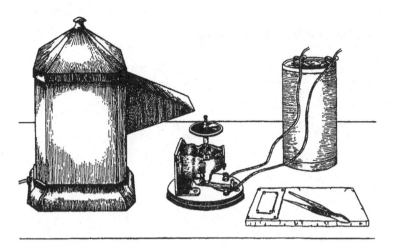

Figure 5.4.3
Color top mounted on Ajax motor with dry-cell connection.

because the light falls less obliquely upon it. Again, top no. VII gives a lighter reading than top no. VIII because it is nearer the source of light. If it is not possible to use diffused daylight, preferably a north light, the worker should employ a constant illumination as near daylight as possible. We employ always a 150-watt nitrogen-filled daylight Mazda lamp with a suitable reflector. Uniformity in speed and movement is attained, as shown in figure 5.4.3, by mounting the top on a little Ajax motor which runs on a single dry cell. The fitting can be obtained at cost on application to the Director.

While these improvements in technique diminish the probable error, there are still many factors which reduce the reliability of the reading. Paper is not an accurate representative of skin and the skin shade is rarely matched with precision by color mixture. It is also probable that color impressions vary with the individual and with his physical and mental condition. This must be borne in mind in every comparison of results.

A Study of Color Combinations

Before putting to the test a comparison of the readings from possible color mixtures, one would naturally inquire into the magnitude of the task.

With four variables (nigrum, red, white, yellow), we first analyze the possibilities as shown in table 5.4.2. Any two may be held constant in definite amount and the other two caused to vary reciprocally. Or any one may be held constant, a second being kept constant until the possible reciprocal variations of the third and fourth are exhausted. Then the amount of the second is changed and the possible reciprocal

Table 5.4.2
The Number of Combinations Possible with Four Variables

Constant ("temporary")	"Changing constant"	Variables
YW		NR
YN		WR
YR		WN
WN		YR
WR		NY
RN		YW
Y	W	NR
Y	R	NW
Y	N	WR
W	R	NY
W	Y	NR
W	N	YR
N	R	YW
N	Y	RW
N	W	RY
R	N	YW
R	Y	NW
R	W	YN

variations of third and fourth again gone through, this process being repeated until the total number of combinations of second, third, and fourth are carried out.

To perform the task suggested by table 5.4.2 is humanly impossible. After a little figuring upon the last array in the table, namely R as "temporary" constant, W as "changing" constant, and Y, N as variables, I find that the total number of combinations for this array alone is 12,341.

As a matter of fact, our purpose is attained by a simple comparison of black (N) and red (R) on two spinning tops. There are twenty sectors on the circumference of the top, each equivalent to 5 per cent of color. Had R been spectrum red, then 10 sectors of N and of R would correspond to 50 per cent of each color. Since the red disc contains 59 per cent of black, 10 sectors of the red disc actually correspond to 29.5 per cent of black and only 20.5 per cent of red. If we start with top I showing only the red disc, it already has 59 per cent of black in it. If we arrange on top II 19.5 sectors of red and 0.5 sector of black, we shall have on top II 60.0 per cent black. This is the difference with which we start. But it is found that the half sector of black on top II results in no appreciable difference in color. It is not until there are 1.5 sectors of black that the distinction becomes apparent. The distinction in color remains obvious if there be a difference of 1.5 sectors of black between the tops until top I has 5.0 sectors of black.

Table 5.4.3 (abridged)
Comparison of N and R on Spinning Tops

Top I (N)		Top II (N)		
Sectors	Per cent	Sectors	Per cent	Noticeable difference
0.0	59.0	0.5	60.0	—
		1.0	61.0	Slight
		1.5	62.1	Apparent
0.5	60.0	1.0	61.0	Slight
		1.5	62.1	Apparent
5.0	69.2	5.5	70.3	—
		6.0	71.3	Slight
		6.5	72.3	Slight
		7.0	73.3	Apparent
7.5	74.4	8.0	75.4	—
		8.5	76.4	Very slight
		9.0	77.4	Very slight
		9.5	78.5	Slight
		10.0	79.5	Apparent
12.0	83.6	12.5	84.6	—
		13.0	85.6	Slight
		13.5	86.7	Slight
		14.0	87.7	Apparent
16.5	92.8	17.0	93.8	—
		17.5	94.9	Very slight
		18.0	95.9	Apparent

It is then necessary to give top II 7.0 sectors of black to demonstrate a color distinction. This significant difference of 2.0 sectors holds until top I has 7.5 sectors of black. With this amount top II must have 10.0 sectors of black to make a difference in color apparent—a relationship which is maintained until top I has 12.0 sectors of black. At this point top II requires only 14.0 sectors of black to register a difference. When top I has 16.5 sectors of black, top II needs but 18.0 sectors of black to permit an apparent difference. Actually the amount of black on top I has been raised by 0.5 sector at a time throughout the range from 0 to 20 sectors and top II pitted against it. A full statement of the results of this comparison is unnecessary. Table 5.4.3 gives all the significant data. The important point for us is that at precisely the range of corrected N figures which is important for negro skins, namely 7.5 to 12.0 sectors on top I, 2.5 sectors of difference are necessary to register a distinction in color. Now 2.5 sectors correspond to 12.5 per cent difference, but one must not imagine that the color top is so vague as this difference indicates. We are pitting N against a red which contains 59 per cent of black. To aid the reader the actual percentages of black have been inserted into table 5.4.3 and it is then apparent that the difference of 2.5 sectors of black actually corresponds to a difference of approximately 5 per cent, not 12.5 per cent. Now 5 per cent is the approximate difference, according to Todd(5), between living and dead skin. It is the limit of probable difference between two observers as determined by Todd and Herskovits working against each other. It also represents a probable extreme range of the personal error (say ±3) as determined by Van Gorder, Beecher, and Blackwood(6). From all these considerations, therefore, we may conclude that a reading for N ± 3 is a fair approximation for comparison of the color-top value for any negro skin.

Practically the same results have been obtained by pitting yellow and white against each other, but in this comparison a difference of 2.5 sectors actually means a necessary difference of 12.5 per cent of color, not 5 per cent as in the comparison of N and R.

The Personal Error

I have stated that a reading for N ± 3 gives a close approximation to color-top value for any skin, but it must be assumed that the observer is practiced, careful, and working in adequate conditions. Inexperience, haste, and difficulties speedily magnify the error.

It is invidious to compare results of different workers without giving some attention to the conditions under which each has made his observations. I am deeply grateful for the opportunity of presenting a validation of the technique through the courtesy of other investigators of pigmentation values in this laboratory. Table 5.4.4 has been drawn up to illustrate what may be expected of the color-top method. There is no doubt that the accumulated knowledge in the laboratory of the practical points to be observed, as well as the greater experience of Miss Blackwood, Mr. Beecher, and myself, accounts for the practical similarity of results in the first three columns. Doctor Van Gorder made the first experiments with these skins (which are always kept in formalin in total darkness) and her results must be read in recognition of the relatively poor conditions under which she made her records. Mr. Foster very graciously allows me to include the first color-top records he ever made, so that we may check up the observations of one to whom the technique is entirely new and strange. It is truly remarkable, having regard to my own early errors, to note with what precision he made his determinations.

The average values at the foot of the table show how reliable the method can be for registering the mean population value for even a small sample. A fuller discussion of this aspect of the problem has already been given by Todd and Lindala(7).

In conclusion, I wish to acknowledge my indebtedness to Doctor Wingate Todd for his encouragement of this work and for his help in arranging the material and presenting the results.

Table 5.4.4

Skin	1928, Bowman	1926, Blaokwood	1927, Beeoher	1920, Van Gorder	1929, Foster
807	93.5	91.0	91.6	91.0	
814	85.3	76.0	85.9	75.0	80.7
822	84.4	87.0	88.7	77.0	79.7
824	92.5	92.5	93.0	88.0	87.5
825	86.2	88.9	87.7	85.0	
835	84.5	85.0	84.9	87.0	82.5
839	72.7	68.0	71.0	71.0	63.2
842	83.5	78.1	85.9	77.0	80.0
845	91.0	91.0	90.8	87.0	86.4
846	82.7	83.0	82.5	82.0	82.8
849	78.0	75.0	82.6	72.0	78.2
851	91.2	92.4	92.4	88.0	85.7
860	81.4	84.0	84.5	81.0	76.4
862	87.6	87.4	85.4	87.0	86.5
866	83.0	83.0	86.8	74.0	78.8
868	89.1	90.0	91.0	87.0	
872	84.7	84.0	82.5	78.0	81.2
874	88.1	90.2	89.4	86.0	85.0
Number	18	18	18	18	15
Average	85.5	84.8	86.5	81.8	80.9

Summary

In conclusion, we may sum up the several points upon which information has been elicited in the course of this study.

1. The technical difficulty of printing reds precludes at present all hope of accurate registration of this range of color and the Bradley Company are to be congratulated upon the practical uniformity which they have attained in the precise shade of the red disc.
2. Several sources of error may be avoided by mounting the color top on an Ajax motor and taking pains to have efficient illumination.
3. A study of color combinations shows that, in the range for N usually required for registration of the pigmentation in negro skins, a difference of 5 per cent N is necessary before the distinction is likely to appear on the color top.
4. The personal error of trained careful workers may be estimated at ± 3 per cent for any single skin and is even better for a population average of very small sample.
5. The color top does not give an absolute evaluation of N, but does give a quantitative determination useful for comparisons.

Literature Cited

1. Davenport, C. B. 1913 Heredity of skin color in Negro-White crosses. Carnegie Inst. Wash., Publ. no. 188, 1–106.

2. ——— 1926 The skin colors of the races of mankind. Natural History, XXVI, 44–49.

3. Harris, R. G. 1926 The San Blas Indians. Am. J. Phys. Anthrop., IX, 17–63.

4. Ridgway, R. 1912 Color standards and color nomenclature. Washington.

5. Todd, T. W., and Van Gorder, L. 1921 The quantitative determination of black pigmentation in the skin of the American Negro. Am. J. Phys. Anthrop., IV, 239–260.

6. Todd, T. W., Blackwood, B., and Beecher, H. 1928 The color top method of recording skin pigmentation. Am. J. Phys. Anthrop., XI, 187–204.

7. Todd, T. W., and Lindala, A. 1928 Dimensions of the body: Whites and American Negroes of both sexes. Am. J. Phys. Anthrop., XII, 35–119.

8. Shaxby, J. H., and Bonnell, H. E. 1928 On skin colour. Man, XXVIII, 42.

9. Sumner, F. B. 1927 Linear and colorimetric measurements of small mammals. J. Mammal., VIII, 177–206.

10. Barnes, I. 1929 The inheritance of pigmentation in the American Negro. Human Biology, I, 321–381.

5.5   "Race and Runners" (1936)

W. Montague Cobb

Introduction

As the physical anthropologist scans the fascinating panorama of contests in simulta-neous progress at a great track meet like the Penn Relay Carnival, he becomes aware of an association between certain types of bodily build and special events. Conspicuous contrasts are the large, heavily muscled, occasionally paunchy athletes who put the shot and throw the hammer farthest, and the tall, lean young men who jump highest. The leading high hurdlers are tall and the stellar distance men of medium to slender build. In the other running and field events distinctive types of bodily build are less apparent. Almost every variety of human form and style of performance competes suc-cessfully in the relay races which endlessly circle the track.

Since athletic accomplishment is jointly dependent upon physical constitu-tion, technique, and the will to achieve, it is obvious that in a few specialized events a particular bodily build may confer advantages which cannot be overbalanced by any amount of training and determination on the part of the less gifted. In the shot put, great bodily weight is an advantage, increasing the impetus imparted to the shot; in the high jump it is a handicap, adding to the load which the muscles must lift from the ground. Similarly, it can be shown that tallness is of advantage to both weight man and high jumper.

Among the sprinters and broad jumpers a diversity of physical types is seen. It is apparent that here determination of the influence of bodily build on performance will be more difficult.

As the anthropologist surveys the striving field in the stadium, he sees noth-ing to suggest an association between race and competition in any particular event. He notices Negro youths in nearly every phase of competition. Their bodily build varies like that of other athletes. The weight men are big fellows, while those topping the bar are more sparely built. However, a number of recent comments in the press upon the current success of American Negro sprinters and broad jumpers have either directly ascribed this success to a longer heel bone or stronger tendon of Achilles than those of their white competitors, or implied that in some way it has been due to racial characteristics. The wide circulation which these suggestions have received warrants a careful appraisal of the facts.

Cobb, W. Montague, "Race and Runners," *The Journal of Health and Physical Education* 7, no. 1 (January 1936): 3–7, 52–56.

Figure 5.5.1
Famous finish of the 100-meter race in the 1932 Olympics at Los Angeles with Ralph Metcalfe and Eddie Tolan breasting the tape together.

Sprinters and Broad Jumpers

In the 1932 Olympics two American Negroes, Eddie Tolan and Ralph Metcalfe, carried off top places in both the 100- and 200-meter dashes, Tolan setting new Olympic records in each event; and another Negro, Ed Gordon, won the broad jump. Since the tenth Olympiad, Negroes have continued to dominate the national field in the sprints and broad jump in the persons of Metcalfe, Jesse Owens, Eulace Peacock, and Ben Johnson.

In 1932, 1933, and 1934, Metcalfe won national A.A.U. championships in both the 100- and 200-meter sprints. In 1935 he won the 200-meter event. With Tolan and Percy Williams of Canada, he is co-holder of the world's record of 10.3 seconds for the 100 meters, and co-holder with Roland Locke of Nebraska of the world's record for the 200 meters, 20.6 seconds. He also holds numerous records at intermediate distances. Jesse Owens, like Metcalfe a consistent performer, won the national A.A.U. broad-jump championship in 1933 and 1934. He has made and equalled various intermediate sprint records. In one afternoon at Ann Arbor, Michigan, in May, 1935, he performed the greatest track feats ever wrought by a single man, breaking three world's records and equalling a fourth. He leaped 26 feet $8\frac{1}{4}$ inches in the broad jump, ran the 220-yard dash in 20.3 seconds, and won the 220-yard low hurdles in 22.6 seconds for new records, besides equalling the world mark of Frank Wykoff of 9.4 seconds for the 100-yard dash. Eulace Peacock was national A.A.U. pentathlon champion in 1933 and 1934. In the 1935 A.A.U. championships he defeated Metcalfe and Owens in the 100 meters to equal the world's record (disallowed because of slight wind), and in the same

Figure 5.5.2
Eulace Peacock, Temple University's famous broad jump and sprinting star.

meet surpassed Owen's remarkable jump of 26 feet $2\frac{1}{2}$ inches with a leap of 26 feet 3 inches to win the broad jump crown. Ben Johnson has had the misfortune to race Metcalfe at peak and twice to suffer injuries when his own condition was very promising. He has beaten Owens, however, to tie the latter's world record for 60 meters and has done the hundred in 9.6 seconds. In the 1935 I.C.A.A.A.A. championships at Cambridge, Massachusetts, he was two and a half yards from the finish of the 100-yard final and leading the field when he pulled a tendon. Only a sophomore at Columbia, his day still may come.

Wide attention has thus come to be focussed on the fact that in the past champion sprinters and broad jumpers have often been Negroes. The first was Howard P. Drew who became national A.A.U. champion at 100 yards in 1912 and 1913 while at Springfield (Mass.) High School. In 1913 he also won the 220-yard title. In 1914 Drew went to the University of Southern California where he became co-holder with Arthur Duffey of Georgetown of the world's record of 9.6 seconds for the 100-yard dash, a mark which stood for many years. Drew also equalled the world's record for

Figure 5.5.3
Howard P. Drew, University of Southern California, one of the first of the famous Negro sprinters to achieve championship records.

the 220-yard sprint when that was 21.2 seconds. His action photographs are displayed today as models of perfect form.

After Drew the A.A.U. list of national track champions shows no Negroes' names among the sprinters until that of Tolan in 1929. For six of the last seven years, however, three Negroes, Tolan, Metcalfe, and Peacock, between them have won both sprints.

The story of the list of broad jump champions is startling. Only three times in the last sixteen years have white men been able to win the national broad jump title. No white man has leaped as far as 26 feet but six white men, Ed Hamm, Bob LeGendre, J. Hill, Lemoine Boyle, Dick Barber, and Al Olson, have officially bettered 25 feet.

In 1920 a parade of six Negro champions began with Sol Butler, Dubuque sprinter and member of the United States Olympic team. He won the national event in that year with a jump of 24 feet 8 inches. Ned Gourdin won the next year. He was then the world-record holder by virtue of a leap of 25 feet 3 inches, which was Harvard's greeting to English invaders the previous spring. Gourdin was the first man known to have jumped 25 feet. He was also national pentathlon champion in 1921 and 1922. Continuing after Gourdin, DeHart Hubbard of Michigan held the national broad-jump title for six consecutive years. In turn the world-record holder, Hubbard set a mark of 25 feet $10\frac{7}{8}$ inches with his last jump in collegiate competition. In 1928 white Eddie Hamm's 25 feet $11\frac{1}{8}$ inches set a new world's record. Then came the fourth Negro, Ed Gordon of Iowa, who took first place in 1929 and 1932. Al Bates won in both the intervening years with leaps of less than twenty-five feet. He has been the only white man except Hamm to win this event in the last sixteen years. In 1933 and 1934 the winner was Owens, and in 1935, Peacock.

Two months after Hamm broke Hubbard's world mark, Silvio Cator, a Haitian Negro, restored Negro supremacy by jumping 26 feet $\frac{1}{8}$ inches. This was the first accredited entry in the 26-foot bracket. In 1928 Hubbard had jumped 26 feet 2 inches, but the take-off was one inch above the level of the landing pit and the record could not be accepted. Chuhei Nambu, the Japanese, jumped 26 feet $2\frac{1}{8}$ inches at Tokyo in 1931 and this record stood until Owens' wonder feat last May.

There have been numerous other stellar Negro performers in the sprints and broad jump like Willis Ward of Michigan and Everett Utterbach of Pittsburgh who never won national or Olympic titles. Particularly noteworthy in this group is James Johnson of Illinois State Normal, who earned a place on the American sprint team for the 1932 Olympics, but for some reason, never clear, was not allowed to run. In 1933 James Johnson set new records in the 100- and 200-meter races of 10.4 and 21.6 seconds respectively, in the junior outdoor A.A.U. championships.

Other Performers

Negro champions have been less frequent among middle-distance runners and field performers, yet in these events the Negro has not failed to produce exceptional men. R. Earle Johnson of the Edgar Thompson Steel Works was national five-mile champion in 1921, 1922, and 1923 and represented this country on the 1920 and 1924 Olympic teams. Gus Moore of St. Bonaventure won fame as a miler. Though never a champion, he several times bettered 4:20. Phil Edwards won the intercollegiate half-mile championship while at New York University and was national champion in 1929. He represented Canada in the 1928 and 1932 Olympics. At the gruelling quarter-mile distance Binga Dismond of Chicago equalled Meredith's world record of 47.4 seconds which withstood all assaults until erased by the great Ben Eastman. Cecil Cooke of Syracuse was intercollegiate champion in this event in 1923 and national champion in 1925. For the past two years the intercollegiate crown has been worn by Jimmy LuValle of the University of California at Los Angeles.

In the high jump Charley Major of St. Bonaventure used to clear 6 feet 4 inches frequently in the early twenties. Owens' present team-mate at Ohio State, Melvin Walker, has cleared the bar at 6 feet 5 inches, and leaps of Temple's Al Threadgill have taken him 6 feet 7 inches into the air. For four consecutive years Cornelius

Johnson of Compton Junior College has won or tied for first in the national high-jump competition. In 1932 he tied with Bob Van Osdel and George Spitz at 6 feet $6\frac{5}{8}$ inches; in 1933 he won with 6 feet 7 inches; in 1934 he tied with Walter Marty, the world's record holder, at 6 feet $8\frac{5}{8}$ inches; and in 1935 he won easily with 6 feet 6 inches. Last May, Leo Williams, a colored Muncie High boy, broke the Indiana interscholastic record with a jump of 6 feet $4\frac{1}{2}$ inches, while in the same meet another Negro lad, Walter Farmer, set a new prep mark in the pole vault of 12 feet 10 inches.

Willis Ward, Charles Drew, and George Beatty have been hurdlers of note. A misfortune snatched from Beatty a place on the 1932 Olympic team. Owens' world's record in the 220-yard low hurdles must be held a tribute to his speed as he is not a finished hurdler and does not intend to specialize in the event.

There is thus no running event and few field events to which Negroes have not contributed some outstanding performer and there is no indication of ineptitude in any event in which no champion has yet appeared. It is to be noted, however, that the sprint and broad jump champions have appeared in a rapid succession, culminating in the present group of contemporaneous performers. For this reason they have been especially conspicuous in the public eye. It is this prominence which has probably stimulated the notion that these stars might owe their success to some physical attributes peculiar to their race.

The Old, Old Story

This sort of suggestion is by no means new. In the days when peerless Paavo Nurmi daily fired every youngster's imagination with new world's records broken in Olympic competition, in the months afterward when more records fell during the memorable duels of Nurmi and his doughty Finnish team-mate, Willie Ritola, while the two toured America, there were reams written on why the Finns seemed to have a permanent corner on supremacy in the distance runs. The historians extolled the conquests of the mighty Hans Kolehmainen in 1913. Geographers showed how Finland's rugged climate bred endurance such that the rest of the world might as well turn in its spikes. Moralizing editors completely effervesced on the subject. But still the ancient records (1904) of England's immortal Al Shrubb for the 6-, 7-, 8-, and 9-mile runs are the world's best. Along have come Kansas' Cunningham, Princeton's Bonthron, and New Zealand's Lovelock to run the mile with "impossible" speed. Who can say whence the next athletic "trust busters" will come, or what records they will attack?

But let us look in other fields. There have been those who felt that continued European victories over Americans in the weight-lifting contests, despite the intensive advertising campaigns of American vendors of bar-bells and other muscle-building and allegedly masculinizing agents, indicated superior inherent European capacity in this line of endeavor. There are people who ask if the fact that professional boxing, once the Irishman's pride, now affords prominence to so many Hebrews, Italians, and Negroes, might not be due to changes in racial physique. Recently the authorities who elucidate the reasons for the meanderings of the Davis Cup have been quite busy.

But to pursue seriously our original inquiry about the relation of the Negro's anatomy to his feats on cinders and pit, track coach and anthropologist must pool their knowledge.

Coach and Anthropologist

To detect and develop athletic talent is the prime function of our track coaches. The track coach is professionally interested only in those qualities of an athlete which make for excellence in performance. He has no concern with the measure in which those qualities may also be characteristic of men of particular occupations or races. These matters are the business of the physical anthropologist.

Let the track coach set down the factors that make a great sprinter and the anthropologist the distinguishing features of the American Negro. If on comparison the two lists have much in common, race may be important; if little, race is of no significance.

Almost at once, however, we are beset with vagaries. The track coach cannot categorically describe the physique and character of the sprint champion, nor can the anthropologist define with useful accuracy the physique and character of the American Negro.

Characteristics of a Sprinter

Most sprinters can broad jump well. If they learn to leave the ground properly after a good run, their inertia will carry them a respectable distance. It used to be thought that participation in one event detracted from ability in the other, but Hubbard, Owens, and Peacock have helped usher this idea into discard. For convenience here, sprinters and broad jumpers are considered together.

It is obvious that superior sprinting and broad-jumping performances require a certain combination of physical proportions, physiological efficiency, and personality which are recognized by the track coach as natural capacity. By methods ably explained in a few manuals on the subject, the coach is able to impart training and technique which convert this potential capacity into the actual ability to perform.

The personal histories and constitutions of our sprinters have not yet been sufficiently analyzed for the formula for the perfect sprinter or jumper to be given. We are not able to say what measure of natural capacity is due to physical proportions, or to physiological efficiency or to forceful personality. Nor can we weight capacity and training scientifically. This does not mean that strongly biased opinions on the subject are non-existent. For instance, it has been said that superior sprinting and jumping ability must be a matter of nine-tenths capacity and one-tenth training because the Negro is not disposed to subject himself to rigorous training.

Despite the fact that adequate data are not available for scientific analysis of sprinting and jumping ability, many useful conclusions may be drawn from a commonsense approach to the problem. We know first of all that the physique, style of performance, and character of our champions have been highly variable.

Physique, Style of Performance, and Character

When the track coach arrays before his mind's eye the galaxy of stars who have done the hundred in 9.6 seconds or better, he notes no uniformity of physique, style of running, or temperament. This group includes the Negro constellation just discussed and Arthur Duffey, Jackson Sholz, Loren Murchison, Charles Paddock, Chester Bowman, Frank Hussey, Charles Borah, Claude Bracey, Emmett Toppino, George Simpson, James Owen, Robert Grieves, Frank Wykoff, Foy Draper, and George Anderson.

Some were tall (Anderson, Metcalfe, Peacock); more were short (Paddock, Tolan, Hubbard, Draper, Grieves, Toppino). Some were slender (Simpson, Anderson); others stocky (Drew, Metcalfe, Paddock, Bracey). Some were well proportioned like Owens, Grieves, and Anderson, but there were a few who could hardly have served as models for the Greeks.

Figure 5.5.4
Ralph Metcalfe, an outstanding sprinter from Marquette University.

For finer distinctions, data of desirable precision are not available but we can say from general inspection that there have been long-legged champions and short-legged ones; some with large calves and some with small. Record-breaking legs have had long Caucasoid calves like those of Paddock and short Negroid ones such as Tolan has.

In the matter of style, there have been fast starters like Hubbard and Simpson and slow ones like Paddock and Metcalfe. We have had "powerhouse" sprinters such as Metcalfe and smooth graceful flashes like Owens whose performances seem without effort. Most of the runners have been mouth breathers; Metcalfe is a nose breather. Owens follows no rule in breathing. The first fifty yards is the faster for some men, the last half the better for others. Paddock started fast, "died" in the middle, and swept to the finish with a final burst, using the orthodox "jump" finish very effectively. Some men use a long stride, others a short one, and so on.

In respect to temperament, again we find no homogeneity. Some maintained calm well before a big meet, others tended to become extremely nervous and required careful handling to be sent off their marks in peak form. Some could ignore efforts by competitors or their sympathizers to gain psychologic advantage; others have been licked before starting a race. Some of rugged constitution could partake of a wide range of edibles and stand a long season, others had to diet carefully and could not long remain at peak. There have been champions of great courage who were undaunted by defeat or misfortune and others who reacted very severely to "bad breaks."

Training and Incentive

We know that general habits, training, and incentive are very important factors in the make-up of the champion. As one studies the records in the *Athletic Almanac* from 1876 to the present, it becomes evident that in every field of athletic endeavor the quality of human achievement has been improving, times are faster, distances farther. To date no calculator of the absolute limits of human capacity seems to have had the correct data for his conclusions. Many have been embarrassed to see men do what had been proved impossible.

The anthropogenist tells us that compensatory improvement in human constitution in a mere sixty years is inconceivable. The fact of our improving performances can mean only that society is learning more and more how to conserve and develop the physical resources which it has had all along. Improved standards of living and two generations of experience have benefited our athletes.

The value of training and incentive is strikingly apparent even in the Negro stars under discussion. If it were true that Negroes had special physical endowments fitting them for sprinting and jumping, it would be expected that from the Negro colleges, which are mostly in the South, there would be a constant crop of good athletes from whom the few champions would emerge. Yet not one of the Negro stars in question came from a Negro college. Both incentive and training facilities are known to be capable of marked improvement in these schools.

Figure 5.5.5
Willis Ward, the University of Michigan's versatile Negro athlete, who was an outstanding football player as well as track athlete.

Some of the Negro stars themselves have declared that the desire to emulate their predecessors and to excel in a white environment have been powerful stimuli to their efforts. Jesse Owens acknowledges the inspiration derived from his high school coach and discoverer, Charles Riley. DeHart Hubbard hitched his wagon to the shining star of Ned Gourdin.

Anthropological Characteristics

We have seen that the variability of the physical, physiological, and personality traits of great sprinters and jumpers, and inadequate scientific data prevent a satisfactory statement as to just what traits are responsible for their success. We have seen also, the importance of training and incentive. Let us now go to the anthropologist. He has to deal with men categorically designated as American Negroes, but they do not look

Figure 5.5.6
Jesse Owens, of Ohio State University, accomplished the remarkable feat of breaking three world's records and equaling a fourth in one afternoon's performance at the 1935 Western Conference Track and Field Meet.

alike. Genetically we know they are not constituted alike. There is not one single physical feature, including skin color, which all of our Negro champions have in common which would identify them as Negroes.

From his photographs Howard Drew is usually taken for a white man by those not in the "know." Gourdin had dark straight hair, no distinctly Negroid features, and a light brown complexion. In a great metropolis he would undoubtedly be often considered a foreigner. Owens and Metcalfe are of rather intermediate physiognomy. Owens is light brown, lighter than Metcalfe. But Owens has somewhat frizzled hair while Metcalfe's is dark, smooth, and wavy. Hubbard, Tolan, Johnson, and Peacock are darker and more definitely Negroid than the others, but not one of them even could be considered a pure Negro according to Herskovits' recent definition.

Extending his view, the anthropologist fails to find racial homogeneity even among the white sprinters. We find blond Nordic and swarthy Mediterranean types and various mixtures. In fact if all our Negro and white champions were lined up indiscriminately for inspection, no one except those conditioned to American attitudes would suspect that race had anything whatever to do with the athletes' ability.

Our situation thus appears hopelessly complicated. Variability is so great in the pertinent characters of both sprinters and Negroes as special groups, that to ferret out and evaluate what the two have in common will be difficult if not impossible.

Test Characters

Another approach to the determination of the influence of race in the making of our Negro sprint champions is the comparison of selected Negroid characteristics which conceivably may affect running capacity, with the conditions found in our Negro stars.

The Negro is long of limb, that is, he has long legs and arms relative to the length of his trunk as compared with the white. In addition, the leg of the Negro is said to be long in proportion to his thigh. Possibly this might be of significance in broad jumping, as leaping animals such as the kangaroo have extremely long shins and very short thighs. The belly of the calf muscle of the Negro is short and the tendon long, whereas in the white, the belly, which produces the prominence of the calf, is long and the tendon short. The size of the pelvis, which is small in the Negro, would also appear of little importance to the runner or jumper.

In one study it was found that the nerve fibers of the Negro are larger in cross section than those of the white. As with electric conductors, the larger the nerve the easier and quicker the passage of the impulse, so this finding would imply better muscular coordination in the Negro.

It would be desirable, of course, to have for comparison the dimensions of all our stellar athletes, Negro and white, but as such data are not available it will be of value to compare the results of an examination of one of the Negro stars with recognized standards. Accordingly, the following data on Jesse Owens are presented and discussed.

Physique of Jesse Owens

Jesse Owens is a tall, slender, well-knit youth of twenty-one years. He stands 5 feet 10 inches and scales 160 pounds. He is muscular, of erect carriage, with broad sloping shoulders, narrow hips, and long legs. The close observer would notice that the legs are better developed than the arms, but there is by no means a disharmony.

Skeletal Proportion

Owens' dimensions will be compared with standard dimensions of whites and Negroes. The standards used are those published by Todd and Lindala based on cadavera at Western Reserve University. As the measurements compared are between subcutaneous bony points, the comparison will be satisfactorily accurate.

The average male white lower limb forms 50.9 per cent of the total height, the limb being measured from the pubic symphysis (mid-point on the front of the pelvic brim). The average male Negro limb forms 52.5 per cent of the standing height. Jesse Owens' lower limb forms 52.7 per cent of his height. Owens thus has a long lower limb relative to his height and in this characteristic he is Negroid.

Now to the proportions of the limb itself, namely, thigh, leg, and foot. As previously mentioned, when nature makes a leaping animal, she shortens the thigh, lengthens the leg, and lengthens the foot, as in the kangaroo. Todd and Lindala found no differences between Negroes and whites in the relative proportions of limb length formed by thigh, leg, and foot. Other investigators, however, have reported that the shin of the Negro is longer in proportion to his thigh than that of the white. Davenport quotes a crural index (length of the tibia expressed as percentage of the length of the femur) of 80.4 for the Japanese, of 84.0 for whites, and 86.0 for Negroes, though certain details of his data are somewhat conflicting. The crural index of Jesse Owens is 84.0. Hence this relationship may be interpreted as being near the white norm, or else supporting the contention of Todd and Lindala that there is no racial difference in this index. It is interesting to note that a member of the group having the lowest crural index, the Japanese, from 1931 until last May held the world's record for the broad jump and still holds the record for the hop, step, and jump. So whatever the significance of the crural index for saltatorial ability in the animal kingdom, it is of no importance to man.

Foot length is proportionate to limb length in both Negroes and whites, the length of the foot being 26 per cent of the length of the limb in both races. Owens' foot, however, is but 24.9 per cent as long as his entire leg. He thus has a short foot, a fact apparent to even the casual observer.

Foot breadth is proportionate to foot length and Owens' foot is on the broad side of the normal in this respect, the breadth being 36.25 per cent of the length. In Todd and Lindala's male whites the ratio was 36.1 and in the male Negroes 35.5.

The proper interpretation of the proportions of the foot itself involves an understanding of certain mechanics of the foot. In walking there is heel and toe contact with the ground. In sprinting the movement is simpler, only the ball of the foot striking the ground. For the present, assuming the foot to be a rigid bar, in sprinting it functions as a lever of the second class, in which the heads of the metatarsals (ball of the foot) are the *fulcrum*, the *force* is the upward pull of the triceps surae (calf muscle) on the hinder end of the calcaneus (heel bone), and the *load* is the weight of the body concentrated on the tibia (shin bone).

Obviously in such an arrangement, the efficiency of the foot as a lever would be increased if the heel bone were prolonged backward. Then the saving in lifting force would give the calf muscle greater driving power, a distinct athletic advantage.

This is the concept which prompted the hypothesis that the calcaneus of the Negro sprinters was exceptionally long. The background for the assumption probably was the widely prevalent impression that the Negro has a projecting heel. Dr. Adolph Schultz has shown that the protuberance when present is a subcutaneous fat pad and not bone.

But there are other faults with the assumption. Why should not all the white champions be long heeled too? If the explanation be that their calf muscles with long bellies and short tendons are more powerful than those of Negroes, it must follow that the short belly and long tendon of the Negro calf gives less muscular power than the muscle of the white and the longer Negro calcaneus would be only compensatory.

For explanatory convenience it was assumed above that the foot as a lever behaves as a rigid bar. This is not true. The foot is arched and the arch bends elastically under pressure. In running and leaping the arch is subjected to severe strains. If on these occasions the supports of the arch are inadequate it will sag and force applied at the heel will be imperfectly transmitted to the heads of the metatarsals. Thus a powerful calf muscle to be effective in giving a runner good drive must act on a foot with a strong or well supported arch.

This emphasizes the importance of the natural supports of the arch. These consist properly of the ligaments which hold the bones of the foot together. But these alone cannot resist the thrust of the body weight. A group of muscles which by their attachments hold the foot bones pressed against each other forms the second line of defense of the arch. Of these the most important are three muscles which arise in the leg beneath the calf muscle. One is a leash for the arch from the inside of the foot (tibialis posterior), another is a sling for the arch from the outside of the foot (peroneus longus), and the third stretches under the inner side of the foot from a pulley on the heel to the great toe (flexor hallucis longus).

It is clear then, that for the calf muscle to be effective in action, the arch of the foot must be strong; and we see that important muscular supports of the arch are not affixed to the heel bone. The proponents of the long heel theory must, therefore, also assume a strong arch for the Negro. The racial anatomy of the arch of the foot and its supports has not been thoroughly investigated but we should like to point out that much has been written about the prevalence of low arches and flat feet in the Negro.

We are now ready to return to Owens. It has been previously noted that he has a relatively short and somewhat broad foot. We cite now that his heel does not conspicuously protrude. On the X-ray picture it may be observed that there is no thick fat pad beneath the skin to produce an effect of protrusion, and that his heel bone is not as long behind the ankle bone as are several bones of approximately the same size belonging to white men. When Owens' roentgenogram was shuffled with those of several male whites of comparable age and size, a fellow anatomist could not pick out Owens' picture, emphasizing that there is nothing unusual about Owens' calcaneus.

His footprints standing and sitting and the conformation of the bones as seen on the roentgenogram reveal an excellent arch. This arch is distinctly a high one.

Our findings thus demonstrate that Owens' foot presents none of the characteristics commonly but often erroneously designated as Negroid. He has a relatively short broad foot with an excellent high arch and a well-formed non-protruding heel. To have what has been described as a typical Negroid foot, the member should be long and narrow with a low arch and projecting heel, the latter being due both to a large fat pad and a long heel bone.

The reader may ask what the average length of the calcaneus relative to the tibia is in Negroes and whites. To our knowledge no such information has been published. But we have tried to show that the calcaneus is only one element in the mechanics of the foot which is far overshadowed in importance for running by the strength of the arch. To single out the calcaneus is like emphasizing the moderate elongation of the kangaroo's heel while completely ignoring the tremendous extension of the foot in front of the shin. This type of lever arm extension can be adapted to give great drive too, and apparently works well in the marsupial.

Calf Muscles

The layman is familiar with the small short calf prominence frequently seen in Negroes as contrasted with the large long calf prominence commonly seen in whites. Dr. George D. Williams found in the dissecting room at Washington University that the tendon of the lateral belly of the gastrocnemius or calf muscle formed 51.0 per cent of the total length of the muscle in whites, but in Negroes this tendon formed 55.3 per cent of the length of the muscle. This gives metrical form to the general observation.

Although it is difficult to measure the proportions of the muscle on the living, the gastrocnemius was measured on Owens. Great care was used and as anatomist we are sure our measurements are useful though as anthropometrist we freely admit the possibility of large error. In Jesse Owens the tendon of the lateral belly of the gastrocnemius formed only 49.2 per cent of the total length of the muscle on the right and 49.9 per cent on the left.

Owens thus has a longer calf than the average white, a feature apparent to the naked eye. In fact, the fine calf development is the most conspicuous feature of Owens' legs. The muscle bellies are long and large. His calf girth standing is 399 mm. or nearly 16 inches. In respect to calf structure then, Owens is of the Caucasoid type rather than the Negroid.

Neuromuscular Coordination

Since the cross-section areas of nerves cannot be measured in the living, Owens was given a tapping test to determine whether in his untrained arms there was any inherent ability to move unusually rapidly. In this test a metal pencil is simply struck alternately on two metal squares about two inches apart as rapidly as the subject is able. Electrical connections make a graph from which is read the number of taps in a given time. In this test Owens was exceptionally fast with his right arm and above the average with his left. He is right handed.

There are other tests of speed of nervous reaction such as the time of the patellar tendon reflex (knee jerk) and the rapidity with which one can run in place which have been studied on athletes. These were not applied to Owens.

Summary and Conclusions

Since man has begun to measure the quality of his athletic performances with stopwatch and tape he has constantly improved. This has been due not to a betterment of human stock but to experience and better nurture.

No particular racial or national group has ever exercised a monopoly or supremacy in a particular kind of event. The popularity of different events with different groups of people has, and probably always will vary, though not necessarily in the same direction.

Negroes have been co-holders but until Owens not single holders of the world's records for the standard sprints. The split-second differences in the performances of the great Negro and white sprinters of past and present are insignificant from an anthropological standpoint. So are the differences in the achievements of the two races in the broad jump.

The physiques of champion Negro and white sprinters in general and of Jesse Owens in particular reveal nothing to indicate that Negroid physical characters are anatomically concerned with the present dominance of Negro athletes in national competition in the short dashes and the broad jump.

There is not a single physical characteristic which all the Negro stars in question have in common which would definitely identify them as Negroes.

Jesse Owens who has run faster and leaped farther than a human being has ever done before does not have what is considered the Negroid type of calf, foot, and heel bone.

Although the world mark for the broad jump has remained the property of Negro athletes for a surprisingly long period, it would seem that the technique of the jump is the only feature involved in the matter of supremacy, for Negro and white sprinters have demonstrated equal speed for the preliminary run.

Chuhei Nambu, the retiring world's champion broad jumper, belongs to a people with an anatomical build the opposite of the Negroid in pertinent features. The Japanese are short of stature, short of limb, long thighed, and short legged. If the view that racial anatomy was important in the Negroes' success were correct, these are just the specifications a jumper should not have. Hence we see no reason why the first man to jump twenty-seven feet should not be a white athlete or the first man to run the mile in four minutes a Negro.

Notes

W. Montague Cobb, M.D., Ph.D.

N.B. A review of pertinent literature will be found in W. M. Cobb, "The Physical Constitution of the American Negro," *Journal of Negro Educ.*, III (1934), 340–88.

6 GLANDULAR DIFFERENCES

Introduction

Evelynn M. Hammonds and Rebecca M. Herzig

It has long been known that the removal or alteration of ovaries and testes influences human behavior in complicated ways. And, as a result, gonadal manipulation has a rich history in western culture, employed at earlier points in time to preserve the vocal range of *castrati* or to ameliorate the "hysteria" of women. Yet only in the nineteenth century did investigators begin considering that the forces governing such behavior might extend beyond the boundaries of the ovaries and testes themselves. Perhaps, they speculated, these influences permeated the entire body in some unseen, subtle manner. In the late nineteenth century, chemists began tracing the production of particular chemical substances to specific groups of cells. In 1905, British physiologists William Bayliss and Ernest Starling developed the term *hormone* (from the Greek "to excite" or "to set in motion") to refer to chemicals produced in the body and transported through the blood stream to influence other organs at a distance. By the 1920s, physiologists had defined the major glands in the body (pituitary, thyroid, adrenals, ovaries, and testes). Furthermore, they claimed that each held a distinct function in the body, chemically controlling specific physiological activities. The study of these functions extended beyond the realm of anatomy, physiology, and psychiatry to include organic chemistry and the new field of endocrinology.

Assessing the importance of hormones meant assessing averages and tendencies; hormone studies focused on *comparative* numbers. Hormonal distinctions between men and women were not absolute (since men were known to possess the so-called "female" hormones just as women secrete "male" hormones), but quantitative and relative. As gland popularizer Louis Berman declared in 1925, "Maleness and femaleness are best understood, in light of our present knowledge, as the expression of chemical influence stimulating or depressing the evolution of the various characters we recognize as belonging to the sexes. One may then logically expect all sorts of combinations of sex characters to occur, which as a matter of fact do occur, and are roughly included in the term 'sex intermediates.'" Berman and others suggested that relative differences might vary even over the course of an individual life: at one period, a person might possess a "feminine" chemistry; at others, "his chemistry will be masculine." The study of sex hormones thus held profound consequences for the conceptualization of sex, opening the possibility of quantitative, fluid transitions between "male" and "female." Post-war research on hormones extended discussions of sexual difference to include questions of social "gender" (which emerged as a new analytical category at this time), including the patient's primary gender identification, the roles attributed by parents and society, sexual desire and practice, and cognitive differences. At the glandular level, sexual difference seemed increasingly layered and dense.

However, statistical sensibilities and the descriptions of sexual "intermediaries" did not necessarily imply a novel view of racial difference. Like earlier accounts, racial studies of hormones were used in competing ways: alternately to affirm absolute typological distinctions between bodies or to demonstrate the plasticity and continuity of an ambiguous spectrum. Hardly exceptional for the period was Louis Berman's conclusion that "the white man's predominance on the planet" was due to "a greater all-around concentration in his blood of the omnipotent hormones. While the Negro is relatively subadrenal, the Mongol is relatively subthyroid. Their relative deficiency in internal secretions constitutes the essence of the White Man's burden."

The first two selections illuminate some of the complexities of emerging accounts of glandular differences, showing the collision of diverse social and physiological forms of deviance and normalcy. The first brief article appeared in a 1920 edition of *Current Opinion*, a New York-based journal of ideas published from 1913–1925. The second was published in *Medical Record* the following year. Both blend descriptions of racial, national, and sexual difference with discussions of disease pathology. As you read, consider how each author evokes concepts such as *nature* or *civilization*. What might these terms imply? What appears to drive evolution in these accounts? And how do the authors use the *race* itself?

These assessments of difference, of course, were situated in particular social contexts, embroiled in tensions over the meaning and import of race, sex, and sexuality. Early studies of "internal secretions," for instance, were tied to guiding questions of the opening decades of the twentieth century. New demands for women's economic independence, political enfranchisement, sexual freedom, and access to education and employment gave glandular politics a particular urgency in the era's anxious quest for sexual definition. Just as "genes" appear in most twentieth-first century social controversies, so, too "glands" once seemed to hold the promise of definitive answers to difficult social issues. By the early 1940s, New York criminologist William Wolf adopted the term *endocrinopathy* to describe those motivated to crime by bad glands. Criminality, sexual depravity, susceptibility to disease, feminism, and labor agitation were all tied to "glandular derangement." The collision of endocrinopathy and race appears vividly in the selections from Berman's best-selling 1921 popular book, *The Glands Regulating Personality*, and in William Wolf's 1000–page textbook, *Endocrinology in Modern Practice*.

As interest in the nature of difference persisted, so, too, did interest in the mutability of that difference, on the limits and possibilities of "endocrine stimulation." If masculinity and femininity, barbarism and civilization were not immutable, inborn qualities, but instead "morphological and psychological attributes that developed under the influence of glandular secretions," then individuals might be feminized or virilized, advanced or degenerated through new surgical, chemical, and nutritional techniques. The Presidential address of the 1922 annual meeting of the Eugenics Research Association made these implications plain. As President Barker suggested in his opening remarks, members of the Association committed to the "improvement of the human race by inheritance" could "scarcely fail to be interested in the discussions that are now going on regarding the glands of internal secretion and their relations to heredity." Barker speculated that eventually the causes of biological diversity (the evolution of man into "types" such as "the Negro, the Mongol and the Caucasian") will be seen to "involve the study of hormonic mechanisms!"

New forms of glandular manipulation seemed to promise machine-like control of these hormonic mechanisms. In the late 1930s, the advent of synthetic hormones made possible cheaper and more potent versions of steroid hormones, produced for a mass pharmaceutical market. Cortisone, for instance, was one of the synthetic mass-produced steroid treatments. Historian David Serlin links this interest in glandular intervention to a larger interest in "self-improvement," through which medical reinvention came to be seen as a purchasable amenity. Like a new frost-free refrigerator, hormonal intervention might become a commodity like any other, a tool for remaking the protean American self. Glands thus revisited old questions about heredity and environment, "constitution" and external influence, while bringing new techniques of medication, sterilization, castration, and injection to the fore. Studies of glandular difference also helped pave the way for altered popular understandings of the body, as the popularity of books such as Berman's attest. The complex interplay of professional and lay concepts of genetics in the twenty-first century was clearly augured by the gland fad of the 1930s.

While reading these selections, you might ask: of what import was a focus on intervention to the development of hormonal studies? How are prospects for manipulation and change described in these selections? What are the various aims of the therapies described? How is race embedded in discussions of self-transformation (and social transformation) through hormonal control, through an effort to render humans—in Berman's memorable words—"less contemptible and more divine"?

Bibliography

Berman, Louis. *The Personal Equation*. New York: The Century Co., 1925.

Hall, Diana Long. "Biology, Sexism, and Sex Hormones in the 1920s." In Marx W. Wartofsky and Carol C. Gould, eds., *Women and Philosophy*. New York: Putnam, 1976.

Oudshoorn, Nelly. *Beyond the Natural Body: An Archeology of Sex Hormones*. London: Routledge, 1994, 163–186.

Rubin, Herman H. *Your Life Is in Your Glands: How Your Endocrine Glands Affect Your Mental, Physical and Sexual Health, Your Appearance, Personality and Behavior*. New York: Stratford House, 1948.

Sengoopta, Chandak. Glandular Politics: Experimental Biology, Clinical Medicine, and Homosexual Emancipation in Fin de Siecle Central Europe." *Isis* 89, no. 3 (1998): 445–474.

Serlin, David. *Replaceable You: Engineering the Body in Postwar America*. Chicago: University of Chicago Press, 2004.

Somerville, Siobhan. "Scientific Racism and the Invention of the Homosexual Body." In Brett Beemyn and Mickey Eliason, eds., *Queer Studies: A Lesbian, Gay, Bisexual, and Transgender Anthology*. New York: New York University Press, 1996.

6.1   "The Gland As a Clue to the Mystery of Human Faces" (1920)

A clue to some of Nature's deepest mysteries—the manner in which she has produced the human races, white or black, yellow or brown, pug-nosed or the reverse—has been afforded by careful study of what to medical men are known as pathological states. An illustration may best bring out the point. When the celebrated Parisian physician, Doctor Pierre Marie, was a young man, first one woman came to him and then another seeking relief for persistent headaches and mentioning incidentally that their hands, feet, faces and bodies had so altered in aspect in recent years that their most familiar friends failed to recognize them. This incident, writes Professor Arthur Keith, of the Royal College of Surgeons, in the London *Mail*, forms the beginning of our knowledge of the quite common disease of growth, as we now know it to be—the so-called "acromegaly." At the same time we came to know that the pituitary gland, attached to the lower or under surface of the brain and in point of size less than a small cherry, is concerned in regulating the size and shape of our features. In some cases where surgeons have succeeded in removing the diseased overgrowths of the pituitary, the patient's features have returned to their normal form.

Next came the discovery that there are certain kinds of giants and dwarfs who are giants or dwarfs because of disease or disorder of the pituitary gland. Then, again, Nature in her disorderly moods produces, by another kind of experiment on the pituitary, a eunuchoid man, long of limb, fat, lethargic, beardless—a veritable human ox. Medical men mark these cruel experiments on the part of Nature and seek for the means to control or circumvent her. For the student of human races these observations are of great interest because they reveal the machinery which controls the "featuring" of our faces and bodies. We can best explain the characteristic appearances of the European face by supposing a predominance in the action of the pituitary gland:

Another series of medical observations has revealed that the thyroid gland, situated in the neck, is also concerned in molding the shape and size of our faces, bodies, and limbs. When this gland becomes the seat of a goitrous disorder in children the growth of the body and brain is profoundly affected.

The experiments carried out by Colonel McCarrison in India throw a new light on the cause of this disorder and lead us to hope that an easy method may be discovered for its prevention. From a study of the bodies and skulls of such goitrous children the anthropologist is led to infer that the characteristic Mongolian features of Chinamen, Japanese, Tibetan, and Burmese can be best explained by a dominance

"The Gland As a Clue to the Mystery of Human Faces," *Current Opinion* 69, no. 2 (August 1920): 207–210.

Figure 6.1.1
The Longest Lady. She achieved fame in the great days of Barnum on account of her size, which is now seen to have been an affair of a gland.

of the thyroid gland in the races of distant Asia. The Mongol is the bulldog of human types.

Our prehistoric ancestors discovered that removal of the reproductive glands brings about a profound alteration in bodily appearance and mental nature. Every farmer knows that the younger the animal on which the operation is performed, the greater the changes that are brought about. The reproductive glands are concerned not only in the bodily and mental characters which differentiate man from woman but influence also other or non-sexual features.

The effect of grafting fresh sex glands on animals and human beings which have been rendered sexless by operation or disease is of recent date. By grafting the male or female gland in the young castrated or sprayed animal the surgeon can call forth the manifestations of the male or female characters of body and mind.

In recent years we have learned that two small glands situated in the loins above the kidney—the adrenal glands—are concerned in the growth of the body, par-

Figure 6.1.2
Barnum's Big Boy. He need not have attained this height if in his time the adrenal and other glands were understood.

ticularly in the manifestation of all the changes which take place as we pass into manhood and womanhood.

On these bodies Nature performs experiments of an exceedingly cruel nature. Cases are known where infants on their mother's lap begin to assume all the disordered manifestations of sexual ripeness; more commonly the subjects of this disease are children just starting at school—"infant Hercules" they become. In all such cases there is a morbid overgrowth, or tumor, of the rind of the adrenal glands.

Disorders of this kind may occur at a later phase of life. Prof. T. R. Elliott has told of the case of a young woman whose sexual life came suddenly to an end and she began to assume the bearded face and some of the bodily attributes of the opposite sex. She was found to be suffering from a tumor of the adrenal gland. With its removal the patient recovered her womanly features.

Figure 6.1.3
The Whiskered Wonder. She got into a circus in her day, but now she would get into a
beauty parlor and have her face made smooth.

Altho the matter is far from proved, we have reason to suppose that the adrenal glands are concerned in determining the color of the skin. Doctor Thomas Addison
discovered years ago that disease or destruction of the adrenal glands is usually accompanied by the deposit of a black pigment in the skin. We know that these glands throw
into the circulating blood a substance which can make pigment-containing cells contract or expand. We have good reason for assuming that these glands are concerned in
regulating the temperature of the body. It is a logical inference that they should have a
direct influence on the deposit of pigment in the skin. Nature will, then, continue to
perform her erratic and often cruel experiments, and if we watch her and have patience
we may learn her secrets.

6.2   "Man and the Endocrines: With Stress on the Female" (1921)

G. K. Dickinson

Buried in the earth of Western Europe for over fifty thousand years lay the remains of primitive man, which, when exhumed, were still in good condition. It is claimed they dated from the period just before the last Ice Age. Being intact, those competent to judge report that the bones were big, the lines of attachment of muscle prominent, the joints large, and from close study conclude that he was not tall, but strong, vigorous, and well adapted for the fight necessary for food and protection, and in appearance somewhat approached the anthropoids.

There were five races: four of them more muscular, heavier, and more cumbersome in body than the fifth. This latter, the Cro-Magnon, on account of greater agility (being more slender and lighter), survived where the four perished, and it is supposed that the present race of man descended from this.

If we study the different nationalities, particularly those from mid-Europe, we will find a type which we are compelled to consider as more closely allied to this primitive state. They have large bones, big joints, plenty of muscle, big chests, wide costal margins, and broad abdomens. Their hands and feet are large, the hair coarse and shaggy, the hair margin on the forehead broad, and the eyebrows heavy, meeting ever the nose and laterally coming down to meet the hair descending from the scalp. They are hirsute all over the body, particularly on the chest and over the sacrum. There is apt to be a fatty prominence at the lower cervical region. The upper jaw is broad with well set, large teeth, there being an interval between the two incisors, and the second incisor often simulating a tusk. The lower jaw is equally well formed.

This anatomical condition being free from nasal defects, leads to good inspiration and well inflated lungs. The abdomen shows an absence of any defect such as imperfect rotation of colon and visceroptosis. They are of those who were brought up in the open. For many generations both men and women worked on the soil. The women were beasts of burden. Even to-day in some countries we see them doing the heaviest work in the field. They were considered of secondary importance and carried the burden with men as their overlords. These people are hypophyseal; the hypophysis in its secretion stimulating all the tissues mentioned and their activity reacting on the hypophysis to stimulate it.

But, as specialization in life, so-called civilization, advanced, when woman came out of the field and remained in the home, when man took the burden, where culture and education predominated, there resulted a lessened call on the hypophysis,

G. K. Dickinson "Man and the Endocrines: With Stress on the Female," *Medical Record* 99 (February 19, 1921): 307–309.

its effect on stimulating growth diminished, and the adrenal, which has charge of energy, became more effective. In people of the adrenal type we have smaller bones and muscle, narrow chests, soft, silky straight hair, a sharp hairline on the forehead, an empty space between the eyebrows, which do not extend far laterally. There is less hair on the body. Being deprived of the full stimulation of the hypophysis, there is greater tendency to defects of development and malformations of both bone and viscera. The upper jaw is prone to be narrow, with a vaulted arch, deviated septum, teeth small, and not so well placed, the chest flatter, the costal margin narrow, and congenital visceroptosis common, together with malformations in the intestinal tract, failure in development of the large gut, mobile colon, dragging cecum, early recession of the appendix, vagotonia, and all the stigmata of imperfect development making for our high-grade defectives and physiological hermaphrodites.

But modifications do occur, largely influenced by climate and diet. In countries where the diet is principally carbohydrate, disturbances due to developmental defects are not as noticeable as in those where high proteins are mainly used.

So much for the anthropological.

Now from the biological viewpoint. For development, sustenance, and protection in the lowest forms of animal life (where there is an arrangement of cells and some fluid content without a nerve system) there exists a hormone, which activates and correlates the several functions. Injuries are also repaired through its influence.

Somewhat higher in life, we find that which is analogous to our vegetative nervous system. It is autonomic, but with it are associated glands which secrete hormones to aid in the functions of circulation, digestion, reproduction, and protection.

Higher still, we come to the cerebrospinal system, which is a late accessory. The vegetative nervous system is fully developed and the several endocrine glands functionate actively, so that man today is cared for by his internal secretions, his autonomic system, and his mentality as expressed through the cerebrospinal. He is an inclusive trinity.

The endocrines are subject to few pathological processes, except through what might be termed the "sins of man." In a man living a normal life, a life approaching the primitive, little pathology occurs to them, but a man living the modern artificial life, far removed from that which was the custom for hundreds of centuries, suffers thereby. The main disturbances of the endocrines are those of function and have a very severe effect upon the constitution.

The autonomic nervous system is more easily affected, being more recent in development. It is disturbed by climate, diet, and mode of life. But even these are not considerable when the nervous system has inherited full growth and vigor. We seldom see what we nowadays call vagotonia in those who are fully hypophyseal in build and physiology. It is more commonly discovered in the adrenal type associated with defective development: the flat chested, the small boned, those with narrow costal margins and visceroptosis. The perverts, morphological and physiological, are the ones who go to make up our so-called neurotics.

Most neurotic pains are preceded by exhaustion of the sympathetic plexus, the stigma of a chronic rather than an acute process. Types of regional neuritis, neuroses, and a neurotic mind are generally associated with lesions of the sclerotic type, where we have vagus expression and mental obsession.

With a life in the open, a life of work, a life of energy, one that stimulates bone, muscle, and gland function, where the only thing taken to inebriate is pure oxygen, we rarely find defective physiology. But in those who for generations have lived in close rooms, have had exercise but spasmodically, who are not earners of their bread by sweat, who are given to lives of leisure, laziness, and intoxication of the toxic type, morbid physiology and a whole vocabulary of symptoms develop. In people who work for a living, who cannot consider their comfort and convenience, whose skin is ever moist, we seldom find a neurosis.

The cerebrospinal system is of the latest development and yet has been the object of study for the longest period. Man's consciousness of himself has led to speculations, diverse theories, and largely unwarrantable conclusions. What this system is anatomically we know pretty well, but what it is physiologically and psychologically we are far from comprehending. To a considerable extent the mind has an influence over the body, though, fortunately, it has none over the sympathetic nor the endocrines. In fact, they are the masters of the mind. Many types of cerebration are practically independent, but character, that is our personality, is dependent upon the endocrines and the autonomic.

With the thyroid juice we are able to metabolize; with the adrenal we give energy to our metabolism; from the liver we obtain glycogen and a something yet unknown to feed the tissue cells of the body and brain and keep their output, be it mental, motor, sensory, or otherwise, at their maximum. It is not inaccurate to say that active mental processes are dependent upon the thyroid, adrenal, and liver, particularly the latter.

But the hypophysis and adrenal are what might be termed "cold" organs. They stimulate for energy. The thyroid and the ovary develop gentleness. They are "warm." The hypophysis and adrenal contract the bloodvessels and raise tension. The thyroid and ovary dilate. We fight with the aid of our adrenals and high tension. We worship, make love, and sympathize through the thyroid, and, in the female, the ovary.

We find that the hypophyseal infect less readily (especially is this true as to tuberculosis), but tend more to tumor formations, while the adrenal type is more susceptible and less resistant to chronicity.

In every gonad there are to be found elements of the other sex, so that a woman way demonstrate male characteristics. Sometimes they are evident anatomically, more often physiologically. When there is a physiological demonstration, there is prone to be soubrette voice, hirsuteness, and sterility. So far as our observations go, the hypophyseal demonstrates this more frequently than the adrenal type.

The endometrium is acted upon by the internal secretion of the ovary, with the resulting production of a periodic menstruation. This is correlated with a temporary enlargement and activation of the thyroid. The hypophysis is also said to enlarge, and through it we have increased tonicity and intermittent contractions of the uterus.

If an ovum should be implanted in the endometrium the hypophysis tends to expel, the thyroid preventing, until such time as the ovarian secretion has enabled the ovum to obtain a placenta. The placenta being formed, it also develops a hormone which passes to the mammary gland, stimulating its growth, eventually with the production of the colostrum. The hormone in the colostrum is absorbed into the general circulation and has an effect similar to that of the pituitary, so that when the tenth œstrus cycle comes around, the pregnant womb is doubly acted upon, with the result that delivery is effected.

If the thyroid secretion fails in its effect upon the endometrium of a pregnant woman, some believe that toxins are formed in the syncytium which disturb the liver function as well as the renal, and may lead to an abortion, pernicious vomiting, or even eclampsia.

As the internal secretion of the ovary recedes, the effect of the hypophysis and adrenal, as well as the thyroid, is not properly balanced, and, periodic bleeding being absent, we often find augmentation of arterial tension. This is not a pathology of the bloodvessels and not so serious as the usual high-tension states.

Not so many years ago, considering the long period that man has existed, education of the fore-brain gradually was instituted, and this for several centuries was the pleasure of the male sex solely. As a result, man became differentiated from woman in strength of mind, breadth of vision, and ability to balance logically opinions on matters brought to his judgment. Woman was still kept either on the farm or in the home. She was not considered man's equal, simply his helpmate and companion in pleasure.

But recently, in the countries where education has advanced, there has been given to her a mind which is observant, a mind whose forebrain is not dwarfed by false speculation, and we have now developed in the female a being who is not as subject to man, who is not so strongly influenced by intuition, and whose thought is clear and unhampered by emotion. The woman of today is not the woman of the past, nor of the recent past. If she be given a hypophyseal body, strong and vigorous, she will have an equally strong, vigorous, and sensible mind. If she be given an adrenal body, persecuted by visceroptosis, intestinal stasis, and diseased functioning of the endocrines, then she will suffer.

We cannot cure malformations which nature has wrought, but with our knowledge of hygiene, diet, and endocrine stimulation we can do much to relieve them. It has long been known that there exists a correlation between the tissues of the different parts of the body, that no one organ can be affected without another responding and sympathizing. We know there is a strong relation between the physiology and pathology of the female pelvic organs and the kidney, liver, stomach, heart, and brain and the different endocrines, particularly the thyroid, adrenal, and pituitary. Let us not forget that, correctly speaking, gynecology means a study of woman. With the specialists it refers simply to the diseases of woman.

The gynecologist whose mind is in the pelvis becomes narrow. He fails to comprehend many existing conditions and misses the great beauty of the Symphony of Nature. To do full duty by woman, to succeed in relieving her of that distress which focuses its symptoms and signs below, one must not only know the unfortunate malformations and diseases which are the result of artificial modern life, but must know woman from all viewpoints, her anthropology, her biology, and her wonderfully perfected psyche.

Notes

G. K. Dickinson, M.D., F. A. C. S., Jersey City, N.J.

Read before the Academy of Medicine of Northern New Jersey, held in Newark, October, 1920.

6.3 The Glands Regulating Personality: A Study of the Glands of Internal Secretion in Relation to the Types of Human Nature (1921)

Louis Berman

THE EFFECT UPON HUMAN EVOLUTION

The ubiquitous and deep-seated influence of the internal secretions upon life and personality comprises but a fraction of what is known and only a hint of what is to become known. There is an endocrine aspect to every human being and every human activity, normal and abnormal, internal process and its external expression, regulated by laws of which we are beginning to catch a glimpse. Their control promises us now a dominion over the most intimate and inaccessible recesses of our lives in a way comparable only to the control we now exercise over the forces and energies once revered as the instruments of the gods—light, heat, magnetism, electricity. We have learned how to control and change our environment. We are now learning, endocrine research is now discovering, how to control and change ourselves.

The story of the evolution of the two types of control has many analogies. When man ceased looking upon his surroundings as inhabited by spirits of good and evil, as he conceived himself, and discovered that they were composed of things malleable and analyzable in his hands, he became their master. When now he drops the old superstitions about himself as a spirit, an emulsion of a spirit of good and spirit of evil, and sees himself more and more clearly as the most complex of chemical reactions, regulated and determined as are the simple and complex chemical reactions around him, he will begin to rule and modify himself as he rules and modifies them. Whether or not he will ultimately come to this final lucidity of thought and action, it behooves us to consider some of the uses to which our present knowledge might be put.

Since every step of the daily routine or adventure, from waking to sleeping, eating, drinking, marrying and giving in marriage, working, idling, fighting, playing, feeling, enjoying, sorrowing, every shade of emotion and nuance of mood, in short, every phase of happiness and unhappiness, are endocrine episodes in the history of the individual; the sphere of applications is as long and broad and deep as life itself. Not only do the internal secretions open up before us the great hope—that Life at last will cease to stumble and grope and blunder, manacled by the iron chains of inexorable cause and effect. They provide tools, concrete and measurable, that can be handled and moved, weighed and seen, for the management of problems of human nature and evolution.

Louis Berman, *The Glands Regulating Personality: A Study of the Glands of Internal Secretion in Relation to the Types of Human Nature* (Garden City, N.Y.: Garden City Publishing Co., 1928 [1921]), 313–329.

Every department of human life, the questions of labor and industry, science and art, education, puericulture, international problems, crime and disease, may be illuminated. War and Sex, those two master interests of mankind, may be understood and handled sympathetically as they have never been before. The reactions of man alone, and man in the crowd, will be clarified. The red thread of individuality which runs through the woof and warp of all human affairs will be unraveled.

Inevitably customs, morals, codes of procedure and practice, institutions, all those expressions of opinion which make conduct, all the currents which contrive the infinite variety of life, will be transmitted into another set of values.

A remolding, a remodeling will take place all along the line. Manifestly an unstable thymocentric should not be treated as a criminal, but treated in a sanitarium. A masculoid woman needs satisfactions not vouchsafed in the "love, honor and obey" home. How absurd it is to found codes of morality upon sermons or even the latest psychologies. During the nineteenth century progress in physics and mechanics overturned traditions thousands of years had painfully toiled to erect. What is to happen when man comes at last to experiment upon himself like a god, dealing not only with the materials without, but also with the very constituents of his innermost being? Will he not then indeed become a god? If he does not destroy himself before, that is surely his destiny. For better or for worse we possess now in the endocrines new instruments for swaying the individual as individual, and as related to other individuals, as a member of a type, family, nation, species, and genus.

The Basis of Variation

The sense of likeness and the sense of unlikeness plays a decisive rôle in the diurnal schedule of the individual. His sense of resemblance to his father and mother, his kin and clan, mark him and them off against the cosmos as an alliance of defense and offense. Yet no matter how closely he is like them and they like him, he differs and varies, they differ and vary, with a sort of mutual forgiveness, because the amount of resemblance overtops the degree of variation. In a paper on the "Rediscovery of the Unique," H. G. Wells emphasized the unique quality of the individual, and how, in spite of the cleverest devices of classification, living things ultimately escaped the classifying net by virtue of their tendency forever to vary.

The individual is unique. Yet when all is said and done the fact remains that between individuals there is a resemblance, and among them variation. What is the reason for their resemblances and what is the cause of their variation?

The conception of a particular chemical make-up of the individual, statable and relatively controllable in terms of the internal secretions, supplies a more rational and satisfactory method of approach to the problem than any so far suggested, as far as vertebrates are concerned at any rate. In effect, the differences between individuals may fundamentally thus be grouped among the differences which distinguish other chemical substances. The difference between water, technically known as hydrogen monoxide, and the antiseptic fluid labeled hydrogen dioxide lies wholly in the possession by the latter of an extra atom of oxygen in its molecules. All the peculiarities and qualities by which hydrogen peroxide is separated from water are referred to that additional quantum of oxygen. So the diversity of constitution and appearance of two brothers, alike in that they have inherited the same internal secretion trends, may be traced to the superiority of the pituitary of the one over the other.

Variation and resemblance are large issues, crucial material of the science of biology upon which much has been thought and written. That the proportion of the endocrines determines variation and resemblance, heredity and evolution is a hypothesis advanced, supported by a large number of facts, and capable of the most interesting experimental verification and observation. If a child resembles particularly either of its parents, grandparents, or relatives, there is good reasoning for believing that it is because their endocrine formulas are very much alike. When people apparently not blood related at all resemble one another the same law must hold. Resemblances may be partial or complete, and the degree will depend upon the amount and ratio of the internal secretions involved.

The same endocrine constitutions will produce corresponding physiques, physiognomies, abilities and characters. Deviations in endocrine type from that of the original stock, more of one endocrine and less of another, is at the bottom of the phenomenon of variation, basic for the origin of new species as well as the extinction of the old. In short, viewing the internal secretions as determinants, by their quantitative variations, of a host of biologic phenomena furnishes a concrete and detailed foundation for Darwin's theory of pangenesis.

Inheritance of Acquired Characters

Darwin's theory of pangenesis was an attempt to harmonize everything known in his time about heredity. It supposed that the various organs of the body gave off into the blood substances, themselves in miniature, which were taken up by the sex cells, and so became responsible for the development of their mother organ in the newly forming individual. Modern knowledge cannot accept all this as a whole. But in a modified version it has become the germ of a theory of heredity of which J. T. Cunningham, of Oxford University, is the chief backer.

Beginning with the traits and qualities which distinguish the sexes, grouped as the secondary sex characters, he showed that they are correlated with the special sexual function of the species in which they occur. These traits appear only when the hormones occur which are present in one sex and that only when the gonads of that sex are mature. In some cases they appear only at the period of the year when reproduction takes place, disappearing again after the breeding season. Their presence makes certain cells develop in excessive numbers at a particular spot in the organism (as in the growth of breasts from a few sweat glands) or causes them to specialize (to make hair on the face in man or to grow antlers on the head of a stag). After castration, the hormones being absent, all these points of contrast between the sexes fail to appear. So by analogy we may explain all somatic and psychic differentiation as functions of the glands of internal secretion. Contemplated from the angle of the effect of environment upon the endocrines, and a reflected action upon the germ cells, we may outline a mechanism of the inheritance of acquired characters at certain times and consequent adaptation. The cycle of events would be as follows:

1. A state of lability of cells at a point because of increased or decreased use.
2. An increased or decreased appropriation by them of the hormone controlling their function.
3. A corresponding increase or decrease in function of the gland of internal secretion and so
4. An increased or decreased representation of it in the reproductive sex cells in the gonads.

   To take a classic illustration, the long neck of the giraffe. The neck of certain animals living in a district populated by trees with high branches would be in a state of instability. If, at the same time, the pituitary, for some reason, was unstable and reacted with an extra supply of its secretion, it would stimulate the neck cells to reproduce themselves. In turn the pituitary would become stabilized in the direction of increased secretion, and hand on the component of increased secretion to the sex cells. That component, in conjunction with other factors, would therefore determine the emergence of a definite species character. In other words, the glands of internal secretion, as intermediaries between the environment and body, and between the body and the reproductive sex cells or germplasm, tender the clue to a phase of the puzzle of heredity, adaptation, and evolution. It is only a dotted outline of an explanation, to be sure, but one certainly capable of being filled in.

The Bearing on Breeding

Since the endocrine glands are so subtly sensitive and responsive to environment, and are at the same time so intimately concerned in the process of inheritance—a law which sums up their influence upon resemblance and variation in animals—there is no need to stress their importance for the practical science and art of good breeding, eugenics. Another mode of approach to its problems is opened up, and fresh enthusiasm instilled into its hopes and aspirations. A method of analysis of the factors involved, together with rules for the prediction of the outcome of certain matings, when finally worked out, will elevate its procedure to the level of the more exact sciences.

   A man's chief gift to his children is his internal secretion composition. The endocrines are truly the matter of breeding as they are of growth. They are the material carriers of the inherited physical and psychic dispositions, powers, abilities and disabilities from the soma to the germplasm and back from the germplasm to the soma. All kinds of questions arise as soon as one attempts to consider the bearing of this underlying principle upon concrete situations. What happens, say, when a pituitocentric mates with a thyrocentric? Or when a pituitocentric marries a pituitocentric? Is there a reinforcement or a cancellation of the dominant endocrine? Is there a quantitative addition of internal glandular tendencies in the germplasm, or a more complex rearrangement dependent upon reactions between all the internal secretions?

The term endocrine dominants brings up the inquiries of Mendelism, and the relation of Mendelian conceptions of dominant recessive to the internal secretion. The Mendelians have emphasized the rôle of the unit factor in heredity, and the conservation of the unit factor as an entity through all the adventures of matings. Also, that when unit factors, say of the color of the eyes, come into conflict, brown or black being mixed with blue or gray, one, the recessive, is submerged and overlaid but not destroyed by the other, the dominant. So brown or black eyes, dark hair, curly hair, dark skin, and so on, are dominant, while blue or gray eyes, light or straight hair, light skin, are recessive. A nervous temperament is dominant to the phlegmatic. A number of psychic qualities have been declared to be Mendelian unit factors: memory, mechanical instinct, mathematical ability, literary ability, musical ability, and even handwriting.

As architects of human qualities the endocrines must be involved in the Mendelian unit factors. Moreover, they seem to act upon a particular locale in different degrees, which is the strongest argument against the resolution of a number of structural traits into Mendelian unit characters. Most characters, somatic or psychic, are the products not of the action of one internal secretion alone, but of the interlinked activities of all of them. The amount of fat deposited under the skin, for instance, is influenced by the pituitary, the thyroid, the pancreas, the liver, the adrenals, and the sex glands. Other qualities, likewise, are resultants of a compromise between all the endocrine factors comprising the equation of the individual. If we are to look more deeply into constitution, we must measure the hormone potentials and their mobilization or suppression.

It will, in all probability, be found that the stability or instability of an endocrine will have a good deal to do with the part played by it in inheritance as well as in the life of the individual. An unstable pituitocentric marrying another unstable pituitocentric will have children either exceptionally small or tall, or abnormally bright or stupid. The instability tends to right itself in the next generation, or that following. Genius as a sport, as well as sudden degeneration of family stock, the whole problem of mutation, may be closely connected with this tendency.

It has been noted that the extinction of species has been preceded by a great increase in their size, for example, the case of the great reptilia of prehistoric time. That possibly represented pituitary stabilization in a changing environment. Indeed, endocrine instability appears the fundamental condition of the tendency to vary, endocrine stability the opposite.

Certain endocrine facts in relation to heredity should be mentioned. The daughters of mothers who menstruated early, themselves menstruate early. Animals fed upon thyroid during pregnancy, comparable to the thyrocentric, give birth to offspring with a very large thymus, comparable to the thymocentric. Women with partial thyroid deficiency, or myxedema, bear cretins. These are suggestive of what the internal secretions may do to an individual in inheritance and development. Inherited endocrine potential is the maximum reaction of which a gland is capable. This matter of potential is comparable to the factor of reserve power or margin of safety demonstrated up to the hilt for such organs as the heart and kidney as varying from individual to individual. A low potential, like instability of an internal secretion gland, may be latent, and not made manifest until the proper stimulus, the maximum amount of stress and strain, like accident, disease, shock or war, arrives.

When the individual is tested, the effects may be purely local because there is always in the organism a point of least resistance. Physical change alone may be prominent. Or because somatic changes are minor, the psychic will dominate the picture. An attack of the "blues," unaccompanied by any demonstrable transformation of the bodily processes, may be the sole symptom of an endocrine failure somewhere in the chain due to hereditary weakness or low potential.

So we may account for family trends and streaks, for varieties and strains among individuals, upon more precise lines based upon endocrine analysis. Finally disturbances of the internal secretions of the extreme sort denominated disease are well known. Indeed, a number of family diseases or predispositions to diseases have been traced to them. Predisposition in any direction will probably be shown to be caused by them within limits. Research here has its opportunity.

The Improvement of Racial Stock

A vast new territory of inquiry and achievement, as yet totally unexplored, is opened by the endocrines to the eugenists, and those idealists whose most earnest aspiration is the improvement of racial stock as a necessary preliminary to improvement of racial life. Beginning with Galton, they have brought to light a great collection of data to prove that human traits and faculties, good and bad, are inherited. Ability has been shown to run in certain families and degeneracy in others. Yet all of the practical net results have been summed up in the term "negative eugenics," the eugenics of prohibition and warnings.

Now the concept of personality, as woven around a system of chemical reflexes, handed on from generation to generation, is bound to change all that, and to create a structure of positive eugenics. It has been said that what radium is to chemistry the internal secretions are to medicine. Just as radium enlightens the chemist about the history of matter, and the integrations and disintegrations constituting the life of an element—the internal secretions illuminate the history of the individual as part of the life of the race, and of its integrations and disintegrations. Seeing the individual as a system of chemical substances interacting will assist enormously to predict the nature, character, and constitution of his descendants, which is essentially what the eugenist is after.

The study of mating, the heart of the matter, will concern itself with the investigation and comparison of the kind of endocrine personality of the offspring. Data bearing upon physique and physiognomy, details of anatomy and function, mind and behavior, will be so coördinated as no eugenist has hitherto succeeded in doing. Laws of endocrine inheritance will emerge that will bring the control of heredity within measurable distance. Standards and norms of a new kind would be obtained.

A beginning of this study of endocrine inheritance, on the pathologic side, has been made. Some of these have been along Mendelian lines. Following up abnormal growth (making giants and dwarfs) and abnormal metabolism (goitre, diabetes, and so on), it has been stated that it would seem that abnormal growth is dominant in the male, and recessive in the female, while abnormal metabolism is dominant in the female and recessive in the male. If an endocrine abnormality like a goitre, or cretinism, or a dwarf or giant appear in a family as a sign of endocrine instability, other members of the family will very likely show internal secretion abnormalities.

If one gland of internal secretion acts as the center of the system and the others as satellites, we should be able to trace what happens to it in the different generations. Does it maintain its supremacy? Or will it be ousted by another member of the group? The time will come when we shall thus be able to advise prospective parents of the consequences of procreation and to forecast the meaning for the race of a particular mariage. Internal glandular analysis may become legally compulsory for those about to mate before the end of the present century.

What are desirable and undesirable matings? The general law followed by nature in her helterskelter way seems to be the production of the greatest number of hybrids and variations possible, whether for good or evil does not matter. Certain endocrine types appear to be specially attracted to others belonging to the same group. Thus thymus-centered types frequently marry. The prepituitary type of male, the strongly masculine, mates often with the postpituitary type of female, the markedly feminine. The children exhibit the lineaments of the pituitary-centered type. The general trend seems to be the establishment of a better balanced, equilibrated type. Yet the children often are apt to segregate into pituitary dominants or pituitary deficients. Happiness and unhappiness in marriage should be examined from the standpoint of endocrine compatibility or incompatibility. Likewise those divorced or about to be divorced.

The correction of endocrine defects, disturbances, imbalances and instabilities, before mating, presents another field. It remains to be seen whether we shall thereby, in one generation, be able to affect at all the germplasm, hitherto revered by all pious biologists as an environment-proof holy of holies. No one can deny, in the face of the multitude of evidence available, that internal secretion disturbances occur in the mother, which, when grave, offer in the infant gross proof of their significance, and therefore when slight must more subtly work upon it. Endocrine disturbances in infancy have been traced to endocrine disturbances in the mother during pregnancy. Pregnant animals fed on thyroid give birth to young with large thymus glands. The diet of the mother has been proved conclusively to influence the development and constitution of the child. As the internal secretions influence the history of the food in the body, they affect development in the womb indirectly as well as directly. Certainly, whether or not we learn how to change the nature of germplasm within a short time, we have in the endocrines the means at hand for affecting *the whole individual that is born and sees the light of day.*

The Control of Mutations

The true physical and intellectual evolution of man depends upon the production of mutations of a desirable kind that can survive. The information furnished by the study of the endocrines concerning the genesis of personality provides the foundations for a positive eugenics, a eugenics of the encouragement of desirable matings, with the proper legal and social procedures. Selective breeding for the production of the best endocrine types should become practicable.

But the physician should be able to go farther. If the eugenist is to limit himself to the method of the animal breeder he will have to rest satisfied with the characters or hereditary factors given, that turn up spontaneously in an individual. But with the internal secretions as the controllable controllers of mutations, the outlook changes. It should become possible to produce new mutations, good and bad, to speed up their production at any rate. The feeding of thyroid to a gifted father before procreation might enhance immeasurably the chances of transmission of his gift as well as of its intensification in his offspring. A field of investigation is opened that would embrace in due time the deliberate control of human evolution.

All the physical traits, stature, color, muscle function, and so on, offer themselves for improvement, as well as brain size, and the intellectual and emotional factors which have dominated man's social evolution. The general prevalence of nervous disorders in civilized countries, visible even in the nervous infants the specialist in children's diseases is called upon to treat, shows that the nervous system of the better part of mankind is in a state of unstable equilibrium. It may be another example of the coincidences that have been called the Fitness of the Environment that the investigation of the endocrines promises to put into our hands the instruments of the control of the future of the nervous system. In general, meanwhile, the eugenist should strive for raising the level of the endocrine potential and discourage its lowering. That means the encouragement of matings in which all the internal secretion activities are reinforced. On the other hand, those internal secretion combinations, generally leading to a deficiency of all of them, which produce types of mental defectives, delinquency and crime, should not be allowed to occur.

The Influence of Environment

What suggestions now are there for the euthentist who would control the influence of environment upon child culture? There are certain pertinent facts and leads that are worth considering.

In analyzing environment, one must distinguish sharply in the jungle, the non-living factors from the living. For while the non-living act upon the endocrines

Table 6.3.1

Countries	Meat per day per capita in grams
Australia	306
U.S. of America	149
Great Britain	130
France	92
Belgium and Holland	86
Austria-Hungary	79
Russia	59
Spain	61
Italy	29
Japan	25

directly, the living act upon the vegetative system as a whole. The non-living factors are those with the intimate scrutiny of which physics and chemistry have busied themselves: food, water, air, light, heat, electricity, magnetism. The living are the animals that prowl all over the planet, the predatories spreading the gospel of fear.

The dietetic habits of a person, for instance, are known to have an influence upon the glands of internal secretion. Meat-eating produces a greater call upon the thyroid than any other form of food. In time this ought to produce a degree of hyperthyroidism in the carniverous populations. Prewar statistics concerning meat-eating in different countries show the greatest meat-eating among the English-speaking groups, who all in all must be admitted the most energetic (Ostertag).

Sea water contains iodine. People living in contact with sea water would be apt to get more iodine in their systems, and so a greater degree of thyroid activity. On the other hand, certain bodies and sources of inland water hold something deleterious to the thyroid, so that whole population of Europe, Asia, and America drinking such water have become goitrous and cretinous, and a large percentage straight imbeciles. Endemic cretinism is the name given to the condition. In parts of Switzerland, Savoy, Tyrol, and the Pyrenees, in America around some of the Great Lakes, there are still such foci. Marco Polo described similar areas he encountered in his travels through Asia.

Certain foods with aphrodisiac qualities may act by stimulating the internal secretion of the sex glands. A type of pituitocentric has an almost uncontrollable craving for sweets. Alcohol and the endocrines remain to be studied.

Light, heat and humidity stand in some special relation to the adrenals. Pigment deposit in the skin as protection against light is controlled by the adrenal cortex. The reaction of the skin blood vessels to heat and humidity is regulated by the adrenal medulla. A change in the adrenal as a response to changes of temperature and humidity in an environment would result in a number of concomitant transformations throughout the body. So variation and adaptation are probably connected. Most Europeans living for a sufficiently long time in the tropics suffer from a combination of symptoms spoken of as "Punjab Head" or "Bengal Head." The condition is probably the result of excessive adrenal stimulation by the excessive heat and light of the tropical sun, followed by a reaction of exhaustion and failure, with the consequent phenomena of a form of neurasthenia. In the section on the pineal gland there was mentioned the relation between light and the pineal gland in growing animals, and how it serves to keep in check the sex-stimulating action of light. The earlier puberty and menstruation of the warmer climates may be explained as due to an earlier regression of the pineal under the pressure of a great amount of light playing upon the skin.

All these, and many more could be cited, are instances of the direct influence of environmental factors upon one or more of the endocrines, and so upon the organism as a whole. Indeed, stimuli may be considered to modify an organism only in so far as they may modify the glands of internal secretion. Consequently, climatic factors will tend to make a population possess certain points of resemblance in common.

Varieties of the human race exist as do varieties of dogs. The pekingese and the fox terrier are as different as the Slav and Latin are different: because of differences in internal secretion make-up. The Slav peasant is definitely subthyroid in his general effect: round head, coarse features, stubby hands, and his stolid, brooding, intellectual and emotional reactions. The Latin shows a pronounced adrenal streak in his coloration, his emotivity, his susceptibility to neurosis and psychosis. H. Laing Gordon, a Scotch physician, reported that of 700 cases he studied, more than twice as many of duplex eyed individuals (brown or black, *i.e.*, adrenal-centered most often) were susceptible to the mental disturbances of war as the simplex (blue or gray-eyed, *i.e.*, thyroid-centered most often). He also pointed out that such individuals tend to have a narrow and abnormally arched palate. The Anglo-Saxon tends to be more sharply pituitarized, his features are more clean-cut, his mentality more stable. The Frenchman is rather a cross between the Anglo-Saxon pituitary-centered and the Italian or Spanish adrenal-centered.

So national resemblances traceable to climatic influences being repeated from generation to generation upon the endocrines may be explained physiologically. The biochemical interpretation of history will indeed be found the broadest, including as complimentary Buckle's climatic theory, Hegel's ideas on the influence of ideas, and Marx's on the superiority of the economic motives and forces.

The Races of Mankind

Arthur Keith, conservator of the Museum of the Royal College of Surgeons of England, was the first to apply the principle of endocrine differentiation to the problem of the color-lines—the lines which have divided mankind crudely into the yellow, the red, the white, and the brown, the Negro, the Mongol, the Caucasian, the copper-tinted American. It has long been recognized by anthropologists that the differences of color match with differences in every comparable trait. Thus the ideal Negro is built upon a pattern in which all the elements are specific and singular. When the looms revolve that make him there is produced a gleaming black skin, kinky, black hair, squat, wide-nostriled nose, thick, protruding lips, large, striking teeth, prominent jaws, and staring eyes. As his upright carriage and bone-muscle-fat proportions are distinctive, so are his musical voice and his easily wrought upon nerves. In contrast, the Caucasian has a good deal of hair on his body, his skin is a pale tan-pink, his lips are thin, and his nose especially has the definite bridge which narrows it. The Mongol, like the Negro, has the hairless body and the beardless face, but unlike him has lank, straight hair on his head, while his features are flattened and fore-shortened.

Upon the basis of these structural, functional, and mental differences, the qualitative and quantitative evolution of which in the race as in the individual is guided by the glands of internal secretion, Keith presents a very good case for the view that the white man is an example of relative excess of the pituitary, thyroid, adrenal, and gonad endocrines. "The sharp and pronounced nasalization of the face, the tendency to strong eyebrow ridges, the prominent chin, the tendency to bulk of body, and height of stature in the majority of Europeans" are the signs of pituitary dominance. Keith is also of the opinion that the "sexual differentiation, the robust manifestations of the male characters, is more emphatic in the Caucasian than in either the Mongol or Negro racial types...in certain Negro types, especially in Nilotic tribes, with their long, stork-like legs, we seem to have a manifestation of abeyance in the action of the interstitial glands." As for the adrenal superiority of the white man, "it is 150 years since John Hunter came to the conclusion...that the original color of man's skin was black, and all the knowledge that we have gathered since his supports the inference he drew. From the fact that pigment begins to collect and thus darken the skin when the adrenal bodies become the seat of a destructive disease we infer that they have to do with the clearing away of pigment, and that we Europeans owe the fairness of our skins to some particular virtue resident in the adrenal bodies." Finally, as regards the thyroid, a comparison of the face of a cretin with that of the Negro or Mongol tells the story. A certain variety of idiocy, Mongolian idiocy, in which the face simulated cretinism so closely as to deceive practised clinical observers, is characterized by a Chinese cast of features and eyes, hence the name. And in the Bushman of South Africa the cretin's face is even more startlingly recalled.

There is every reason, then, for believing that the white man possesses more pituitary, adrenal, gonad, and thyroid internal secretions as compared with the yellow man or black man. And since these endocrines control not only physique and physiolognomy, anatomic and functional minutiae, but also mind and behavior, we are justified in putting down the white man's predominance on the planet to a greater all-around concentration in his blood of the omnipotent hormones. While the Negro is relatively subadrenal, the Mongol is relatively subthyroid. Their relative deficiency in internal secretions constitutes the essence of the White Man's burden.

Man's Attitude toward Himself

A last, but by no means least, application we may consider of the developing knowledge of the internal secretions in relation to human evolution is its effect upon Man's attitude toward himself and so toward his fellow men. Whatever else he is, man is a land animal with ideas. That makes him the last word of mind working upon materials. In a word, he is the last word of mind working upon matter. But persistently he has refused to recognize himself as matter and as subject to the laws, to the physics and chemistry of matter.

History consists of the protocols that record the high lights of the interactions of materials and ideas which is the adventure of man in time and space. Materials and ideas have reacted, the record shows; materials come upon have begotten strange fantasies. Ideas that flashed from nowhere into a consciousness have transformed utterly the face of the earth. The herd-brute, agglutinated with his fellows by a magnetism beyond his ken, could be infected without, and so cast in the heroic mold. The possibility of communion—that possibility of possibilities, for without it none other could be possible—has rendered man the heir of divine destiny. For the progressive education of the race, a single discoverer here, an inventor there, and thinkers everywhere have been inspired. In due time their inspiration becomes the possession of even the lowest brain but capable of grasping it.

Man's attitude toward himself, his self-consciousness, and his attitude toward his fellow creatures has grown and varied and evolved with his education about himself. According to the theory he formulated concerning his being, his why and wherefore, he directed and governed, punished and mutilated himself and them. But the pressure of his curiosity, and the inexorable quality of the truth would not let him stand still. The poetic genius within him, as Blake called it, struggled on from one dogma concerning his nature to another. Behavior malignant or beneficial, horrible in its tragedy and pitiable in its comedy, flowed inevitably on. Witchcraft, trials, and the tortures of the Spanish Inquisition belong among the more mentionable consequences of some of man's theories about his own nature and its requirements.

Heretofore the imaginative spirit has had its day in the matter. And, curiously enough, an obsession to subjugate the natural has made it exalt the supernatural. Visions, dreams, portents, revelations, all symptomatic of an order of things above nature, are the stuff of what more than ninety-nine per cent of the millions of the race believe about themselves and their fate. Man's cruelty to man, through the ages, is a comment upon how vast and ramifying may be the consequences of a delusion.

But now for a couple of centuries the critical spirit, which is the spirit of science, has been invading the affairs of men. Humble but persistent corrosive of delusion, it has infiltrated the furthest bounds of ignorance and superstition. It has not dared to assert the supremacy of its fundamental views upon the everyday problems of human life because it was without concrete means of vindicating its claims. That lack is now supplied by the growing understanding of the chemical factors as the controllers and dictators of all the legion aspects of life.

The profoundest achievement of the biochemist will be the change his teachings and discoveries will bring about in man's attitude toward himself. When he comes to realize himself as a chemical machine that can, within limits, be remodeled, overhauled and repaired, as an automobile can be, within limits, when he becomes saturated with the significance of his endocrine-vegetative system at every turn and move of his life, and when sympathy and pity informed by knowledge and understanding will come to regulate his relationships with the lowest and most despised of the men, women, and children about him, the era of the first real civilization will properly be said to be born.

Morality, as society's code of conduct for its members, will change in the direction of a greater flexibility with the establishment of organic differences in human types. There is nothing that is more emphasized to the pathologist than that one man's meat is another man's poison. In the family, as nature's laboratory for the manufacture of fresh combinations of the internal secretions, allowances will be made of divergences in capacity and deportment from a new angle altogether. Schools will function as the developers, stimulators, and inhibitors of the endocrines, as well as investigators of the individuals who have not enough or too much of one or some of them. Prisons will have the same function, only they will be named detention hospitals. The raising of the general level of intelligence by the judicious use of endocrine extracts will mean a good deal to the sincere statesmen. The average duration of life will be prolonged for an enormous mass of the population. If the prevention of war depends upon the burning into the imagination of the electorates what the consequences of war are, a high intelligence quotient and revaluation of life will count for a good deal.

Man is the animal that wants Utopia. So long as human nature was looked upon as fixed constant in the ebb and flow of life, a Utopia of fine minds could be conceived only by the dreamer and poet. The desire for such a Utopia could only be regarded as a tragic aspiration for an impossibility. The physiology of the internal secretions teaches that human nature does change and can be changed. A relative control of its properties is already in view. The absolute control will come.

For need any one fear that the science of the internal secretions in its maturity will signify the abolition of the marvelous differences between human beings that create the unique personalities of history? A derangement of the endocrines has been responsible for masterpieces of the human species in the past and will be responsible for them in the future. The equality of Utopia can be the equality of the highest and fullest development possible for each of its inhabitants. The applications of endocrine control will not necessarily interfere with the life of the individual. There will be breeding of the best mixtures of glands of internal secretion possible. And there will be treatment for those born with a handicap, or who have become handicapped in the life struggle. There will be a stimulation of capacity to the limit.

The internal secretions are the most hopeful and promising of the reagents for control yet come upon by the human mind. They open up limitless prospects for the improvement of the race. A few hundreds of investigators are engaged upon their study throughout the world. That is one of the ironies of our contemporary civilization. A concerted effort at the task of understanding them, backed by the labors of tens of thousands of workers, would, without a doubt, accomplish much more for humanity than the vast armies and navies that consume the substance of mankind. If we could not obtain Utopia then, we might, at least, by abolishing the subnormals and abnormals who constitute the slaves and careerists of society, render the human race less contemptible and more divine.

6.4   "The Relation of the Endocrine Glands to Heredity and Development" (1922)

Lewellys F. Barker

Since the object of the Eugenics Research Association is the advancement of knowledge that will contribute to the improvement of the human race by inheritance, its members can scarcely fail to be interested in the discussions that are now going on regarding the glands of internal secretion and their relations to heredity. As a medical man, deeply interested in the problems of constitution and of condition and profoundly impressed with the recognizable influences of internal secretions upon form and function in both normal and pathological states, I welcomed the suggestion of Dr. Davenport that I deal in my presidential address with the topic announced. The progress of research in endocrine domains and in heredity has of late been so rapid that no single person can keep pace with its strides. My remarks, therefore, will make no pretence to completeness of discussion of the reciprocal relations of heredity and endocrinology. They are intended rather to direct the attention of the members of the association to some of the more important facts that have been established and to stimulate interest in some of the newer problems that are emerging and clamoring for solution.

The Endocrine Organs and Their Products

It is only comparatively recently that the significance of the so-called ductless glands and of the substances they manufacture has become recognized, but, in a very short time, a considerable body of knowledge concerning their structure, their functions and their inter-relations has been accumulated. At the moment, studies of the internal secretions, or, as many now call them, the "incretions," are, on account of their astonishing and novel revelations, attracting the attention not only of scientific workers in biology and medicine but, and perhaps to too great an extent, also of the laity. Important as a knowledge of these incretions is for an understanding of bodily and mental states, there is some danger, I think, of over-emphasis and of disproportionate prominence. Popular articles and treatises on endocrine subjects too often assume what is mere conjecture, or wild speculation, to be established as fact and reveal a tendency to exploitation that must sooner or later be followed by disappointment and disillusionment. There is, I fear, some danger that even scientific endocrinology may, temporarily at least, be brought into undeserved discredit. It would seem especially desirable, therefore, that those who write or speak upon the subject should discriminate carefully between fact and fancy. Every effort should be made rigidly to control hypotheses by

Lewellys F. Barker, "The Relation of the Endocrine Glands to Heredity and Development," *Science* 45, no. 1435 (June 30, 1922): 685–690.

accurate observation and careful experiment, for only thus can an orderly advance in knowledge be assured.

Though an incretory function has been ascribed to many organs of the body, the principal incretory organs, those whose function is best understood, are seven in number: (1) the thyroid gland, (2) the parathyroid glands, (3) the hypophysis cerebri, or pituitary gland, (4) the epiphysis cerebri, or pineal gland, (5) the suprarenals (consisting of two parts of entirely different functions, (*a*) the medulla or chromaffine portion and (*b*) the cortex or interrenal portion), (6) the islands of Langerhans of the pancreas, and (7) the interstitial tissue of the gonads (ovaries and testicles) or so-called "puberty gland."

There is evidence that each of these organs yields an internal secretion that, distributed through the blood, exerts important chemical influences upon other, more or less distant, organs and tissues. Some of these influences have been definitely determined, but it will doubtless be a long time before all of them will be well understood. The knowledge that has been gained concerning the thyroid, the pituitary, and the suprarenals gives promise, however, that steady research will gradually enlarge our information regarding the influences exerted by each of the incretory glands.

The chemical substances contained in the incretions have been called "hormones" and the determination of the precise chemical constitution of these hormones sets fascinating tasks for the biochemist. That the chemical constitution of some endocrine products may be closely approached, if not definitely established, has been shown by researches upon epinephrin (from the medulla of the suprarenal gland) and upon iodothyrin and thyroxin (from the thyroid gland). Studies of concentrated functionally potent extracts from other glands may before long reveal the chemical nature of other hormones; I have in mind, especially, studies of so-called "pituitrin" (hypophyseal extract) and of so-called insulin (extract of the islands of Langerhans of the pancreas). Clues as to the chemical nature of the hormones of the parathyroids, the pineal body, the interrenals and the gonads will probably be more difficult to obtain. Biochemical researches to establish the precise nature of the single hormones are extraordinarily important and should be vigorously prosecuted in order that experimental studies of hormone influences may be more systematically, exactly and intelligently pursued.

The Better-Known Endocrinopathies

Our knowledge of endocrine functions has been variously derived, partly through keen clinical-pathological observations, partly through experimental work upon animals (surgical removal of single organs; organ transplantations; injections of organ extracts or of isolated hormones). Before discussing the relations of the endocrine organs to heredity and development, it may be helpful briefly to refer to a few of the classical clinical syndromes that are now justifiably believed to be endocrinopathic in origin. Time will not permit me to refer to more than a few of these, but those chosen will serve as illustrative paradigms.

I may cite first two characteristic clinical syndromes met with in association with disease of the thyroid gland, namely, exophthalmic goitre and myxœdema.

In the former, known also as Graves' disease or Basedow's disease, we observe, in typical instances, a markedly enlarged pulsating thyroid gland (goitre) in the neck, a persistently accelerated pulse rate (say 150 or more to the minute instead of the normal rate of 72), marked nervous symptoms including fine tremor of the fingers, outspoken protrusion of the eye-balls (exophthalmos), a tendency to profuse sweats and to watery diarrhœa, sensitiveness to heat, a peculiar psychic over-alertness and apprehensiveness, and a tendency to rapid emaciation (despite an abundant food intake) associated with demonstrable acceleration of the rate of the basal metabolism. Since similar symptoms can be produced by feeding thyroid gland extract, it is believed that there is a hyperfunction of the thyroid gland (hyperthyroidism) in exophthalmic goitre.

In the idiopathic form of myxœdema (or Gull's disease) the clinical conditions are diametrically opposite to those in exophthalmic goitre. The thyroid gland is small, the pulse-rate is usually slow, the eyes look sunken (enophthalmos), the lid-slits are narrow, the bodily movements are slow and clumsy, the patient is mentally dull, forgetful and apathetic, there is sensitiveness to cold and a tendency to constipation, the hairs fall out, the skin is dry, thick and wrinkled and there is a tendency to obesity (despite a restricted food intake) associated with demonstrable retardation of the rate of the basal metabolism. Since patients with idiopathic myxœdema rapidly improve if they are fed the thyroid gland of the sheep, and since a condition precisely similar to it occurs if the thyroid gland be surgically removed (cachexia thyreopriva), it is believed that myxœdema is due to a hypofunction of the thyroid gland (hypothyroidism).

Two similarly contrasting clinical syndromes due to disorders of the hypophysis cerebri or pituitary gland may next be mentioned, namely, (1) gigantism and acromegaly, due to over-function, and (2) Froehlich's syndrome of obesity with genital dystrophy, due to under-function.

When there is overfunction of the pituitary gland in early life before the epiphyses of the long bones have united with the shafts of those bones there is overstimulation of bony growth and the patient becomes excessively tall (gigantism). When the overfunction of the pituitary gland occurs in later life (after epiphyseal union), bony overgrowth is still stimulated but manifests itself in enlargement of certain parts of the skull and of the hands and feet (acromegaly). There is also enlargement of the tongue and of the internal organs (splanchnomegaly). The victim presents a very characteristic appearance. The face is hexagonal, the nose is broad, the chin is prominent and curved so as to bend sharply upward, the cheek bones are outstanding and the arches above the eyes are prominent. Looked at from the side, the face resembles that of Punch (nut-cracker profile). The hands are spade-like, the fingers are sausage-shaped, and the feet are huge.

On the other hand, when there is under-function of the pituitary gland during development a condition (Froehlich's syndrome) in marked contrast to gigantism and acromegaly results. The skeletal development is defective, the growth of bone being less than normal. The patient is short in stature, the face remains child-like and the hands and feet are small (acromikria). The subcutaneous fat is markedly increased (obesity), and is distributed in an uneven way over the body, being most abundant on the abdomen, over the buttocks, and in the proximal portions of the extremities. The secondary sex characters either fail to develop or develop in a faulty way. The pubic and axillary hairs do not appear or are scanty. The external genitals remain in an infantile state. In young men the voice is high pitched and there is a lack of normal virility. In young women, the menstrual flow is scanty or absent.

Next, let us contrast two clinical pictures believed to depend upon disorders of the suprarenal capsules, (1) Addison's disease, met with in destruction of the suprarenals (hyposuprarenalism), and (2) pseudo-hermaphrodism, premature puberty, and hirsutism, met with in association with hyperplasias of the suprarenals (hypersuprarenalism).

In Addison's disease there is great weakness and prostration, associated with low blood pressure, diarrhœa and other digestive disturbances, chronic anæmia and often a peculiar bronzing of the skin (melanoderma).

On the other hand, in cases in which there is believed to be overfunction of the suprarenals, the clinical picture is markedly different though it varies somewhat with the time of onset of the assumed hyperfunction. Should this occur during fœtal life, a pseudo-hermaphrodite appears, the person presenting the external genital appearances of one sex while possessing the internal sex organs of the other sex. When the overactivity exists soon after birth rather than before birth, puberty appears prematurely, a little girl of three or four menstruating regularly and exhibiting the bodily and mental attributes (sexually) of an adolescent, or a boy of seven presenting the external genitals and the secondary sex characters of an adult. Should the overactivity of the suprarenals not occur until adult life, it may reveal itself in a woman of middle age by the rapid development of hairiness over the body (hirsutism) and by the exhibition of masculine characteristics (virilism).

Other examples of clinical pictures might be mentioned but these few will suffice to illustrate the extraordinary mental and physical changes that may become manifest when there are disturbances of function of the endocrine organs.

Constitution and the Endocrine Organs

Biologically considered, a developed human being, like all developed higher organisms, must be looked upon as the resultant of a long series of reactions between the zygote (fertilized ovum) and its environment. The germinal type or genotype, reacting with the surroundings, becomes the developed type or phenotype, in the case of human beings, the "realized person." The germ plasm provides the determining factors, the environment the realizing factors. Everything in the phenotype attributable to inheritance may be spoken of as "constitution," everything attributable to environment as "condition." Medical men as well as biologists must, then, when studying a person or a single organism, be interested in differentiating, when they can, what is "constitutional" from what is "conditional" in origin. In experiments upon animals and plants such a differentiation may be relatively easy; in studies of human beings it is always extremely difficult and, as regards many features, as yet wholly impossible.

The importance of constitution will need no emphasis among biologists who are predominantly students of heredity. Among medical men, too, throughout the centuries, especially among practitioners, there have always been those who have been fully aware of the significance of constitution and of its relation to disease-disposition. During the past fifty years, however, under the spell of bacterial and proto-zoan etiology, medical men have been so absorbed by studies of influences arising in the environment that they have, too often, forgotten to continue their investigation of influences of endogenous origin. For a time, it was almost taboo to speak of "consti-tution," or of "disposition," owing to a justifiable reaction, perhaps, against the earlier prevalent tendency to use these words as a mask for ignorance. Recently, however, there has been a welcome revival of studies of constitution. Now that facts that supply a scientific basis for a general pathology of constitution have been accumulated, we may look forward to a greatly increased interest among physicians in the part played by inheritance in disease. Indeed, during the past five years, several treatises upon this and allied subjects have been published; and we may expect, I think, during the period just ahead of us, many attempts to present, more systematically than hitherto, the rôle played by constitutional disposition in the pathogenesis of a whole series of diseases.

The chemical consideration of endocrine disorders, has in my opinion, given a strong impetus to this movement toward a revival of studies of the physiology and the pathology of constitution. For though the endocrine organs are, in some instances, accessible to trauma and to poisons and parasites that reach them through the blood-stream, diseases of these organs, especially those "idiopathic" chronic diseases that develop insidiously and give rise to the classical endocrine syndromes, appear to be, usually, of endogenous rather than of exogenous origin, that is to say, they develop as the results of special anomalies of constitution. This accounts for the fact that endocri-nopathies tend to run in families, and the interrelationships that exist among the different endocrine organs may explain why a disease of the thyroid (exophthalmic goitre) may appear in one member of a family, a disease of the pancreas (diabetes mel-litus) in another, a disease of the hypophysis (dystrophia adiposogenitalis) in a third, or a pluriglandular disorder in a fourth member of the same family. The experienced clinician can now often recognize phenotypes in which there are anomalies of consti-tution that predispose to endocrine disorders; and as a result of this recognition he may, sometimes, be able to institute a rational prophylaxis. The thyreotoxic constitu-tion, the hypothyreotic constitution, the hypoparathyreotic constitution, the hyperpi-tuitary constitution, the hypopituitary constitution, the hypergenital constitution and the hypogenital constitution are instances in point. Unfortunately we have not learned as yet how effactually to intervene in a prophylactic way in all of these anomalies of constitution, but rewarding experiences with the hypothyreotic and with the hypopar-athyreotic constitution give us hope that, with widening knowledge, suitable preven-tive measures will be discovered.

Studies of the symptoms of endocrine disorders and studies of partial anomalies of constitution affecting the endocrine organs are thus throwing much light not only upon (1) the mode of action of the incretions, but also upon (2) inheritance as a determining cause of endocrinopathic phenotypes. The incretions may affect distant parts directly, being carried to them by the blood; or they may affect those parts indirectly through the intermediation of the autonomic nervous system, which they sensitize. When they act directly, they may influence the substances and processes in the localities that they reach (chemical correlation; regulation of metabolism) or they may supply materials for incorporation by the cells (nutritive and formative influences). When they act indirectly through the vegetative nervous system they may exert profound effects through the secretory activity of glands, through the contraction of smooth muscle, or through modifications of those neural mechanisms that have to do with the emotions and the will. During the developmental period, it is clear that the incretions are in part responsible for the dimensions and proportions of the skeletal apparatus and the soft parts. A normal functioning of the incretory organs is essential for the shaping of parts and for the maturing of functions in the right place and at the right time. Through correlative differentiation (due in part at least to the action of the incretions), the developing organism gradually comes to exhibit the characteristics of its species, its age and its sex. Even the anthropologists now maintain that the solution of the problem of how mankind has been demarcated into types so diverse as the Negro, the Mongol and the Caucasian will involve the study of hormonic mechanisms!

Can Hormones Modify Unfertilized Germ-Cells So As to Influence Inheritance

Thus far in our discussion of the relation of the endocrine glands to heredity and development we have confined our attention to (1) the genotypic determination of endocrine functions in developing organisms, (2) the rôle played by the incretions in normal and pathological ontogeny, and (3) the fact that there exist heredo-familial anomalies of body make-up that predispose to endocrine disorders. But we must, for a few moments at least, consider the possibility that hormones, reaching unfertilized germ-cells, may modify the germ plasm in such a way as to give rise to new inheritance factors that will be transmitted from generation to generation.

Experiments upon the influence of incretory substances upon the development of cold-blooded animals have yielded such striking results upon cells of the soma that many have wondered whether incretions circulating in the blood might not also permanently alter the germ-cells so as to account in animals for the origin of mutations and new biotypes. You will recall the experiments to which I refer (1) the acceleration of tadpole metamorphosis by feeding thyroid substance and (2) the retardation of the same process by feeding thymus substance.

In endocrine diseases of either endogenous or exogenous origin, the cells of the soma are also markedly altered; and the question has naturally been asked, May not the germ-cells be simultaneously profoundly changed?

Since 1895, a number of investigators have suggested that the influence of specific internal secretions might easily be used for the explanation of the inheritance of acquired characters. Last year, an English evolutionist published a volume on "Hormones and Heredity" and suggested that environmental influences influencing an organ, or part, of the mother may set free chemical substances (hormones) that, carried through the blood to the ovaries, may affect the ova in such a way as to lead to similar changes in the same organ, or part, of the offspring. By such a mechanism he would attempt to account for a progressive evolution in the animal series. His theory would seem practically to be a modification of the pangenesis theory of Darwin with the substitution of "hormones" for Darwin's "gemmules."

Many physicians, too, have leaned toward Lamarckian or neo-Lamarckian theories that assume the inheritance of acquired characters and some of these have suggested that in such inheritance the incretions must be concerned. Those who have been trained in the methods of modern biology, however, usually reject Lamarckism, and attempt to explain the apparent inheritance of "acquired characters" for a generation or two by assuming either a "germinal injury" (in the sense of Forel's "blastophthoria") or a "parallel induction."

The consensus of biological opinion in this country is strongly opposed to the inheritance of acquired characters. Mendelian studies lend no support to the view that conditional influences can affect inheritance factors. Mendelism is, however, difficult if not impossible to apply to man. As some one has put it, "the propagation of man consists of a continual crossing of polyhybrid heterozygote bastards," not susceptible to analysis by Mendelian methods such as can be applied to the study of the propagation of plants and experimental animals. But if inheritance of acquired characters really occurred, why should there not be, as Conklin emphasizes, an abundance of positive evidence to prove it? When one plant or animal is grafted on another, there is no evidence that the influence of the stock changes the constitution of the graft. When an ovary is transplanted, the foster mother does not affect the hereditary potencies of the ova. Until more proof has been brought than had hitherto been advanced, we shall not be justified, so far as I can see, in accepting the theory that conditional influences change hereditary factors. There are, moreover, aside from the problem of the inheritance of acquired characters, enough relationships of the endocrine organs to heredity and development to long keep us rewardingly occupied.

Conclusion

Let me summarize in a few words the situation as I see it. The endocrine organs are of the greatest importance in normal development, their incretions exerting profound formative and correlative influences. In pathological development, the abnormal plenotypes that appear often point decisively to partial anomalies of constitution involving especially the ductless glands and their functions. Whether or not under normal or pathological conditions, hormones arising in the soma can so change the germ plasm of ova or sperm-cells as to account for certain mutations or for germ-cell injury is a question that deserves consideration and merits experimental test. Finally, the conjecture that conditional influences upon the soma can through hormonal production and transportation to parental gametes so modify the germ-plasm as to result in the inheritance of the conditioned modification seems, as yet, to have but little, if any, evidence to support it.

Note

Presidential address at the tenth annual meeting of the Eugenics Research Association, held at Cold Spring Harbor, Long Island, June 10, 1922.

6.5   *Endocrinology in Modern Practice* (1939)

William Wolf

Our next question is, What makes glands go wrong? The answer is in part indicated above in the enumeration of the conditions upon which normal gland activity depends. *Hereditary factors* play an extremely important rôle, but while it is true that some endocrinopathies are transmitted as such from one of the ancestors through the mother to the offspring, it seems that more often the child is born with an endocrine weakness or susceptibility to a glandular disorder rather than with an overt endocrine disease. The congenital susceptibility may affect the same gland as that affecting the ancestor or a different gland entirely. Thus a thyrotoxic parent may have a child who later develops acromegaly. Still more frequent is the occurrence of an overfunction of a gland in the parent and an underactivity of the same gland in the offspring. Any variety of combinations may, however, present itself. The diagnostician should not be deceived thereby and should by no means discount hereditary influences because the glandular disturbances in the parent and child seem widely different. It is also possible that a gene defect occurred during intra-uterine life as a result of injury through trauma, *x*-rays, or compression by a tumor, or as a result of such severe infections as lues and perhaps tuberculosis.

Then there are *toxemias* of various kinds. Thus we are reminded of the degenerative influences upon the adrenal cortex and thyroid by acute infections, the baneful effect of infection upon diabetes, etc. The resulting toxic absorption not only causes glandular dysfunction, but perceptibly hinders recovery even when endocrine therapy would otherwise be adequate. Under this heading focal infections should not be forgotten, although their removal should be performed with great discretion. The necessary operative procedure may unduly strain the weakened organism or the loss of vital organs may prove more detrimental to the patient than his endocrinopathy. Chronic toxemias due to hepatic disease, protein sensitivity and the like also belong to this group of causative factors.

Abnormal action of *other glands* must of necessity also affect endocrine function. For example, an underactive pituitary as seen in Simmonds' disease depresses the adrenals, thyroid, parathyroid, gonads and other glands, while an overactive adrenal cortex causes gonadal and probably pituitary hyperfunction.

Diseases of the *blood vessels* and *nervous system*, by inadequately supplying the gland tissue with raw materials or allowing only incomplete removal of hormones, also hinder proper function. A notable example here is the arteriosclerosis seen in diabetes

William Wolf, *Endocrinology in Modern Practice*, 2nd edition (Philadelphia: W. B. Saunders Company, 1939), 22–25.

and the infiltration of the skin with mucoid material in myxedema. Again, continued emotional stress places great burdens upon the adrenals, pancreatic islands, thyroid, etc.

Tumor formation may affect a gland in two different ways. If it compresses or invades the gland structure so as to destroy its functioning epithelium the net result will, of course, be a decrease in hormone output. But the endocrines are unique in that they may also develop adenomata which are composed of secreting tissue, with the result that not only is the hormone output not diminished, but it is actually greatly increased. Examples of this type of tumor are the eosinophile adenoma of the pituitary which produces acromegaly, the parathyroid tumors causing von Recklinghausen's disease, adenomata in the pancreas resulting in hyperinsulinism, or those of the adrenal cortex producing the adrenogenital syndrome.

Food deficiencies, notably a lack of certain *vitamins* and *minerals*, also affect endocrines adversely. Vitamin A deficiency is frequently associated with adrenal atrophy, poor lactation, diabetes and injuries of the reproductive system in both male and female, and contributes to thyroid atrophy in males and thyroid hypertrophy in females. There is also some similarity in the effects of this vitamin and the parathyroid hormone. Vitamin B deficiency is a factor in thymic atrophy, in hypothyroidism, in gonadal hypofunction, probably resulting from lack of the pituitary gonadotropic hormone, and also in adrenal hypertrophy. Vitamin C is an important constituent of the adrenal cortex, the pituitary and the corpus luteum, and is also found in the pancreas, testis, ovary and thyroid. When it is ingested in inadequate quantities the well-known signs of vitamin C deficiency, such as scurvy and hemorrhages, result, and in addition the stores of the vitamin in the adrenal cortex and other tissues are lowered and the secretion of adrenalin is reduced. Vitamin D is directly related to calcium metabolism (fixation) and the parathyroid glands as well as to the sex function. Vitamin E deficiency affects gonadal function, causing castration-like changes in the pituitary, atrophy in the testes, and degenerative changes in the developing tissues if pregnancy does take place.

Other food constituents besides the vitamins also have a relation to glandular function. It is well known that calcium and phosphorus are directly associated with parathyroid function, and iodine with that of the thyroid. In this connection one may also mention the geographic factor, the lack of iodine in certain localities accompanied by the presence of goitrogenic factors contained in certain common foods, causing the establishment of a genuine goiter belt with a high incidence of thyroid disease.

Racial and constitutional factors play a most important rôle in the susceptibility to endocrine diseases, but very little definite knowledge is available which would furnish us with useful data in the diagnosis or treatment of disease. Perhaps the greatest advantage in recognizing this fact is that many apparently endocrine diseases are racial characteristics and consequently are not amenable to therapy. By recognizing this, the physician and patient may be spared many disappointing experiences. However, use can be made of the knowledge of so-called "definite types," which seem to have a predisposition to exhaustion of certain glands. Thus the so-called "adrenal type," with the dark skin, low hair line, coarse hair, broad body build and hypersexuality may be distinguished from the thyroid type with his fine delicate skin, thin silky hair and a nervous trembling temperament, or from the thymic type with the angelic, more or less child-like, infantile appearance. While no clear-cut conclusions can be drawn from this sort of inspection it may help in evaluating presenting features, since as an empiric fact we know that in these individuals it is the indicated gland rather than others which is apt to be affected.

The conditions which conduce to endocrine disorder are so widespread and the endocrine disabilities themselves lead to symptoms of such wide implications, extending into so many of the bodily systems, that it is advisable to check upon the endocrines whenever a clinical diagnosis is uncertain and not particularly well supported. Aside from such well-marked endocrine syndromes as Addison's disease, Graves' disease and diabetes, such obscure syndromes as neurasthenia, nervousness, headaches, obesity, menstrual disorders without pelvic pathology, growth disorders, hypertension, bone conditions, retardation of development, whether physical or mental, and other similar states may be manifestations of endocrine malfunction. In addition, even where a non-endocrine clinical diagnosis has been verified and is accepted as correct, if expected results do not follow adequate treatment the endocrine situation will often bear watching.

In diagnosing endocrine disorders, however, one meets an immediate problem in that rarely indeed is only a single gland affected. Graves' disease is not a disorder of the thyroid alone nor Addison's disease of the adrenals. In the less well-defined endocrine syndromes the many-sided aspect is even more pronounced. Therefore the problem of diagnosis necessitates an attempt to discover which one of the chain started the process in the wrong direction. This is often an impossible task, although a well-taken history is of inestimable help. After this search has been exhausted we must attack the next point, namely, attempting to determine which of the glands is most profoundly affected, or better, which of the hormones secreted abnormally is apt to produce the most undesirable manifestations in the patient and which effects can be most readily corrected. This is important for several reasons. First, because severe damage to the organism may thus be prevented; secondly, because the picture generally becomes clearer as one or two of the glands begin to function more normally; and thirdly, because the chain of events implicating one gland after another is broken and many of the manifestations correct themselves spontaneously.

While the statement is simple and uncontroversial, the solution of the problem is by no means as easy as it might appear. It is for this reason that a considerable portion of this volume is devoted to the various methods at our disposal in arriving at a useful diagnosis. A carefully taken history is extremely important, special emphasis being placed upon any factor which may have an endocrine bearing. Similarly a general physical examination with any hereditary or hormonal influences in mind is a *sine qua non*. Only then may we have recourse to special aids, such as anthropometry, x-ray examination, laboratory analyses of the blood, urine and other body fluids, etc. The tendency so prevalent at the present time to proceed in the reverse order should be assiduously avoided. While in unusually experienced hands it may occasionally furnish a short cut to a diagnosis it is bound sooner or later to lead to disaster.

A plan of procedure found useful by the author and others is to keep the following questions in mind: What are the principal presenting symptoms? Are they likely to be of endocrine origin? If so, which gland or glands are likely to produce such symptoms? If my surmise is correct, what other symptoms should I expect to find? Are any of them present? What other glands may produce similar manifestations? If the second guess was correct, what else should I expect to find in corroboration? Is there more than one gland fundamentally affected? If so, do the effects of one neutralize or enhance those of the others? Which of the glands causes most damage to the organism? Which is apt to be most easily corrected? What is the probable cause of the endocrinopathy? Which organs in the body are principally responsible for the presenting symptoms? Is it possible that the affected organ is incapable of response to a normally functioning endocrine gland? Can I determine this by direct examination of the organ or can I deduce this possibility by the fact that only certain organs, which are under the control of the particular, suspected gland, are affected while other organs which are under the influence of the same gland function normally? How are the affected glands influenced by other organs in the body, such as the heart, kidneys, blood, spleen, liver, etc.?

In order to answer all these questions we frequently must go through all the procedures mentioned in the preceding paragraph, and then we may be rewarded with the thrill of a definite, correct and useful diagnosis. This naturally brings us to the next problem as to what we can do to normalize our patient.

7 HYBRIDITY AND ADMIXTURE

Introduction

Evelynn M. Hammonds and Rebecca M. Herzig

Taxonomic approaches to race segregate individuals into groups and subgroups based on specifiable characteristics—characteristics which might range from hair form or skin color to passivity, trustworthiness, or predisposition to specific diseases. Such classifications necessarily rely on "typical" or "ideal" characteristics, ignoring variation *within* a given group in favor of differences across groups. As we have seen, challenges to this typological view of race have appeared throughout the history of racial science. Some investigators have emphasized the mutability of racial traits under the influences of climate, education, religious discipline, diet, or labor. Others questioned the numbers behind the quantitative and statistical research that served to normalize "typical" attributes, demoting the concept of averages that governed inherited taxonomic approaches. Still others sought to insist on the fundamental hybridity of race. "Between [the] whitest of men and the blackest negro," wrote Xavier Mayne in 1908, "stretches out a vast line of intermediary races as to their colours: brown, olive, red tawny, yellow." Against long-standing cultural preoccupations with purity of types, Mayne argued for the inevitability of intermediacy: "Nature abhors the absolute, delights in the fractional." By the 1920s, nineteenth-century efforts to identify and assess stable, distinct racial categories ran headlong into new studies emphasizing racial mutability and blending. The studies emerged in the context of a massive influx of new immigrants from Southern and Eastern Europe in the first decades of the century and increasing African American migration from southern to northern states. Amidst increased public discussion about the prospect of racial "passing," scientists sought to identify the characteristic physiological, intellectual, and emotional characteristics of the "hybrid," and to develop new techniques for discerning racial identity where ancestral mixing or "admixture" had occurred. In his 1920 anthropometry manual (a field reference guide which would go through multiple editions), Aleš Hrdlička described a "skin test for mixed-bloods":

On the pectoral parts of the chest may also be made certain tests developed by the author which in many instances of doubtful mixtures between Whites and Indians or other Yellow-browns, and between Whites and other colored races, will help us arrive at a conclusion. They are tests for the blood reaction of the skin. In a full-blood individual of the Yellow-brown or other dark races, if the chest is exposed and the observer makes three or four vertical lines over the pectoral parts by drawing his finger nail over the skin with a certain amount of pressure, there will be little or no visible reaction; but if there is any mixture with Whites the lines will show as fairly broad red marks, and the flush will be of some duration—both features being the more marked the more white blood is present in the individual under examination, provided he is in the ordinary state of health. In malarial, anaemic and phthisical subjects, where the condition and supply of blood are much altered, the value of this test does not hold good.

Of course, these were not the first scientific efforts to discern racial mixture. Well before the term *miscegenation* entered the American lexicon in 1863, natural historians and physicians endeavored to delineate the relative effects of varying ancestral lines on individual qualities and capacities. Elaborate taxonomies were developed to classify and name various "amalgamations" of the ideal-typic races African, Indian, and Causcasian. Identifiers such as *mestee, mulatto,* and *octoroon* were used to specify mixed racial status. Generally based on ideas of percentages of racialized blood ("blood fractions"), nineteenth-century studies reinvigorated the notion of pure racial types even while studying multiple generations of interracial marital and extramarital heterosexual relations. The notorious "one-drop rule" of hypodescent, a standard of mixture that applied only to individuals of African descent, laid bare the white supremacist underpinnings of these nineteenth-century discussions.

Received interest in the prevalence and consequences of racial mixing was given fresh force by the work of German biologist August Weismann (1834–1914), who established a novel analysis of heredity, based on a "theory of the germplasm," in 1885, eventually publishing his results in 1893. Weismann's studies determined that unlike the mortal cells of the body ("soma"), the reproductive cells comprising the germplasm were essentially immortal—passing largely unchanging from generation to generation. With this theory, and with the naming of the gene as the united of germplasm transmission, studies of racial admixture acquired a new language. (It is worth noting that the phrase *melting pot* entered American parlance around the same time as *germplasm*.) In this 1917 selection from the *Proceedings of the American Philosophical Society*, Charles Davenport, perhaps the most influential eugenicist in America of the early twentieth century, offers a definition of race based on genes. How does this definition appear to inform his assessment of race "intermingling"? What are the consequences of this intermingling, according to Davenport? What kinds of evidence does he provide for his claims?

The next three selections are also noteworthy for this stress on Mendelian genetics. In the first, Harvard zoologist William Ernest Castle (1867–1962) discusses human "race-crossing" in the context of his genetic studies of rabbits. It is noteworthy that Castle spells out an opposition between the "biological" and "social" attributes of race—an opposition still dominant in debates over the category. What are the contents of each term here? On which side of the equation to institutions such as marriage or slavery appear to fall? How might such categories give rise to intra-professional battles between various scientific authorities, and inter-professional battles between scientists and lawyers, physicians, legislators, and other kinds of experts?

The subsequent selection by physiologist S. F. Cook addresses the topics pertaining to "racial fusion" on the Pacific Coast (a region only gestured toward in Castle's discussion). Here note the persistence of the blood quantum ("full," "quarter," etc.) in the discussion of "social" and "genetic" amalgamation. How does Cook calculate percentages of "Indian blood"? What prediction does he tender with respect to the futures of the California Indians and the Nevada Indians? The final selection, from Louis Wirth and Herbert Goldhamer's monograph "The Hybrid and the Problem of Miscegenation," reveals the interpenetration of themes in the sciences of race, from anthropometry to disease susceptibility. The essay also highlights ongoing intra-professional debate around racial identity and difference. How do the coauthors navigate these disputes here?

Bibliography

Fields, Barbara J. "Of Rogues and Geldings." *American Historical Review* 108, no. 5 (December 2003): 1397–1405.

Fullwiley, Duana. "The Biologistical Construction of Race: 'Admixture' Technology and the New Genetic Medicine." *Social Studies of Science* (forthcoming).

Gleason, Philip. "The Melting Pot: Symbol of Fusion or Confusion?" *American Quarterly* 16 (Spring 1964): 20–46.

Greenberg, Julie A. "Definitional Dilemmas: Male or Female? Black or White? The Law's Failure to Recognize Intersexuals and Multiracials." In Tomi Lester, ed., *Gender Nonconformity, Race, and Sexuality: Charting the Connections.* Madison: University of Wisconsin Press, 2002, 102–120.

Hodes, Martha, ed. *Sex, Love, Race: Crossing Boundaries in North American History.* New York: New York University Press, 1999.

Hollinger, David A. "Amalgamation and Hypodescent: The Question of Ethnoracial Mixture in the History of the United States." *American Historical Review* 108, no. 5 (December 2003): 1363–1390.

Hrdlička, Aleš. *Anthropometry.* Philadelphia: The Wistar Institute of Anatomy and Biology, 1920.

Kennedy, Randall. *Interracial Intimacies: Sex, Marriage, Identity, and Adoption.* New York: Vintage Books, 2003.

Lemire, Elise Virginia. *"Miscegenation": Making Race in America.* Philadelphia: University of Pennsylvania Press, 2002.

Mayne, Xavier [Edward Irenaeus Prime Stevenson]. *The Intersexes: A History of Similsexualism as a Problem in Social Life.* New York: Arno Press, 1975 [1908].

Pascoe, Peggy. "Miscegenation Law, Court Cases, and Ideologies of 'Race' in Twentieth-Century America." *Journal of American History* 83 (1996): 44–69.

Tapper, Melbourne. *In the Blood: Sickle Cell Anemia and the Politics of Race.* Philadelphia: University of Pennsylvania Press, 1999.

Williamson, Joel. *New People: Miscegnation and Mulattoes in the United States.* New York: New York University Press, 1984.

7.1   "Effects of Race Intermingling" (1917)

C. B. Davenport

The problem of the effects of race intermingling may well interest us of America, when a single state, like New York, of 9,000,000 inhabitants contains 840,000 Russians and Finns, 720,000 Italians, 1,000,000 Germans, 880,000 Irish, 470,000 Austro-Hungarians, 310,000 of Great Britain, 125,000 Canadians (largely French), and 90,000 Scandinavians. All figures include those born abroad or born of two foreign-born parents. Nearly two thirds of the population of New York State is foreign-born or of foreign or mixed parentage. Even in a state like Connecticut it is doubtful if 2 per cent. of the population are of pure Anglo-Saxon stock for six generations of ancestors in all lines. Clearly a mixture of European races is going on in America on a colossal scale.

Before proceeding further let us inquire into the meaning of "race." The modern geneticists' definition differs from that of the systematist or old fashioned breeder. A race is a more or less pure bred "group" of individuals that differs from other groups by at least one character, or, strictly, a genetically connected group whose germ plasm is characterized by a difference, in one or more genes, from other groups. Thus a blue-eyed Scotchman belongs to a different race from some of the dark Scotch. Strictly, as the term is employed by geneticists they may be said to belong to different elementary species.

Defining race in this sense of elementary species we have to consider our problem: What are the results of race intermingling, or miscegenation? To this question no general answer can be given. A specific answer can, however, be given to questions involving specific characters. For example, if the question be framed: what are the results of hybridization between a blue-eyed race (say Swede) and a brown-eyed race (say South Italian)? The answer is that, since brown eye is dominant over blue eye, all the children will have brown eyes; and if two such children inter-marry brown and blue eyes will appear among their children in the ratio of 3 to 1.

Again, if one parent be white and the other a full-blooded negro then the skin color of the children will be about half as dark as that of the darker parent; and the progeny of two such mulattoes will be white, $\frac{1}{4}$, $\frac{1}{2}$, $\frac{3}{4}$ and full black in the ratio of $1:4:6:4:1$.

Again, if one parent belong to a tall race—like the Scotch or some Irish—and the other to a short race, like the South Italians, then all the progeny will tend to be intermediate in stature. If two such intermediates intermarry then very short, short, medium, tall and very tall offspring may result in proportions that can not be precisely

C. B. Davenport, "Effects of Race Intermingling," *Proceedings of the American Philosophical Society* 130 (1917): 364–368.

given, but about which one can say that the mediums are the commonest and the more extreme classes are less frequented, the more they depart from mediocrity. In this case of stature we do not have to do with merely one factor as in eye color, or two as in negro skin color, but probably many. That is why all statures seem to form a continuous curve of frequency with only one modal point, that of the median class.

What is true of physical traits is no less true of mental. The offspring of an intellectually well developed man of good stock and a mentally somewhat inferior woman will tend to show a fair to good mentality; but the progeny of the intermarriage of two such will be normal and feeble-minded in the proportion of about 3 to 1. If one parent be of a strain that is highly excitable and liable to outbursts of temper while the other is calm then probably all the children will be excitable, or half of them, if the excitable parent is not of pure excitable stock. Thus, in the intellectual and emotional spheres the traits are no less "inherited" than in the physical sphere.

But I am aware that I have not yet considered the main problem of the consequence of race intermixture, considering races as differing by a number of characters. First, I have to say that this subject has not been sufficiently investigated; but we may, by inference from studies that have been made, draw certain conclusions. Any well-established abundant race is probably well adjusted to its conditions and its parts and functions are harmoniously adjusted. Take the case of the Leghorn hen. Its function is to lay eggs all the year through and never to waste time in becoming broody. The brooding instinct is, indeed, absent; and for egg farms and those in which incubators are used such birds are the best type. The Brahma fowl, on the other hand, is only a fair layer; it becomes broody two or three times a year and makes an excellent mother. It is well adapted for farms which have no incubators or artificial brooders. Now I have crossed these two races; the progeny were intermediate in size. The hens laid fairly well for a time and then became broody and in time hatched some chicks. For a day or two they mothered the chicks, and then began to roost at night in the trees and in a few days began to lay again, while the chicks perished at night of cold and neglect. The hybrid was a failure both as egg layer and as a brooder of chicks. The instincts and functions of the hybrids were not harmoniously adjusted to each other.

Turning to man, we have races of large tall men, like the Scotch, which are long-lived and whose internal organs are well adapted to care for the large frames. In the South Italians, on the other hand, we have small short bodies, but these, too, have well adjusted viscera. But the hybrids of these or similar two races may be expected to yield, in the second generation, besides the parental types also children with large frame and inadequate viscera—children of whom it is said every inch over 5' 10" is an inch of danger; children of insufficient circulation. On the other hand, there may appear children of short stature with too large circulatory apparatus. Despite the great capacity that the body has for self adjustment it fails to overcome the bad hereditary combinations.

Again it seems probable, as dentists with whom I have spoken on the subject agree, that many cases of overcrowding or wide separation of teeth are due to a lack of harmony between size of jaw and size of teeth—probably due to a union of a large-jawed, large-toothed race and a small-jawed, small-toothed race. Nothing is more striking than the regular dental arcades commonly seen in the skulls of inbred native races and the irregular dentations of many children of the tremendously hybridized American.

Not only physical but also mental and temperamental incompatibilities may be a consequence of hybridization. For example, one often sees in mulattoes an ambition and push combined with intellectual inadequacy which makes the unhappy hybrid dissatisfied with his lot and a nuisance to others.

To sum up, then, miscegenation commonly spells disharmony—disharmony of physical, mental and temperamental qualities and this means also disharmony with environment. A hybridized people are a badly put together people and a dissatisfied, restless, ineffective people. One wonders how much of the exceptionally high death rate in middle life in this country is due to such bodily maladjustments; and how much of our crime and insanity is due to mental and temperamental friction.

This country is in for hybridization on the greatest scale that the world has ever seen.

May we predict its consequences? At least we may hazard a prediction and suggest a way of diminishing the evil. Professor Flinders-Petrie in his essay on "Revolutions of Civilization" suggests that the rise and fall of nations is to be accounted for in this fashion. He observes that the countries that developed the highest type of civilization occur on peninsulas—Egypt surrounded on two sides by water and on two sides by the desert and by tropical heat, Greece, and Rome on the Italian peninsula. It is conceded that such peninsulas are centers of inbreeding. Flinders-Petrie concluded that a period of prolonged inbreeding leads to social stratification. In such a period a social harmony is developed, the arts and sciences flourish but certain consequences of inbreeding follow, particularly, the spread of feeble-mindedness, epilepsy, melancholia and sterility. These weaken the nation, which then succumbs to the pressure of stronger, but less civilized, neighbors. Foreign hordes sweep in; miscegenation takes place, disharmonies appear, the arts and sciences languish, physical and mental vigor are increased in one part of the population and diminished in another part and finally after selection has done its beneficent work a hardier, more vigorous people results. In them social stratification in time follows and a high culture reappears; and so on in cycles. The suggestion is an interesting one and there is no evident biological objection to it. Indeed the result of hybridization after two or three generations is great variability. This means that some new combinations will be formed that are better than the old ones; also others that are worse. If selective annihilation is permitted to do its beneficent work, then the worse combinations will tend to die off early. If now new intermixing is stopped and eugenical mating ensues, consciously or unconsciously, especially in the presence of inbreeding, strains may arise that are superior to any that existed in the unhybridized races. This, then, is the hope for our country; if immigration is restricted, if selective elimination is permitted, if the principle of the inequality of generating strains be accepted and if eugenical ideals prevail in mating, then strains with new and better combinations of traits may arise and our nation take front rank in culture among the nations of ancient and modern times.

Note

Read April 13, 1917.

7.2 "Biological and Social Consequences of Race-Crossing" (1926)

W. E. Castle

What constitute the essential differences between human races seems to be a question difficult for anthropologists to agree upon but from a biologist's point of view those appear to be on safe ground who base racial distinctions on easily recognizable and measureable differences perpetuated by heredity irrespective of the environment. See Hooton, 1926. It is still a moot question how races originate, not merely in man, but also among the lower animals and plants. At one time natural selection was thought to be an all-sufficient explanation of the matter, but the more carefully the question is studied and the more exact and experimental in character the data which enter into its solution, the more fully do we become convinced that forms of life are rarely static, that organic change is the rule rather than the exception. Change is inevitable and is not limited to useful or adaptive variations. Natural selection undoubtedly determines the survival of decidedly useful variations, which arise for any reason, and also the extinction of those which are positively harmful, but a host of other variations fall in neither of these categories and survive among the descendants as a matter of course, quite unaffected by natural selection.

The experimental study of evolution indicates that genetic (hereditary) variations are all the time arising, and with especial frequency in such organisms as are bisexual and cross-fertilized.

In a state of nature no species can for long be separated by geographical barriers into non-interbreeding groups, without the origin of specific or racial differences between such groups. This is because new variations are from time to time originating in each group, and if chance is an element in the origin of variations, it would be a rare event for the same variation to appear simultaneously in two geographically separated groups. Hence such groups become different irrespective of the action of natural selection. Hence the maple, sassafras, chestnut and oak trees of Asia have become specifically different from those of North America since the land connection between the two continents disappeared, although the species found in one continent will grow perfectly well in the other. Also the reindeer of Eurasia is different from the caribou of Alaska, although the two are still enough alike to interbreed and produce fertile hybrids. For a like reason the North American Indians are racially distinct from the Mongolians, their nearest of kin among human races. Time and isolation have made them different.

W. E. Castle, "Biological and Social Consequences of Race-Crossing," *American Journal of Physical Anthropology* (Cambridge, Mass., Harvard University, Bussey Institution), 9, no. 2 (April–June 1926): 145–156.

When isolated groups of flowering plants have become specifically distinct, they often show a tendency to remain distinct even if subsequently they are brought into the same territory. One may have become earlier or later than the other in its time of flowering, or structural or physiological differences may have arisen which make cross fertilization between the two difficult. Similarly in the higher animals (particularly among birds and mammals) a psychological element enters into the maintenance of group differences. The individual prefers to mate only in his own group and with his own kind, but circumstances may overcome racial antipathy and the overpowering impulse of sex bring about mixed unions when mates of the same race are not available. Thousands of mules are produced annually by matings between a mare and a jackass, but it often requires considerable finesse on the breeder's part to bring them about, and if asses and horses of both sexes were turned loose together on a range, it is doubtful whether a mule would be produced once in a century.

In mankind, where the race differences are less profound, so far as the physiology of reproduction is concerned, the psychological element in the maintenance of racial differences is even greater. In a population mixed in its racial composition, differences in language, religion, dress, or social customs, often keep the racially different elements distinct for centuries. The castes of India are a case in point. Since there are no biological obstacles to crossing between the most diverse human races, when such crossing does occur, it is in disregard of social conventions, race pride and race prejudice. Naturally therefore it occurs between antisocial and outcast specimens of the respective races, or else between conquerors and slaves. The social status of the children is thus bound to be low, their educational opportunities poor, their moral background bad.

There is a school of writers who insist that mixed races are inferior just because they are mixed. They cite the poor cultural attainments of the mixed races of the West Indies and of certain South American countries, maintaining that the half-breeds have all the vices of both parent races but the virtues of neither. They compare the cultural attainments of the southern U.S. with those of the northern U.S., much to the disadvantage of the former, and ascribe the difference wholly to the presence of the mulatto. They overlook or ignore a number of other factors which enter into the question, such as the kind of individuals who contracted the mixed matings and the character of the physical inheritance of their offspring, the conditions under which the children of mixed race were reared, the nature of their intellectual and moral education, the character of their economic and industrial opportunities, their ability to share in the equal protection of the law. Does the half-breed, in any community of the world in which he is numerous, have an equal chance to make a man of himself, as compared with the sons of the dominant race? I think not. Can we then fairly consider him racially inferior just because his racial attainments are less? Attainments imply opportunities as well as abilities.

Writers who appeal to race prejudice are very much in vogue. Their task is easy. We inherit from a long line of animal ancestors the group instinct, loyalty to the herd against the rest of creation. It is not difficult to persuade us that our group of races is the best group, our particular race the best race and all others inferior. There was a time when divine revelation was relied upon to establish the claim to the status of "chosen people," but now it is sufficient to write "science says." Would it not be well to inquire into the credentials of a "science," which so confidently proclaims one race superior and another inferior, and all mixtures worse than either? Is it really *science*, truth established by adequate evidence, or is it *assumption* backed up by loud voiced assertion? I share the views in this connection of Dr. Hooton as recently expressed in Science. He says (p. 76):

A third group of writers on racial subjects, usually not professional anthropologists, associates cultural and psychological characteristics with physical types on wholly insufficient evidence. These race propagandists commonly attribute to the physical subdivision of mankind to which they imagine that they themselves belong all or most of the superior qualities of mankind, physical, mental and moral. They talk of the psychological characteristics of this or that race as if they were objective tangible properties, scientifically demonstrated. Starting from an *a priori* assumption that physical types have psychological correlates, they attempt to refer every manifestation of the psychological qualities assumed to be the exclusive property of this or that race to the physical type in question. Great men of whatever period are claimed to be members of the favored race on the basis of their achievements and sometimes with a total disregard of physical criteria. In no case has any serious effort been made by such ethnomaniacs to isolate a pure racial type and to study either its mental qualities or its material culture. The fact that most if not all peoples are racially mixed is consistently ignored. While some of the conclusions of such writers may be correct, none of them have been scientifically established.

A commendable attempt to obtain experimental evidence on the effects of race crossing was made a few years ago by Dr. Alfred Mjoen who crossed dissimilar races of rabbits. His general conclusion was that racial crossing tends to produce physical deterioration both in rabbits and in humans. He admitted the impracticability of investigating the question critically in human populations and for that reason resorted to experiments with rabbits for critical evidence. He offers the results of two sets of experiments in one of which two different races of rabbits were crossed, in the other three. The evidences which he observes of physical deterioration are: 1. Increased size in F_1 (first hybrid generation), greater than that of either parent. This is regarded as a "weakness" because "abnormal." 2. Decreased size of some individuals in F_4 (fourth hybrid generation), which are smaller than either ancestor of pure race. Other individuals of F_4 are intermediate in size between the uncrossed ancestors. 3. Diminished fertility and increased mortality in the young in F_4 as compared with earlier generations. 4. Failure of the sexual instinct in many F_4 individuals. 5. Asymmetrical carriage of the ears, one erect, one pendant, among cross-breeds between lop-eared and albino rabbits.

The increased size of F_1 individuals is a phenomenon familiar to animal and plant breeders and frequently utilized by them. It is regularly attended by unusual vigor of growth and resistance to disease as well as by high fecundity. I have never before seen it mentioned as an evidence of physical deterioration. If it is "deterioration" to be "abnormal," all superior individuals have "deteriorated," because they are "abnormal." The races with which Dr. Mjoen started were "abnormal" as compared with the ancestral wild rabbit of Europe, all being medium to large sized (3400–4300 grams). The weights given for the F_1 individuals are 4160 and 4645 grams respectively. In F_4 weights are given for eight individuals, ranging from 2560 to 3850 grams. The smallest of these is well above the average weight of wild rabbits and so "abnormal." Is it an evidence of "deterioration" that some of these are less "abnormal" than their immediate ancestors, or that the group is more variable than the F_1 generation?

The diminished fertility of F_4 individuals and the increased mortality of their young more probably resulted from unsanitary environment than from the mixed racial nature of the parents. The animals are reported to have been kept in this generation one or two males in a common hutch with eight females. Rabbits cannot be bred successfully in such crowded quarters and it is not surprising that only one litter of young was obtained in a period of six months. Failure of the sexual instinct and inability to produce viable young are well known consequences of an inadequate or unbalanced diet but not of race crossing in any species of animal that ever I heard of.

The asymmetrical carriage of the ears which Dr. Mjoen regards as "the most distinct outward sign of a disharmonic crossing that can well be imagined," and which he observes among three of his cross-bred rabbits, is a feature not confined to cross-breeds but of frequent occurrence among rabbits of large size, irrespective of race. Ear size in rabbits is closely correlated with body size, as I have shown elsewhere. When the ears are long, the muscles at their base are often unable to hold the ears erect, and they may lop over both to the same side, or one may lop over while the other remains upright. Ossification at the base of the ear adapts itself to this abnormal relation, as observed by Darwin (*Animals and Plants*) and the condition thus becomes permanent. The purest races of large rabbits, such as Flemish Giants and pure bred lop-eared rabbits, often show this asymmetrical ear carriage. Breeders naturally consider it a defect and in lop-eared rabbits seek to correct it in the young by mechanical means, such as manipulation with the hands to separate the connective tissue beneath the skin which joins the ears together. Books on rabbit-keeping figure leather caps to be placed over the top of the head of the young lop-eared rabbits to hold the ears apart and down. It is evident, therefore, that the asymmetrical ear carriage of Dr. Mjoen's rabbits was not due to their cross-bred state, since this same condition is found in uncrossed individuals of one, at least, of his "pure races."

From an experience of more than twenty years in the breeding of rabbits, in the course of which I have crossed nearly all known breeds, some of which differed much more in size and other characters than did those used by Dr. Mjoen, I am satisfied that there are no breeds of domestic rabbits so distinct racially that their hybrids show the slightest diminution in fertility or vigor, as compared with the uncrossed races. Breeds of rabbits show no more racial distinctness than breeds of cattle, which are so frequently crossed in the most enlightened agricultural practice, without any indications of diminished fertility being observed.

Dr. Mjoen's conclusions rest on insufficient and uncritical observations. It would not be necessary to point this out to an experienced geneticist, but the sociologist is perhaps entitled to a biological rating of these observations.

Organic Misfits

Dr. Mjoen's argument, if I understand him rightly, assumes that all inheritance in rabbits and in men is Mendelian, and that if this is so all possible recombinations of the inherited characters will occur in F_2 and later generations. Among these recombinations, he thinks, are sure to be many organic misfits, such as small legs on large bodies. It might be supposed that in the evolution of existing races organic misfits had been eliminated by natural selection, and therefore, that surviving types are superior types which could only be made worse by intercrossing, since the frequency of organic misfits must be increased by such crossing.

The equation of the production of skeletal misfits in crossing the largest with the smallest known races of rabbits, I have subjected to an extensive and intensive experimental investigation, but I have failed to observe any indication of the occurrence of misfits either in F_1 or F_2. There is a remarkable constancy in the degree of correlation between part and part within the body, quite irrespective of size. The genetic agencies which control the size of particular parts are identical with the agencies which control the size of the body as a whole. From an intimate study of the subject I am able to deny categorically Dr. Mjoen's assumption that there is inheritance, independent of general body size, of types of bone structure which regulate "the way or mode of jumping and holding—carrying—the body."

Why, it might be asked, if nature abhors race crossing, does she do so much of it? Why is it that distinct races of the same species of animal occur only where geographical isolation exists? Why does she go to such pains to ensure cross fertilization rather than self fertilization or close fertilization?

Are there such thing as "harmonic" and "dishormonic" race crossings? It is assumed in Dr. Mjoen's argument that some combinations of inherited characters are better than others, have greater survival value, and for that reason are found in existing races. As race crossing brings about recombination of inherited characters, it is to be expected on genetic principles that mixed races will be more variable than unmixed races. Is such variability a disadvantage? Yes, if all *new* combinations are inferior to those which previously existed. This Dr. Mjoen seems to assume to be true in certain cases, as in Norwegian-Lapp crosses, which he regards as "dishormonic." From his viewpoint any infusion of Lapp characters into the Norwegian complex is deterioration. Perhaps the Lapp might reasonably take a similar view of the situation. Race pride and race prejudice narrow down to just that view of all alien stocks. But to an outside observer it is conceivable that *some* inherited characters of the Lapp might be combined with other inherited characters of the Norwegian to produce meritorious racial combinations, which would be viewed with satisfaction both by the intelligent Lapp and by the intelligent Norwegian. When these combinations had gained such recognition, Dr. Mjoen would doubtless designate them "harmonic race crossings."

Now is there any way, other than trial and error, by which harmonic can be distinguished from disharmonic race crossing? I doubt it. I doubt whether there is any race of human beings whose genetic qualities are all inherently bad. I doubt whether there is any human race so "superior" that it is incapable of improvement. Dr. Mjoen is looking for some simple "blood test" chemical or serological which will show whether a proposed mating, either inter-racial or intra-racial, is "harmonic" or "disharmonic." I doubt whether he finds it. The methods of genetic analysis of inherited qualities are far in advance of chemical knowledge of the material determiners of those inherited qualities. We may reasonably expect to learn more from a study of the genetic qualities of races and individuals and their mode of inheritance than from blood tests.

Race Crossing and Social Inheritance

I doubt whether there are any race combinations which are, so far as biological qualities are concerned, inherently either harmonic or disharmonic, that is productive of better or worse genetic combinations. Both better and worse should theoretically result, if all inherited characters follow Mendel's law in transmission. A more variable population would then result, which should be on the whole more adaptable to a new or changing environment either physical or social. Is it not possible that the racially mixed character of the populations of France, Germany, England and the United States have been one factor in their adaptability to social and economic changes?

If all inheritance of human traits were simple Mendelian inheritance, and natural selection were unlimited in its action among human populations, then unrestricted racial intercrossing might be recommended. But in the light of our present knowledge, few would recommend it. For, in the first place, much that is best in human existence is a matter of social inheritance, not of biological inheritance. Race crossing disturbs social inheritance. That is one of its worst features. And, limiting our attention to biological characters only, few of them follow the simple Mendeliam law, with presence or absence of single characters, dominance or elimination. Most inherited characters are blending (the Mendelian interpretation of which is in terms of multiple factors). When parents differ in a trait, the offspring commonly possess an intermediate degree of it. This is true of stature, weight, and, I think, of general mental powers. Neither parent is devoid of stature or weight or is without mental ability. The children as a rule are intermediate between their parents as regards such traits. It is so in racial crosses, except for the complication of hybrid vigor or "heterosis" in the F_1 generation, a well known occurrence both in animal and human crosses. When two races cross which differ in stature, the children may surpass either parent in this respect, as Dr. Mjoen has observed. But the "overgrowth," as he well calls it, does not persist into later generations. It disappears, as heterosis disappears, and the population of later generations will be intermediate in character, though probably more variable than either uncrossed race. This is the outcome in numerous careful experimental investigations among animals, and may confidently be predicted as the result with similar characters in the crossing of human races.

When traits blend in human crosses, deterioration is not to be expected as a consequence, but rather an intermediate degree of the characters involved. Whether from a purely biological standpoint a particular race cross is considered desirable or undesirable will depend on whether a greater or less degree of the characters under consideration is desired.

Race Crossing in the United States

Consider for a moment the physical (not social) consequences in the United States of a cross between African black races and European whites, an experiment which has been made on a considerable scale. The white race has less skin pigment and more intelligence. The first difference will not be disputed, the second can be claimed at least on the basis of past racial accomplishment. As regards skin color the F_1 hybrids are intermediate; as regards intelligence it is not so easy to judge, since their environment has commonly been that of the blacks, but it will be generally admitted that they are superior in this respect to the blacks and that this has been a factor in their social advancement which has been more rapid than that of the blacks. Repeated back-crosses with whites, if permitted, might be expected to result in an approximation to the skin color and level of intelligence of the whites in a few generations. Similarly back-crosses with the blacks would naturally result in an approximation to their physical and mental standards. Matings of F_1 individuals *inter se* would continue indefinitely a race varying about intermediate standards, but varying more widely than either uncrossed race.

So far as biological considerations are concerned, there is no race problem in the United States. If social considerations were not much more powerful than biological ones, the future population of the United States would certainly be highly variable in skin color and intelligence, passing by scarcely perceptible gradations from a pure black type of the present "black belt" to a pure white type such as would result from a mixing of European races. But the social considerations *are* of much more importance than biological ones in this connection, and the racial future of the United States cannot be predicted from the latter alone.

Mixed Races from Inferior Stocks

Dr Mjoen would like to believe that the mixed race constituent of Norway's population will die out of itself, because he finds that it coexists with bad physical and social states of the population. He seeks biological support of this hope in animal experiments, but will not find it, I think, if those experiments are made critical and interpreted without bias. He should investigate also the social environment under which race crossing occurs and in which the hybrids are forced to live. In these, if I mistake not, rather than in any mysterious biological disharmonies, will be found the explanation of the alleged greater prevalence of tuberculosis, drunkenness, theft and other social evils among the mixed population. He should inquire what sort of individuals contract mixed marriages, and under what conditions. Are they the best or the worst of their respective races? Do those who contract such marriages do so from deliberate choice or only because they can find no eligible mates among their own people? Are they individuals of force of character with passions well under control, or are they of the feebler sort, yielding readily to impulse and with unbridled passions? Is it to be expected that a cross between poor specimens of two races will result in anything but poor offspring? It is illogical to ascribe the poor quality of a mixed race to the fact that it is a mixture, provided that the original ingredients are poor. How could it well be otherwise?

Consider also the social environment in which race crossing usually occurs and in which the hybrids grow up. Crossing occurs clandestinely or, if in legalized wedlock, between individuals lost to shame. For parties to such matings are despised by both races and their children are social outcasts. Their social opportunities are decidedly limited. Is it any wonder that their social attainments are limited and that they show lack of the ordinary social inhibitions? It is not necessary to invoke biological disharmonies in order to explain the poor results of many race mixtures. Social agencies afford a sufficient explanation.

Outlook for the Mulattoes

Let us consider further, in this connection, the black-white race mixture in the United States. According to Willcox, about nine-tenths of the present population of the United States consists of whites without admixture of African blood, the other tenth consisting of blacks or black-white hybrids, known as mulattoes. If there were free intercrossing of all elements of the population, the proportion of mixed bloods should steadily increase, but this has not been the case in the past and is not likely to be in the future. At the first United States Census in 1790, according to Willcox, the negroes and mulattoes constituted about one-fifth of the total population, or twice the proportion they now represent. Instead of the increase which random matings would produce, there has been a steady decrease in the proportion of blacks and mulattoes. This has been due in part to white immigration, in part to a lower rate of increase among the blacks, but chiefly to a strong social prejudice among the whites against mixed marriages, which in many States has found expression in legislation against miscegenation, and in all States takes the effective form of a strong public sentiment against it. This same public sentiment insists on classifying as black every individual who is known to have or is suspected of having any trace of negro blood in his veins. The consequence is that marriages between whites and blacks or mulattoes are at present extremely rare and clandestine unions are uncommon.

So far as back-crossing of mulattoes with blacks is concerned, this probably does not occur with random frequency, since pure-blooded negroes on one hand and mulattoes on the other, have each a degree of group consciousness which tends to keep them apart. The mulattoes as a rule are more intelligent and have enjoyed better educational advantages so that they find more ready employment in urban life as porters, janitors, or even in clerical or professional occupations. But with urban life goes a reduced birth rate among blacks as well as whites. The prospect is that, if things go on as they now are, the mulattoes will not amalgamate either with the whites or with the blacks, but will form a separate but diminishing proportion of the total population. The blacks are holding their own in certain rural sections of the South, but elsewhere are going back numerically. No complete amalgamation of blacks with whites is to be anticipated, simply because of social impediments, though no biological barrier whatever is discoverable.

Indian-White Crosses

Another distant racial cross which has been made on a considerable scale in North America is that between European whites and North American Indians. To be sure, the number of hybrids resulting from this cross is insignificant compared with that of the mulattoes, but it is sufficient to be instructive as a biological and social experiment. The early English colonists kept clsoe to the coast and steadfastly refused to associate with the "savages," but the French in Canada were more disposed to roam the woods. Their young men explored the interior of the continent, lived with the Indians as trappers and hunters and often took Indian wives. Thus a half-breed population grew up of hardy adventurous frontiersmen. It would be difficult to find in them evidences of physical or intellectual degeneracy, other than those entailed by the introduced vices of the white race.

Within the United States, the settlement of the Mississippi Valley took place so rapidly that it amounted to a complete dispossession of the Indian tribes found there. These moved bodily westward to "reservations" beyond the great river. Accordingly there was little opportunity for race mixture. Nevertheless, renegade whites, who had reason to lose their identity temporarily, often joined the Indians on their reservations. As the reservation lands became valuable through the occupation of the surrounding territory by whites, the "squaw men" and their half-breed children found it an economic advantage to be members of the tribe. So when later the wild Indians were domesticated and "given lands in severalty," the individuals of mixed race often found themselves wealthy land owners. This gave them social advantages which resulted in frequent marriage alliances with the whites. For there is no strong social prejudice against the red man such as exists against the black man, recently a slave. Consequently the pureblooded Indians are a rapidly vanishing element of the population of the United States, and those of mixed race are being rapidly assimilated in the white population, frequently attaining positions of influence and authority. The difference in results following crossing with the black and with the red races in the United States are not referable to any biological harmonies or disharmonies existing in the respective cases but wholly to the social attitude of the whites, which is hostile in one case, indifferent in the other.

A further illustration of the surpassing importance of social over biological considerations in race-crossing is seen in the attitude of the Pacific Coast States towards Chinese and Japanese intermixture. No one questions the virility of these races or their biological fitness. Their cultural attainments are very high and antedate our own. Hybrids between these races and white races, so far as our information goes, are of high quality physically and intellectually. Yet public opinion is unalterably opposed to Oriental immigration or race mixture, not on biological grounds, but purely on social, economic, or political grounds.

So far as a biologist can see, human race problems are not biological problems any more than rabbit crosses are social problems. The rabbit breeder does not cross his selected races of rabbits unless he desires to improve upon what he has. The sociologist who is satisfied with human society as now constituted may reasonably decry race crossing. But let him do so on social grounds only. He will wait in vain, if he waits to see mixed races vanish from any biological unfitness.

Note

This paper, prepared at the suggestion of Dr. Hrdlička, is based largely on an article published under the same title in The Journal of Heredity, Vol. 15, Sept. 1924. Thanks are due to the editor of that journal for permission to use the material here.

Literature Cited

1. Hooton, E. A. Methods of Racial Analysis. *Science*, 63 pp. 75–81, Jan. 22, 1926.

2. Mjoen, J. H. Harmonic and Disharmonic Race Crossings. *Eugenics in Race and State*, pp. 41–61. Baltimore, 1923.

3. Willcox, W. F. Distribution and Increase of Negroes in the United States. *Ibid.*, pp. 166–174. 1923.

4. Hoffman, F. L. The Problem of Negro-White Intermixture and Intermarriage. *Ibid.*, pp. 175–188. 1923.

5. Castle, W. E. Genetic Studies of Rabbits and Rats. *Publication No.* 320, Carnegie Institution of Washington. 1922.

7.3 "Racial Fusion Among the California and Nevada Indians" (1943)

S. F. Cook

I

In a series of recent papers the course of racial adjustment by Pacific Coast Indians has been discussed from the standpoint of environmental adaptation.[1] It has been shown that these aboriginal peoples have made considerable progress during the past century with respect to dietary habits, urbanization and certain social customs such as marriage and divorce. Ecological changes, however, are likely to be associated with alteration in genetic constitution, with varying degrees of racial fusion. Hence it becomes of significance to learn if possible to what extent there has been admixture of Caucasian with Indian blood.

II

The pertinent information is available in the form of several quite adequate censuses. The first, and most comprehensive, is that of 1928 in which the attempt was made to enumerate all the Indians in California.[2] In conjunction with other data the degree of blood for each individual was tabulated. In addition there are several censuses taken by resident agents at intervals during the past forty years.[3] Not all of these are complete but those for Hupa Valley in 1910, 1930, 1940, for Carson Agency for 1940, and for the Mission Agency in 1940 state degree of blood for each person enumerated. In the 1928 comprehensive census and in all others taken within recent years the degree of blood is expressed as an exact fraction—$\frac{1}{2}$, $\frac{1}{4}$, etc. Since such notation is cumbersome for purposes of calculation I have converted it to terms of simple percentage, calling full Indian blood 100 per cent, half blood 50 per cent, etc.

The accuracy with which a white man can determine the precise degree of blood in a given Indian has always been subject to considerable debate. Where records are available they may serve as a valid criterion. Otherwise the word of the Indian, his relatives and friends, must be accepted. Granting that the individual may be ignorant of his ancestry or may be prone to falsification, nevertheless the enumerators have been consistently careful to utilize all sources of evidence and arrive at a reasonably close approximation. In particular the agency employees who customarily perform this task know their Indians personally and are not likely to be misled seriously by false statements. Moreover, in the statistical sense errors which are introduced are random. I know of no generalized tendency for Indians to represent themselves as possessing

S. F. Cook, "Racial Fusion among the California and Nevada Indians," *Human Biology* 15 (1943): 153–165.

more or less native blood than is actually the case. Finally we are forced to accept the data of the censuses for what they are worth since there is no better source of information available. On the whole they may be regarded as quite reliable.

The extent of miscegenation between any two racial stocks obviously depends upon a host of factors, physical and social, which are so interrelated as to defy any final and complete analysis. We can at best select a few and test them with respect to the factual material at hand.

1. Perhaps the least common denominator applicable to all such situations is physical propinquity. This may be expressed in terms of population density for it is obvious that if both races are present in large numbers over a small area more crossings will occur than if either one or both is sparsely represented. There is no numerically significant relation between degree of blood and white density alone, but an association between degree of blood and the ratio of Indian to white density based on county populations yields a chi-square value of 9.84. Since this is moderately significant, one may conclude that density as such is operating to increase racial fusion.

2. Irrespective of the crude densities, if the Indian population is in any way segregated from the adjacent whites the normal principle of random contact will no longer hold with undiminished validity. The reservation system constitutes such a segregation.[4] Furthermore, numerous small bodies of natives may be settled in geographically isolated spots within otherwise heavily settled areas, and thus fail to make the contacts which would be assumed from the basic density data.

The effect of isolation may be tested by separating the California tribes into three arbitrary divisions. In the first may be placed those tribes which have been closely under federal supervision—on and off actual reservations—even though they were located in thickly inhabited regions.[5] The second will include those tribes geographically isolated, whether or not under federal supervision and the third will consist of the remainder, those who have been isolated in neither the political nor geographical sense. The mean degree of blood in 1928 for the three divisions was: first 67.0 per cent, second 76.9 per cent and third 51.4 per cent. There appears, therefore, a rough but distinct correspondence between degree of isolation and extent of racial mixture.

3. A third possible factor in determining fusion is the social condition of the individual tribe. If an aboriginal unit is still moderately intact, without serious losses in number, with many of the primitive institutions in force we would expect to find in it less dilution with Caucasian blood than in one which had been violently disrupted and dispersed. The most convenient quantitative index to such disruption is the per cent of the aboriginal population now surviving. Associating the latter with degree of blood a chi-square value of 5.89 is obtained, which indicates a just significant relationship.

III

Although analysis of contemporary conditions can yield some information concerning the factors responsible for racial mixture, it is equally clear that the process must be considered essentially from the historical point of view. Environmental changes of an adaptive nature may, under sufficient pressure, occur with great rapidity—genetic alteration cannot. One must, therefore, attempt to reconstruct the temporal course of racial fusion among the California Indians.

It is both convenient and logical to subdivide the Indian population into five historical categories, rather than treat it as a composite. The first segment (Group I) consists of those tribes which were untouched by white civilization prior to 1848, but were at that time subjected to an inundation of pioneers, particularly gold miners. From 1850 to 1860 they passed through a catastrophic period of physical, economic and social upheaval, subsequent to which the survivors were permitted to settle down in the newly established communities and have existed under fairly stable conditions ever since. This group includes most of the tribes in northern and central California: Tolowa, Yurok, Hupa, Wiyot, Chimariko, Karok, Yuki, other Athabascans, Wylackie, Pomo, Shasta, Maidu, Miwok, Wintun, Yokuts, Mono, and Tulatulabal of Kern River. Collectively they comprise approximately one half the Indian population of the State.

The second segment (Group II) consists of the Atsugewi (Pit Rivers), Washo, Northern Paiute, Eastern Mono (commonly called Paiute), other Shoshoneans, and Yuma. These peoples inhabited the fringes of the Great Basin east of the Cascades, Sierra Nevada, and Colorado desert. They did not meet the white race till 1850, were relatively little disturbed physically and have never had an opportunity to mix freely with large masses of white men.

The third and fourth segments, in contradistinction to the others, were subjected to the Ibero-American civilization of the missions as early as 1770. The coastal natives, and many from the interior, became what the administrators of the system were pleased to call "reduced" to Christianity. Despite huge losses from disease and social causes the survivors by 1850 had become thoroughly adapted to the pastoral civilization of the time. Subsequent to 1850 the remnants from the missions as far south as San Gabriel at Los Angeles merged unnoticed with the general population and have never been treated as wards by the Indian Service. Their history has thus been in many respects similar to that of other racial groups which did not originate in California but migrated thence. This segment constitutes Group III. Group IV comprises the Indians from the three southern establishments of San Luis Rey, San Juan Capistrano, and San Diego to whom are added the partially missionized Cahuilla. Large in numbers, this group received much attention from the Federal Government which indeed placed most of them on reservations and set up a special agency for them. Thus, unlike Group III, instead of merging easily with the surrounding white population they have been kept rather isolated.

The fifth segment (Group V) is composed of the natives of northern and western Nevada-Paiute, Washo, and Shoshone—who have been under the jurisdiction of the Carson Agency. The history of these Indians resembles that of Group II in that their contact with whites has been limited. Indeed, their isolation in all senses has been even more complete than that of the aborigines living over the line in California.

The genetic differences in the five natural groups or subdivisions can be demonstrated by three methods of analysis. The first consists in breaking down the data to show the percentage of each group having different degrees of blood—see Table 7.3.1. The geographically remote groups (II and V) still possess over 75 per cent full bloods, whereas only 16 per cent are found in Group III and 26 per cent in Group I. That the process of mixing has been proceeding for a long time in Group III is indicated by the fact that 5 per cent of these persons are of the fifth mixed generation (1/32 Indian blood) and 5 per cent of the sixth generation (1/64 Indian blood) with a few individuals reporting only 1/128 Indian blood.

The tendency to intermix with the white race is also shown by the distribution within a given generation. Thus if we examine the individuals having 25 and 75 per cent ($\frac{1}{4}$ and $\frac{3}{4}$) Indian blood we find that the former predominate in Groups I, III and IV, the latter in Groups II and V. Otherwise expressed, in the isolated segments of the population half-breeds tend to marry full-bloods rather than whites, whereas among the former mission Indians half-breeds are more likely to marry persons whiter than themselves. This tendency is likewise manifest in subsequent generations. For instance in the third mixed generation the progression in Group III is: 1/8 blood 11.17 per cent, 3/8 blood 5.38 per cent, 5/8 blood 0.49 per cent, and 7/8 blood 0.39 per cent. In Group V the corresponding values are 0.0 per cent, 0.58 per cent, 1.59 per cent, 3.22 per cent. Indeed a general principle is demonstrated, viz. the lower the level of Indian blood in the individual, the greater is the probability of his or her marrying into the white race.

However, in order for this principle to become operative in fact there must be free physical opportunity for the social urge to fulfill itself. The coastal mission Indians of Groups III and IV probably followed identical courses until the late 19th century at which time the former group went on reservations. This threw those persons of low Indian blood into an environment of relatively pure Indian stock and thus retarded their mixture with the whites. Among the non-reservation Indians of Group III no such environmental inhibition was present and they have proceeded to merge more rapidly with the white population. The same considerations apply to Groups II and V although in lesser degree. The tendency to backcross into the full-blooded stock has been somewhat greater in Group V. This is doubtless associated with the more stringent confinement on reservations and the greater geographic isolation to which they have been subjected.

The second method of analysis is ideally the most direct and satisfactory. This is to secure a set of complete censuses for a single group, taken at frequent intervals since the beginning of white settlement. Unfortunately such a set of data does not exist. We possess merely a few fragments. The earliest agency census I have found is that for Hupa Valley in 1890, but not until 1910 was degree of blood stated and then only as "full" and "quarter plus". The other agencies did not record exact degree of blood until some ten years ago. In spite of these restrictions, nevertheless, we can assemble a few items pertaining to the per cent of full bloods and their median ages. These are

Table 7.3.1
Per Cent of Total Population Corresponding to Different Degrees of Mixture

Per cent Indian blood	0	1	2			3					4								
	100	50	25	75	Total	12.5	37.5	62.5	87.5	Total	6.2	18.7	31.2	43.7	56.2	68.7	81.2	93.7	Total
Group I	26.46	23.51	15.83	10.46	26.29	9.19	4.64	4.07	2.83	20.73	0.92	0.54	0.29	0.43	0.19	0.27	0.25	0.04	2.93
Group II	77.84	9.79	3.59	5.30	8.89	0.65	0.31	0.42	1.85	3.18	0.11	...	0.02	0.02	0.08	...	...	...	0.23
Group III	16.66	23.55	22.45	6.48	28.93	11.17	5.38	0.39	0.49	17.43	2.99	0.59	0.01	0.01	...	...	...	...	3.60
Group IV	42.36	16.41	11.35	7.52	18.87	8.23	3.50	1.84	1.53	15.10	1.77	1.82	0.96	0.46	0.11	0.61	0.07	0.11	5.91
Group V	75.66	9.93	1.81	6.84	8.65	...	0.58	1.59	3.22	5.39	...	...	0.07	...	0.10	0.02	0.10	0.02	0.31

Per cent Indian blood	5									6			7	
	3.1	9.3	21.8	28.1	34.3	46.8	53.1	78.1	Total	1.5	10.9	Total	0.7	Total
Group I	...	0.03	...	0.01	...	...	0.02	0.02	0.10	...	...	0.00	...	0.00
Group II	...	...	...	...	...	...	...	...	0.00	...	...	0.00	...	0.00
Group III	4.19	...	...	...	...	...	...	...	4.19	4.99	...	4.99	0.39	0.39
Group IV	0.23	0.35	0.16	...	0.47	0.02	...	...	1.23	...	0.02	0.02	...	0.00
Group V	...	...	...	...	...	...	...	...	0.00	...	...	0.00	...	0.00

Generations from pure Indian blood

Table 7.3.2

Data for Certain Tribes Taken from Successive Censuses

Jurisdiction and date	Per cent of full bloods	Median age of full bloods minus median age of total population
Hupa Res., 1910	70.5	7.65
Hupa Res., 1930	37.9	13.25
Hupa Res., 1940	29.3	14.04
Mission Tribes, 1928	53.7	8.36
Mission Agency, 1940	47.1	12.15
Paiute, Mono, and Inyo Cos., 1928	71.2	7.41
Same, Carson Agency Census, 1940	61.2	7.52

presented in Table 7.3.2. In all three cases, the Hupa, Paiute (Western Mono) and Southern Mission, it is clear that the relative number of full bloods is steadily diminishing. Furthermore, in two out of the three cases their median age with respect to that of the entire population is increasing. Although no further refinements in calculation are justified these data tend to support the results obtained previously.

The third approach to the problem of historical changes is by analysis of degree of blood with respect to age. The relations are most clearly expressed by determining the mean degree of blood for each age group using five-year intervals. The results are then plotted graphically (see Figs. 7.3.1 and 7.3.2).

If a pure-blooded population has been exposed to consistent hybridization over a reasonably long period of time, the mean degree of aboriginal blood should be at a maximum among the oldest persons and at a minimum in the youngest. The velocity with which the process has occurred should be indicated by the slope of the line in the plot of age against degree of blood, whereas the vertical distance of the points from the base line is a rough index to the relative extent of hybridization at any given epoch.

In Fig. 7.3.1 are plotted the data, using five-year intervals for Groups I to IV. Group V is omitted since it substantially coincides with Group II. Consider Group II, which by other methods has been shown to have undergone the least change. The oldest age category to possess 100 per cent Indian blood is that centering around 67.5 years. Since the census was taken in 1928 the inference may be drawn that miscegenation began approximately 67.5 years earlier or, say 1860. This is wholly in conformity with the known fact that the Great Basin and Colorado desert tribes made no material contact with the whites prior to the silver rush of 1860 and the Civil War. Subsequent change has been steady and consistent but relatively slow for the mean decrease in Indian blood has been 2.2 per cent per decade.

The history of Group I is strikingly indicated. The youngest category showing full Indian blood is that centering around 82.5 years of age which puts first contact

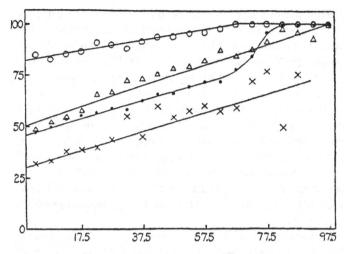

Figure 7.3.1
Degree of Indian blood with respect to age ordinate degree of Indian blood in per cent. Abscissa age. The points are placed at five year intervals from 2.5 to 97.5 years. Symbols: solid dots for Group I, circles for Group II, crosses for Group III ,and triangles for Group IV.

with white men at 1845. In persons between 62.5 and 82.5 years old—i.e. born between 1845 and 1865—there is a tremendous increase in white blood. This corresponds precisely to the pioneer and mining era during which these tribes were nearly destroyed by the inrush of white men. The maximum rate of hybridization seems to have been approximately 18 per cent per decade. After the restoration of stable conditions the rate fell to a constant value which has been maintained ever since—a decrease in Indian blood of 4.7 per cent per decade.

Using a similar index Group IV began to show racial mixture somewhere near 1830. This is puzzling if we remember that the territory was first entered by the Spanish in 1770. However, although intermarriage must have occurred in the subsequent 60 years it does not appear to have reached quantitative importance. This is probably due to several factors. In the first place, the missions south of San Gabriel did not encounter the great mass of Luiseño and Cahuilla until very late in the mission period. Secondly, mission administration as a rule was better than in the north with the result that fewer illicit unions were consummated. Thirdly, the Diegueño and to a lesser extent the Cahuilla were wild, warlike, and highly refractory to mission rule. The real social demoralization of these tribes began with secularization in 1834 and since that time, except for the reservation system there has been little inherent obstacle to miscegenation. As the graph suggests, this process has been effective for the past century, for the rate of decrease in Indian blood with Group IV has been on the average 5.0 per cent per decade.

In Group III the oldest persons may show white blood. Indeed, the categories over 67.5 years of age contain so few individuals and are so erratic that they are of little use in determining the exact trend. However, it is not unreasonable to extrapolate the line backward, since its slope in the past 60 years has been fairly uniform. If this is done, the line crosses the 100 per cent level at the hypothetical age category centering around 152.5 years. The earliest racial mixture would hence appear to have been at or near the year 1775 which is quite in conformity with probability. The whitening of Indian blood among the central and northern missions, therefore, must have begun immediately upon settlement and have continued to the present in a generally uniform manner. By contrast to Group IV these Indians were always peaceable, tractable and easily assimilated socially. Secularization and its attendant disturbance consequently entailed little more than a possible slight acceleration of an already well established process. The mean rate of decrease of Indian blood in recent times, and probably earlier, has been 4.5 per cent per decade.

It will be noted that the mean decennial decrease in per cent Indian blood is very nearly the same in Groups I, III, and IV (4.7, 5.0, 4.5) whereas it is markedly less in Group II (2.2). The difference undoubtedly is referable to the more intense interracial contact which has existed in the coastal and valley strip of California as opposed to the desert region east of the mountains. Thus, since 1860, the coastal groups have reacted almost identically, irrespective of their previous history.

Fig. 7.3.2 is introduced in order to examine more closely the situation in Group IV. Here are two selected stocks: (1) the Gabrieleño, Fernandino, Juaneño, and

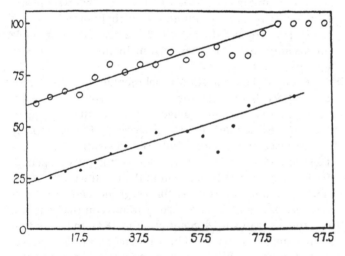

Figure 7.3.2
Degree of Indian blood with respect to age ordinates as in figure 7.3.1. Solid dots represent the Luiseño, Juaneño, Gabrieleño, and Fernandino of Los Angeles and Orange Counties; circles represent the Luiseño and Diegeño of San Diego County.

Luiseño living in Los Angeles and Orange Counties, and (2) the Luiseño and Diegueño living in San Diego County. The former were highly missionized and exposed to heavy social contact from an early date. The latter were particularly intractable and unassimilable during the mission era although they were exposed to Caucasian influence for as long a period. The different reaction is demonstrated in the graph by the much earlier appearance of miscegenation among the Los Angeles County Indians. Indeed, not until nearly 1850 did the San Diego County natives begin to yield to the pressure of the white race.

It is very tempting to extrapolate these data into the future, however dangerous the process, in order to make some prediction concerning the final racial amalgamation of the California Indian. The principal difficulty lies in the assumption, tacit or otherwise, that the present rates will continue unaltered as practical social assimilation is achieved. Furthermore in the ultimate stages the decrease in per cent Indian blood cannot be a linear function of time; it must become exponential and approach the zero value as a limit. In order to simplify, however, we may consider that when an individual has 1/32 Indian and 31/32 white blood, he is in the social sense fully white. Making the further gratuitous assumption that the rates of change in degree of blood will continue as in the past three generations, we may extrapolate the graphs. Starting from the 1928 levels the number of years it will take the four California groups to attain a mean of 3.1 per cent Indian blood for children under 5 years of age is respectively 55, 230, 30 and 65. To achieve reasonable social amalgamation as well as genetic, however, we should use the epoch at which the oldest members reached the designated mean degree of blood. Here we obtain the values, respectively, of 100, 450, 70, and 115 years subsequent to 1928.

Even though the above calculation is admittedly very crude, it is reasonably safe to conclude that if existing conditions do not alter, the cis-montane California Indians will have merged fully with the white race in approximately one century, while the Great Basin and Nevada Indians will not reach that point for nearly five hundred years.

At the beginning of this paper mention was made of progress achieved by the California and Nevada aborigines with respect to various environmental and cultural factors. It was suggested that degree of miscegenation might be concerned in the entire behavior complex. We may now compare certain data in such a way as to emphasize the correspondence. In Table 7.3.3 are brought together the primary results of four distinct lines of investigation, those pertaining to (1) hybridization, (2) food habits, (3) migration, and (4) marriage. The tribes and groups cannot be completely identical nor can the numerical values be correlated in any rigid statistical sense. Nevertheless the overall picture emerges clearly and the conclusion is justified that there is a broad parallelism between adoption of Caucasian customs and mean degree of Indian blood. One cannot be dogmatic with respect to causes but it is possible to affirm that there has been a definite interrelation between environmental and genetic factors in determining the rapidity with which the Indian is assimilated.

Table 7.3.3
Comparison of Racial and Social Trends, Derived from Various Sources

Racial mixture (degree of blood, expressed as per cent Indian blood)		Food habits (per cent of total diet which consists of aboriginal food stuffs)		Migration (per cent of the stock who had migrated from the aboriginal territory by 1928)		Marriage (per cent of marriages which were legal rather than Indian custom for the period 1880–1940)	
Group I	37.7	Maidu	0.3	Group I	15.4	Round Valley Res., Yurok, Karok, Hupa	32.2
Group II	57.4	Nevada and California Paiute and Washo	6.0	Group II	6.3	...	...
		Atsugewi (Pit Rivers)	5.0			...	...
Group III	32.6	...		Group III	40.1	...	...
Group IV	42.7	...		Group IV	6.5	Old Mission and Cahuilla*	55.2
Group V	57.9	...		...		Nevada Paiute and Washo	5.2

*The high value here is due to the influence of the Missions and more recently the individual efforts of the Catholic clergy.

Notes

1. Cook, S. F. The mechanism and extent of dietary adaptation among certain groups of California and Nevada Indians. *Univ. Calif. Publ., Ibero-Americana*, no. 18, 1941.

————. The conflict between the California Indian and white civilization. Part IV. Trends in marriage and divorce since 1850. *Univ. Calif. Publ., Ibero-Americana*. In press.

————. Migration and urbanization of the Indians in California. Human Biology. 15: 33–45.

2. This census is described in detail in an earlier paper (Migration and urbanization, etc.) to which reference may be made.

3. These records were kindly put at the disposal of the author by the agents in charge of the Hupa Valley, Sacramento, Carson, and Mission Agencies.

4. The reservation system implies more than the mere enclosure of an Indian group upon a plot of land with a fence around it. Rather it embodies all forms of paternal oversight on the part of the government to the end that the needs of the Indians shall be so satisfied that physical and social contact with the whites becomes unnecessary.

5. The first group includes: Yurok, Hupa, Pomo, Yokuts, Western Mono, Cahuilla and descendants of Missions Santa Ynez, San Luis Rey, San Juan Capistrano, and San Diego.
    The second group includes: Karok, Achomawi, Paiute, Washo, Shoshone, and Yuma.
    The third includes: Tolowa, Wiyot, Athabascan remnants, Chimariko, Wylackie, Yuki, Shasta, Maidu, Miwok, Wintun, Yana, and the Mission Indians from Sonoma to Los Angeles.

7.4 "The Physical Characteristics of the Hybrid" (1944)

Louis Wirth and Herbert Goldhamer

The study of the physical characteristics of the Negro-white mixed blood is not simply a matter of anthropometric interest. The visibility of racial characteristics and the important role that physical type plays in the social stratification of the Negro community profoundly affect social relations. It is also pertinent to inquire into the validity of stereotyped notions that mixed-blood persons exhibit a variety of inferior physical traits.

One of the outstanding studies of the American mixed-blood Negro is that of Day, which was carried out under the supervision of Hooton at Harvard University.[135] Day studied a large number of mixed-blood families for whom unusually complete genealogical records were available. These families comprised 2,537 adults over 14 years of age. She divided this series into two groups. Group I is largely an ante-bellum group and comprises 958 deceased persons and 194 living persons over 68 years of age; Group II, for whom, of course, much fuller information is available, comprises persons born since 1860; these number 1,385, of whom 1,347 were living at the time of the study.[136] In an extraordinary number of instances Day was able to present relatively complete genealogies of these families in photographic form. It should be emphasized that the collection and publication of these photographs of persons whose racial antecedents are known form one of the major contributions of the study, a contribution which, unfortunately, cannot be reproduced here.

Anthropometric measures of such traits as skin color and lip thickness cannot in themselves provide a picture of "how Negro" Negroes with given amounts of white blood appear. The general appearance of Negroes with specified amounts of white blood can be ascertained adequately only from seeing these characteristics *in situ* in the living person or from photographic or pictorial representations. Since it is not possible to reproduce Day's photographs here, it will be well, before discussing the individual anthropometric measures, to present her own verbal descriptions of the different mixed-blood types:

The Mulatto   The dominant type of mulatto is occasionally fair enough to be mistaken for a swarthy European, and usually has brown hair with deep waves and medium features. The intermediate type is apt to be of a golden, yellowish color, with heavier features, frizzly hair, and sometimes gray-brown or greenish brown eyes. The recessive

Louis Wirth and Herbert Goldhamer, "The Physical Characteristics of the Hybrid," in Otto Klineberg, ed., *Characteristics of the American Negro* (New York and London: Harper & Brothers, 1944), 320–329.

type is light brown or tan, and is similar to one type of Hawaiian, with features varying in their degree of heaviness. A recessive mulatto is frequently the same type as a person of $\frac{3}{4}$ Negro blood and $\frac{1}{4}$ white blood, while a dominant mulatto looks the same as the recessive or intermediate type of the next group which we call the $\frac{5}{8}$ group.[137]

Five-eighths White ($\frac{3}{8}N \frac{5}{8}W$)   The surprising feature of this group is the fact that although just $\frac{1}{8}$ removed from mulattoes we get a large number of persons who look non-negroid and who are phenotypes of other nationalities.

Quadroons ($\frac{2}{8}N \frac{6}{8}W$)   When the quadroon group is reached we fail to find the combination of all three Negro factors, namely swarthy skin, frizzly hair, and heavy features, which occasionally occurs among the recessives of the $\frac{5}{8}$ group. A fair appearance of uniformity has not been achieved, and very rarely is there an instance of a combination of major Negro characteristics; that is to say, we may find frizzly hair, but it is apt to be light in color and coupled with gray or blue eyes and white or ruddy complexion. Or again there may be a sallow complexion and a slightly heavy cast to the face, but the hair is apt to be as straight as that of an Indian. Again, we may find an everted, thick lower lip, as an isolated indication, or, as in one instance, very meager calves of the legs of an otherwise well developed girl.

Octoroons ($\frac{1}{8}N \frac{7}{8}W$)   In the few examples of octoroons which I have studied I have been able so far to see no traces whatever of Negro admixture.[138]

Day partially summarizes these accounts with the statement that "dominant mulattoes and $\frac{5}{8}$ individuals are frequently mistaken for foreigners of various nationalities, or for White Americans, and ... I know of no case of a quadroon who could not easily 'pass for White.'"[139]

In a separate chapter Hooton discusses the results of the anthropometric measures made by Day. Turning first to bodily size or proportions, the results show little difference between pure-blood Negroes and those with up to $\frac{6}{8}$ white blood except in arm length, sitting height, face length, nose breadth, and nasal index. Arm length, facial height, nasal breadth, and nasal index decrease with addition of white blood; sitting height increases with such additions. However, only in the case of nasal breadth and nasal index are the changes sufficiently pronounced to be taken as reliable. Hooton observes that "most of the bodily dimensions of Day's series of Negroids, regardless of the amount of White blood, tend to cluster around Negro means rather than the values observed in Whites."[140] He points out that the probable errors are large, but believes that "it may be an anthropological fact that, with the exceptions noted, Negroids tend to preserve Negro proportions even when White blood becomes predominant."[141] However, as Hooton points out, marked differences according to amount of white blood are not to be expected since in many of these bodily characteristics no significant differences exist between even fullblood Negroes and whites.

A fairly close correlation between skin color and the amount of Negro blood was observed. When the thirty skin colors of the Von Luschan scale were grouped into seven categories, the calculation of mean square contingencies showed coefficients of .73 for the male series and .71 for the female series. With the exception of one male, no $\frac{5}{8}$N males or females or persons with less Negro blood were found among the darker colors of the scale. White skin color characterized 12.5 per cent of the $\frac{1}{2}$N (mulatto) males and 20.6 per cent of the $\frac{1}{2}$N females. This percentage reached 85.7 in $\frac{1}{4}$N (quadroon) females. "The evidence," Hooton concludes, "indicates a Mendelian segregation of skin pigment with multiple factors. There is no clear indication of dominance."[142]

Straight hair appears for the first time in the $\frac{1}{2}$N (mulatto) group (in 8.8 per cent of the cases). Straight hair occurs in 31.6 per cent of the $\frac{3}{8}$N group. Hooton concludes: "As far as our data carry us we may conclude that $\frac{1}{4}$N males, $\frac{1}{2}$N females, and even $\frac{3}{4}$N females may exhibit the entire range of hair curvatures generally recognized, but that, if Mrs. Day's information is valid, distinctively Negroid forms of hair, such as frizzly and woolly, do not appear unless there is at least $\frac{3}{8}$ of Negro blood in the individual."[143]

Herskovits's study[144] of the anthropometry of the American Negro was carried out with the use of more refined measuring techniques than those employed by Day. His measurements of a considerable number of physical characteristics do not enable one, however, to classify individuals readily into physical types or constellations of characteristics as these exist in the actual person.

In his study of pigmentation Herskovits matched skin color against a color wheel with variable segments of red, white, black, and yellow. Table 7.4.1 presents the skin color of his different ancestry groups expressed in terms of the percentage of black in the color wheel required to match their skin colors.

Although the relation between black skin pigmentation and amount of white ancestry is quite clear, Herskovits emphasizes that in pigmentation there is so much overlapping between the various ancestry groups that the chances of estimating a person's ancestry accurately from skin color alone are slight.[145] He does not share Hooton's belief that pigmentation in the Negro indicates Mendelian segregation with multiple factors. He states:

Table 7.4.1
Per Cent Black Pigmentation*

Ancestry class	Upper outer arm	
	Number of cases	Average per cent black
Negro	109	75.5
More Negro than white	127	68.2
Same amount of Negro and white	94	61.2
More white than Negro	30	48.7

*Herskovits, *The Anthropometry of the American Negro*, p. 225. The data are for persons without Indian ancestry.

Miss Barnes' analysis of the heredity of pigmentation in the families of the series of this study, while it showed that simple Mendelian heredity, or even multiple Mendelian heredity with two or three factors is certainly not operative, also demonstrated that simple blending is also not the process by which the present form of pigmentation of the American Negro has developed.[146]

In his study of Negro criminals Hooton finds confirmation for his statement, based on Day's body measurements of mixed bloods, that the mixed blood is much closer to the Negro racial type than to the white type. He points out that in the case of pigmentation only .2 per cent of 3,325 Negroids had white skins, only 1.4 per cent had olive skins, and 31.6 per cent had dark-brown skins.[147] However, Hooton himself states that there were few quadroons and octoroons in this series of criminals.[148] Such a specially selected series perhaps ought not to be utilized in any case for general statements concerning the mixed blood. It is possible also that any attempt to get a fair sample of Negro physical types might be thwarted by the fact that many of the lightest Negroes may be presumed to be "passing" and therefore lost to view.

Having reviewed some of the findings on specific physical characteristics, let us turn to an examination of what may be taken as the major conclusion derived by Herskovits from his study. It has been customary to assume that the physical characteristics of hybrid groups show greater variability than the corresponding characteristics of pure or relatively pure races. Herskovits, however, contends on the basis of his findings that the American Negroes are forming a definite new physical type with low variability:

In trait after trait, if one measures them and computes their variabilities, and if one then compares these with the variability in the same traits of unmixed African, European, or American Indian populations, one will find that in most of the traits measured the variability of the greatly mixed American Negroes I have measured is as low as, or lower than, that of the unmixed populations from which it has been derived.[149]

The foregoing statement perhaps lacks necessary qualifications. In his more extended work Herskovits states his case as strongly but on the grounds that "in ten of the traits these adult male American Negroes are at about the top of the list [in amount of variability], that in seven, they are about at the center, while in about six they are among the least variable."[150] He also points to the study of Davenport and Love, which showed that in the majority of traits the Negroes were less variable than the whites; and to Todd's comparison of Negro and white cadavera which showed Negroes more variable in 36 traits, whites more variable in 36 other traits, and the two groups possessing equal variability in 3 more traits.[151]

From these data Herskovits concludes that:

What is apparently happening, therefore, it may again be repeated, is that this group of Negro-Indian-White hybrids, so greatly mixed racially, are inbreeding to form a type, the general variability of which in numerous traits is not only less than that of an unselected sample of American Whites, but in many instances no greater than the unmixed European, African, and American Indian peoples who have contributed to its ancestry.[152]

Herskovits's conclusions have been subject to some questioning, despite the fact that a few other investigations[153] also have shown relatively low variabilities in hybrid populations. The major point of criticism lies in the fact that in discussing variable traits Herskovits does not distinguish adequately between traits in which there is little original difference between the parent groups (and in which therefore little variability is to be expected) and traits in which marked differences exist in the parent races and in which variability might more reasonably be anticipated. This criticism has been made more particularly by Davenport and Steggerda.[154] As Cobb[155] points out, when one takes this criticism into account it is still difficult to come to any definite conclusions, for if one selects for special examination those physical characteristics in which the parent racial groups show considerable differences, three of them (nasal breadth, lip thickness, and skin pigmentation) show greater variability in the Negro; and four of them (hip width, interpupillary distance, ear height, total facial height) do not show an increase in variability or are less variable than in the parent stocks. On the other hand, in favor of Herskovits's contention would appear to be the fact that

...the lowest summated average variability is not that of the unmixed Negro group, where it would be expected were low variability and lack of racial mixture as closely correlated as biologists have assumed, but rather in the group whose descent is more Negro than white, mixed with Indian. On the other hand, the greatest variability is not at the point of greatest mixture (the NW) but rather in that group which is composed of persons who are of preponderantly White ancestry with some Negro mixture.[156]

It would appear that Herskovits's conclusions concerning the lower variability of the Negro and consequently the formation of a relatively uniform Negro type in the United States cannot be accepted without qualification in view of the conflicting evidence derived from an examination of different physical characteristics. The fact that some physical traits show lesser variability than would ordinarily be assumed in view of the differences in those traits in which this low variability is not found prevents an unequivocal conclusion from being drawn.

Herskovits explains the low variability of Negro physical traits in terms of the inbreeding (since the period of extensive miscegenation ended) of the American Negro—inbreeding on a large scale, to be sure, but within a socially isolated population; and in terms of the role of skin color in Negro marriage selection which tends to bring about biological homogeneity.[157] It is possible that these factors account for the low variability in the few physical traits of the Negro in which the variability is significantly low.

Hybridization has been claimed to result in both unfavorable and beneficial traits. On the whole, the unfavorable consequences have been given the greater emphasis, especially in popular literature. A variety of biological weaknesses are attributed to the mixed blood.

The confusion that exists concerning the biological consequences of miscegenation is well illustrated in the discussion of the fecundity of the hybrid, for here diametrically opposed opinions are to be found. Some authorities maintain that hybrids show a lowered fecundity and sexual impulse while others, impressed by the opposite phenomenon, attribute "hybrid vigor" to the mixed blood as one of his principal characteristics. Low fecundity is apparently often imputed to the human mixed blood on the basis of its existence, or alleged existence, in other species. The mule, which is incapable of reproduction, is perhaps the best known and most striking instance of the effect of hybridization on the fertility of a hybrid organism. The work of Mjøen in crossing dissimilar races of rabbits is much quoted by those who contend that hybrids show a diminished reproductive capacity. Mjøen reported beginning with the fourth hybrid generation "a diminished fertility and increased mortality in sexual instinct."[158] Castle, who has subjected Mjøen's work to detailed analysis and criticism, believes this result was due to the particular physical conditions, especially insanitary environment, to which the experimental animals were subjected.[159] In any case, it scarcely seems profitable to draw conclusions concerning the human organism from the study of nonhuman organisms in view of the highly variable results in the latter field.[160]

A number of "observers" have noted the lowered fecundity of the mulatto in the United States, but other "observers" have noted precisely the opposite condition.[161] These "observations" are unaccompanied in either case by convincing supporting data. Boas has shown, in connection with half-breed Indians, that

... the average number of children of five hundred and seventy-seven Indian women and of one hundred and forty-one half-blood women more than forty years old is 5.9 children for the former and 7.9 for the latter. It is instructive to compare the number of children for each woman in the two groups. While about ten per cent of the Indian women have no children, only about 3.5 per cent of the half bloods are childless.[162]

While family size can scarcely be taken as an adequate index of reproductive capacity, it is worth noting that Fischer's study of the Hottentot-Boer hybrids showed 7.7 children per family several generations after the original crossing;[163] and that Shapiro concludes from his study of the fecundity of the offspring of the mutineers of the *Bounty* and Polynesian women that there is no loss of reproductive capacity, but rather "the crossing of two fairly divergent groups leads to a physical vigor and exuberance which equals if not surpasses either parent stock."[164]

The conflicting nature of the available evidence is apparent from the conclusion drawn by Davenport:

This survey of the results of race crossing leads to the conclusion that there is no single rule that applies to all racial hybrids. Some of them, like the French Canadian-Indian hybrids and the Chinese-Hawaiian seem to show hybrid vigor; others, like the Eurasians, show an enfeeblement.[165]

Concerning the American mulatto specifically there appears to be even less evidence than in the case of many less accessible hybrid populations. Certainly it can at least be said that there is no satisfactory evidence that the mulatto shows a lowered reproductive capacity.[166]

Apart from diminished reproductive capacity the principal alleged dysgenic consequence of hybridization is a disharmony of parts in the hybrid organism. Here again there is frequent reference in the literature to Mjøen's work on rabbits and his finding that asymmetrical carriage of the ears was "the most distinct outward sign of a disharmonic crossing that can well be imagined."[167] Castle, however, states that this is by no means confined to cross-bred rabbits but is of frequent occurrence among rabbits of large size.[168] More pertinent to the present inquiry is the study of race crossing in Jamaica by Davenport and Steggerda. They found that "some of the Jamaican browns have the long legs of the Negro and the short arms of the white which would put them at a disadvantage in picking things up from the ground."[169] Castle, however, points out that the "long legs of the Negro" turn out to be .5 of a centimeter longer than those of the whites; the legs of the brown skinned would therefore, presumably, show an even smaller excess than this negligible difference. Since Davenport and Steggerda's conclusions have been given some prominence in proving disharmony of parts, it may be well to quote the later remarks of Pearson:

The only thing that is apparent in the whole of this lengthy treatise is that the samples are too small and drawn from too heterogeneous a population to provide any trustworthy conclusions at all. There are sound biometricians and anthropologists in the United States, and it would have seemed worth the while of the Carnegie Institution of Washington to have placed the manuscript of the work before them before authorizing its publication.[170]

Susceptibility to disease, particularly tuberculosis, also has been mentioned as of frequent occurrence among hybrid groups. The Negro, of course, is well known to have a high tuberculosis rate, but no evidence has been adduced that the rate is not so high among those of pure Negro ancestry.

One further aspect of Negro-white miscegenation probably requires some comment in view of the persistent misconception at least in the lay mind on this particular point, namely, the possibility that a white person mating with an individual who passes for white but has some Negro ancestry may produce a child darker than the mixed-blood partner. It should be pointed out in this connection that, while two parties with Negro blood may occasionally have an offspring with somewhat more Negroid features than themselves, it is not possible for a white person and a person with some Negro ancestry to have an offspring more Negroid than the partner with Negro blood.[171] East's comment on the widely believed danger of "relapse" into the original physical traits of the Negro ancestor on the part of the offspring of Negro-white crossing is as follows:

A favourite short-story plot with which melodramatic artists seek to harrow the feelings of their readers is one where the distinguished scion of an aristocratic family marries the beautiful girl with tell-tale shadows on the half-moons of her nails, and in due time is presented with a coal-black son. It is a good framework, and carries a thrill. One waits shiveringly, even breathlessly, for the first squeal of the dingy infant. There is only this slight imperfection—or is it the advantage?—it could not possibly happen on the stage as set by the author. The most casual examination of the genetic formulae given above demonstrates its absurdity. If there ever was a basis for the plot in real life, the explanation lies in a fracture of the seventh commandment, or in a tinge of negro blood in the aristocrat as dark as that in his wife.[172]

And Hooton says on this same point:

There is no reversion to the Negro type in the offspring of mixed parents which would support the traditional notion of seemingly white couples producing fully Negroid infants, but there is no doubt that by a combination of features from both parents an occasional child may intensify the Negroid appearance not particularly obvious in either of his progenitors. In other words, a Negroid child may look more like a Negro than either of his parents, *if both of them carry Negro blood*. This is theoretically impossible if one parent is pure white, and I do not believe that it occurs. Negroid features seem to be attenuated, rather than intensified, by successive generations of inbreeding of mixed types, even when approximately identical proportions of blood are maintained. White features seem to gain upon Negroid features. I am convinced that some sort of Mendelian inheritance, involving many factors, is concerned in this process.[173]

Notes

135. C. B. Day, op. cit.

136. Ibid., p. 6.

137. Ibid., p. 10. In the above passage it is important to note that mulatto is used in the strict sense of the word, as the direct offspring of a white person and a fullblood Negro. The terms "dominant," "intermediate," and "recessive" are not used by Day in their technical biological sense, but simply to indicate a threefold division of mulattoes according to the prominence of white characteristics.

138. Ibid.

139. Ibid.

140. E. A. Hooton, in C. B. Day, op. cit., pp. 80–81.

141. Ibid.

142. Ibid., pp. 83–84.

143. Ibid., pp. 84–85.

144. M. J. Herskovits, *The Anthropometry of the American Negro*. A more popular account of his findings is contained in his book, *The American Negro*.

145. Ibid., p. 227.

146. Ibid., p. 228. See also pp. 81, 89, 91, 93.

147. E. A. Hooton, *Crime and the Man*, p. 306.

148. Ibid.

149. M. J. Herskovits, *The American Negro*, pp. 21–22.

150. M. J. Herskovits, *The Anthropometry of the American Negro*, pp. 250–251.

151. Ibid., pp. 41–42.

152. Ibid., p. 251.

153. K. Wagner mentions in this connection, in addition to Herskovits's study, Sullivan's study of half-blood Sioux, Williams's study of Maya-Spanish crosses in Yucatan, and Fischer's work on the Rehoboth Bastaards. K. Wagner, "The Variability of Hybrid Populations," *American Journal of Physical Anthropology*, 16: 306 (1932).

154. C. B. Davenport and M. S. Steggerda, "Race Crossing in Jamaica," Washington, 1929.

155. *Journal of Negro Education*, July, 1934, pp. 354–355.

156. M. J. Herskovits, *The Anthropometry of the American Negro*, p. 275.

157. M. J. Herskovits, *The American Negro*, pp. 30–31, 261.

158. W. E. Castle, "Biological and Sociological Consequences of Race Crossing," *American Journal of Physical Anthropology*, 9: 148–149 (1926).

159. Ibid., p. 149.

160. Gates points out that in the case of plants some species that are so closely similar that they can barely be distinguished show complete sterility, and that sterility as a criterion of species has almost completely broken down. (R. R. Gates, *Heredity in Man*, p. 302.)

161. H. G. Duncan (*Race and Population Problems*, New York, 1929, pp. 103–104) has brought together a number of these statements. See also S. J. Holmes, *The Trend of the Race*, 1921, pp. 253–254.

162. Quoted in Duncan, op. cit., p. 104.

163. Otto Klineberg, *Race Differences*, New York, 1935, p. 216.

164. Quoted from Herskovits, "Critical Discussion of the Mulatto Problem," *Journal of Negro Education*, July, 1934, p. 397.

165. C. B. Davenport, in E. V. Cowdry (ed.), *Human Biology and Racial Welfare*, New York, 1930, p. 564.

166. In a study by Roberts (op. cit.), 92 Negro-white couples showed an average of 1.5 children per couple; 66 of the 92 couples had been married 10 years or more, but without taking into account the ages of the women the data are not readily usable. In any case, such sources of information for the solution of the present problem are open to the objection that in the original crossing there may be an attempt, for obvious social reasons, to restrict the appearance of children; 29 of the 66 couples married over 10 years were childless.

167. W. E. Castle, op. cit., p. 149.

168. Ibid.

169. Quoted from C. Dover. *Half-caste*, p. 32.

170. Ibid., pp. 33–34.

171. Cf. Edward M. East, *Heredity and Human Affairs*, New York, 1929, p. 100.

172. Ibid., pp. 99–100.

173. E. A. Hooton, in C. B. Day, op. cit., p. 107.

8 TOWARD GENETICS

Introduction

Evelynn M. Hammonds and Rebecca M. Herzig

Questions of admixture necessarily raised questions of ancestry, and ancestry necessarily entailed larger histories of forced and voluntary migrations, obstacles and encouragements to marriage, and other dimensions of human history. The theory of germplasm and the rediscovery of Mendelian genetics also increased interested in ascertaining the microscopic mechanisms of heredity. Over the course of the 1930s and 1940s, the fundamentals of a new field of genetics were laid out which combined these two interests. This project displaced the individual as the central unit of analysis in favor of the population—all the members of a species who can and do interbreed with one another.

The first selection by the Russian émigré Theodosius Dobzhansky highlights the importance of the population concept in sciences of race. Dobzhansky, who arrived to the U.S. in 1927 to work in T. H. Morgan's famous "fly room" *Drosophila* laboratory at Columbia University, is remembered for providing one of the first significant syntheses of Darwinian evolutionary theory and Mendelian genetics. In this essay, published in *Scientific Monthly*, Dobzhansky explains the emerging understanding of race and population, stressing the influence of adaptive pressures on the process of "raciation."

Dobzhanksy also introduces the concept of "clines," a term of genetic analysis which dovetailed with the emphasis on flows and circulations emerging from Boasian anthropology. The rise of population-based approaches emphasized the importance of frequencies. A *cline* is a continuous gradation over space in the form or frequency of a trait, a mapped pattern of variation. (Any number expressed as a fraction of the whole population is called a frequency. The frequency of a particular allele in the population is called its gene frequency.) The concept of clinal variation was introduced in 1939, and gained speed around the time of Frank B. Livingstone's 1962 publication of "On the Non-Existence of Human Races," also reproduced here. Livingstone studied mathematics as an undergraduate at Harvard before shifting to the study of anthropology at the University of Michigan, where he devoted the remainder of his career. In this influential essay, Livingstone lobbied for the abandonment of the concept of race, noting that anthropologists were moving toward a "genetic definition" of populations.

The concept of gene frequency altered the scientific study of race. Under the definitions crafted in population genetics, *race* typically refers to a geographic subdivision of a species, one distinguished from others by the allele frequencies of some number of genes. Frequencies of certain genes, then, became more important than skin color, eye shape, hair type, or other morphological features. Races in this definition are groups of similar populations whose boundaries are necessarily permeable.

More importantly, only a population might have a gene frequency. In sharp contrast with most eugenic research, the focus of these studies shifted from the reproductive habits of individuals to the characteristics of the whole breeding group. This shift in focus from the individual to the aggregate is apparent in William C. Boyd's "Rh and the Races of Man," which maps the occurrence of the Rh- gene across the globe. Working with his wife, Lyle, the Missouri-born Boyd compiled a vast survey of blood data derived in part from transfusion centers used in the First World War. His used his analyses to challenge existing typological definitions of race, insisting on the importance of allelic variations in blood groups and the irrelevance of the sorts of superficial characteristics used to segregate human beings in Jim Crow America. His views on race also led him to collaborate with the science fiction writer Isaac Asimov on several occasions, including on their 1955 book, *Races and People*.

In the final selection, Bentley Glass extends the population approach to race, noting its development among geneticists such as Sewall Wright and J. B. S. Haldane. Glass was born in China, the son of Baptist missionaries from Texas. His long career included a number of influential positions, including editorships of the journals *Science* and *Quarterly Review of Biology*, presidencies of the American Association for the Advancement of Science and the Phi Beta Kappa academic honor society, and several governmental advisory posts. In Glass's essay we again see the importance of the theory of natural selection for explanations of difference, and the slippery relations between "population" and "race" as analytical categories. What future for human racial variation does Glass portend?

Bibliography

Boyd, William C., and Isaac Asimov. *Races and People*. New York: Abelard-Schuman, 1955.

Bulmer, Michael. *Francis Galton: Pioneer of Heredity and Biometry*. Baltimore: Johns Hopkins University Press, 2003.

Carlson, Elof Axel. *The Gene: A Critical History*. Philadelphia: Saunders, 1966.

Dobzhansky, Theodosius. *Genetics and the Origin of Species*. New York: Columbia University Press, 1982 [1937].

Dunn, L. C. *A Short History of Genetics: The Development of Some of the Main Lines of Thought, 1864–1939*. New York: McGraw-Hill, 1965.

Jacob, François. *The Logic of Life: A History of Heredity*. Translated by Betty E. Spillmann. Princeton: Princeton University Press, 1993.

Ludmerer, Kenneth M. *Genetics and American Society: A Historical Appraisal*. Baltimore: Johns Hopkins University Press, 1972.

Reardon, Jenny. *Race to the Finish: Identity and Governance in an Age of Genomics*. Princeton: Princeton University Press, 2005.

8.1 "The Race Concept in Biology" (1941)

Theodosius Dobzhansky

The perennial discussion of the nature of races, particularly of those in man, has become especially lively, frequently acrimonious and notoriously inconclusive during the last decade. Although the problem obviously is in part a biological one, biologists have, with few exceptions, disdained to take part in the debate. An apparently good reason for this forbearance is that the debate on the "race problem" is not conducted on a scientific plane at all. Yet biologists can not escape a part of the blame for the disrepute in which the race problem has fallen. The plain fact is that in biology itself no clear definition of what constitutes a race has been evolved. The existing concepts are either fundamentally unsound or so ambiguous as to be of little use for rigorous thinking. The refined analytical methods of modern genetics may permit a better insight into this problem to be gained than was possible in the past, but the work in this field is now barely begun. The purpose of the present article is to outline the salient features of the situation.

Most taxonomists and anthropologists cling perforce to the habit of describing races in terms of averages of morphological and sometimes of physiological and psychological characters. We are told that the average Eskimo has such-and-such a height, cephalic index and intelligence quotient, while different sets of figures are given for the average German or Hottentot. This method is, because of its simplicity, undeniably convenient for a rough description of the observed variety of humans or of other living beings. The trouble is that it leads to a hopeless confusion when an analysis of the underlying causes of this variety is attempted.

A race defined as a system of averages or modal points is a concept that belongs to the pre-Mendelian era, when the hereditary materials were pictured as a continuum subject to a diffuse and gradual modification. Genetics has established that the hereditary material, the germ-plasm, is not a perfect continuum, but rather a sum of discrete particles, genes, which change one by one by mutation. This is no trifling distinction, and its corollaries must be appreciated. If germ-plasms could blend with each other as a water-soluble dye commingles with water, every interbreeding population would soon reach a reasonable uniformity, and every individual would in a very real sense be a child not only of its parents but of its race as well. A "pure race" would be formed in each locality occupied by the species. With the germ-plasm being particulate, the variety of genes present in a population tends to be preserved intact indefinitely; the genetic constitution of an individual does not necessarily lie midway between those of its parents; some of the genes of an individual may resemble those

Theodosius Dobzhansky. "The Race Concept in Biology," *Scientific Monthly* 52, no. 2 (February 1941): 161–165.

commonly present in the population from which it sprung, while other genes may be identical with those usually found in representatives of another race. Except in asexually reproducing organisms, pure races can be formed only under very exceptional circumstances (a long-continued inbreeding of close relatives). Since the germ-plasm is particulate, the variation within a population can adequately be described only in terms of the frequencies of the variable gene alleles and of their combinations. Differences between populations must likewise be stated in terms of the differences in the frequencies of genes present in them.

A geneticist can define races as populations that differ from each other in the frequencies of certain genes. The obvious flaw in such a definition is that differences in gene frequencies may be quantitatively as well as qualitatively of diverse orders. The statement that two populations are racially distinct really conveys very little information regarding the extent of the distinction. This can be made evident by a series of examples illustrating the different degrees of racial separation. The examples given below concern mostly lower organisms, and particularly the small flies belonging to the genus Drosophila. The reader may be inclined to question the applicability of the conclusions reached through studies on this material to organisms in general, and particularly to man. Although in dealing with man the complications resulting from his social organization must not be lost sight of, the laws of heredity are the most universally valid ones among the biological regularities yet discovered. The mechanisms of inheritance in man, in the Drosophila flies, in plants and even in the unicelluars are fundamentally the same. The race concept is very widely applicable, at least among the sexually reproducing forms of both the animal and the plant kingdoms. It can be elucidated most effectively through use of a favorable material, which is, for technical reasons, readily amenable to the application of the experimental and quantitative methods of modern genetics.

The fly *Drosophila pseudoobscura* is a species widely distributed in western North America. Although its representatives from any part of its geographic range appear to be similar externally, genetic analysis reveals a considerable variabitity under the guise of external uniformity. The variability concerns both the gene arrangement and the gene contents of the chromosomes. Genic variability is displayed in the occurrence of mutant genes that affect the external structures, viability, development rate and other characters. Most of the mutants are recessive to the "normal" condition, and rare enough so that only heterozygotes occur in natural populations.

None of the populations of *Drosophila pseudoobscura* so far examined proved genetically uniform; in every one of them some individuals carried chromosome structures and mutant genes not present in others. Every population may be characterized by the incidence of the genetic variants present in it. Comparison of populations from different localities usually shows them to be unlike, since some of the genetic variants present in one either do not occur at all or occur with different frequencies in others. It is astonishing that even contiguous localities may harbor different populations. In forms which can move only very slowly, such as land snails, differences of this kind have been known for many years. Yet similar differences are observed in the much more mobile Drosophila. In one instance a statistically significant "racial" difference has been observed between populations of localities about 100 meters apart, although the intervening terrain contains no obvious barriers that could impede the migration of the flies. Mobility of an individual organism does not always prevent an extremely fine subdivision of the population of a species into local races. Studies of Dahlberg and others suggest that such a subdivision may occur also in man, since the incidence of certain genes may be different in populations of neighboring villages.

More unexpected still is the fact that the genetic composition of a population of *Drosophila pseudoobscura* does not remain constant with time. In certain populations from California the incidence of various chromosome structures has been observed to change not only from year to year but from month to month. The causes of such alterations are as yet not clear. The most probable conjecture is that the food sources are unevenly distributed in the territory inhabited by the flies, and that a single or a few individuals which first reach and monopolize an abundant food supply leave an offspring large enough to impress their individual characteristics on the population of the surrounding area. As shown by Sewall Wright on basis of theoretical considerations, both temporal changes and a gradual drifting apart of the genetic composition of local colonies are expected to occur where the effective sizes of local populations are limited. It seems, then, that local populations may be effectively small even in species possessing as good locomotion means as Drosophila. Perhaps an analogous situation in man is the occurrence of villages in which some family name, the heritage of a prolific early settler, is much more frequent than in the population of the surrounding territory. However that may be, changes in the racial composition of local populations may be observed in nature well within a human lifetime. Evolutionary changes in nature are not too slow to be observed directly.

The drifting apart and the consequent racial differentiation of local populations is a process which, by itself, can not be regarded as an adaptation to the environment. It is rather the forerunner of an adaptive differentiation. Genetic changes which arise in a species are subject to natural selection which eliminates the unfit and preserves the valuable variants and the populations in which such variants become frequent. Since the environment is seldom uniform throughout the distribution area, the species differentiates into local races that are adjusted each to the environment prevailing in its particular habitat. Such local races, termed by Turesson ecotypes, do not, as a rule, form continuous populations over large parts of the species area. They recur wherever the proper environment is available, while the intervening localities are occupied by different ecotypes or not inhabited by the species at all. Ecotypic differentiation has been described in many plants by Turesson, J. Clausen, Gregor and others; among animals this phenomenon seems less wide-spread, possibly because the mobility of most animals makes them less dependent than the plants are upon the micro-environment of their habitat. Nevertheless, Dice and Blossom have shown that in the mammals of the North American desert the coat color becomes darker or lighter, depending upon the prevailing shade of the soil on which they live. Dark ecotypes occur on the outcroppings of lava, and light ones on stretches of light sand. An important fact is that ecotypic differentiation does not, as a rule, involve the entire mass of individuals residing in any particular habitat. Thus, the light average shade of the coat color in mammals inhabiting light sand is due merely to a greater frequency of lightly colored specimens in sandy localities, although the darkest individuals on light soil may be much darker than the lightest ones on dark soil.

While the exigencies of adaptation to the strictly local and recurring conditions of the habitat lead to the formation of ecotypes, adaptation to more general variations in the environment results in formation of geographic races (otherwise known as subspecies or ecospecies) which occupy more or less continuously definite parts of the species area. Taxonomists are well aware of the fact that the differences between geographic races are slight in some cases and much more striking in others. Since the environment changes more or less gradually as one passes from one region to another, the changes in the appearance of the species population may be correspondingly gradual. Where a definite geographic boundary between races is discernible, the races are nevertheless found to merge into each other in at least a narrow boundary zone. This situation is described by taxonomists as "overlapping" or as presence of intermediates between the races. This is a very misleading way of stating the observed facts, for it implies the notion of "pure races" the intermediates between which are sometimes formed. It is more accurate to say that the frequencies of the variable genes change more or less gradually or abruptly during the passage from one portion of the species area to another. If the characters distinguishing races are examined one by one, geographically graded series or, to use the term recently proposed by Huxley, "clines" are encountered. The clines in gene frequencies are what cause the appearance of clines in the outwardly visible characters. The naive concept of pure races connected by intermediates must be replaced by the more authentic one of the varying incidence of definite genes. The idea of a pure race is not even a legitimate abstraction: it is a subterfuge used to cloak one's ignorance of the nature of the phenomenon of racial variation.

As a general rule, the further two populations are removed geographically the greater are the genetic differences between them. In *Drosophila pseudoobscura*, this rule is infringed upon chiefly where very small distances are involved, since the fluctuations in the composition of the population in any one locality (see above) may be large enough to obscure the more general geographical trends. Thus, the variations in the local populations on Mount San Jacinto, California, appear to be haphazard. The localities from which these population samples were taken are from 100 meters to about 25 kilometers apart. Populations from Mount San Jacinto and from the Death Valley region, a distance of about 400 kilometers, are difficult to distinguish if only small samples are available. The difficulty is alleviated if a number of large samples from several localities in each region are studied. Comparing the data for the eastern and the western parts of the Death Valley region, Mount San Jacinto, Mount Wilson, San Rafael

Mountains and San Lucia Mountains, California, we find pronounced east to west racial clines. The populations inhabiting Texas are, however, so different from those of California that a single small sample can be determined as coming from one or the other of these regions. Nevertheless, the ability to distinguish groups of individuals, populations, does not necessarily imply that every individual may be classified as a representative of one or the other races. Thus, some individuals from Texas are identical in chromosome structure with those from California, although the Texas and California races as groups are undoubtedly distinct.

While the process of "raciation" must be regarded as predominantly an adaptive one, it does not follow that every difference in the gene frequencies is a direct result of natural selection. Some of the characteristics distinguishing races appear to be adaptively neutral. Without going into the details of this very perplexing problem, one may say that the racial subdivisions of a species are a product not only of the environment now existing but also of a long historical process of evolutionary development. A race inhabiting a country is what it is, not only because it lives there but also because it came from a definite source, following a definite distribution path or paths.

Superior adaptive types, having originated in different parts of the species area, may spread and finally confront each other across a more or less narrow boundary zone. When this stage is reached, the races may develop isolating mechanisms that would prevent them from interbreeding and hence from exchanging genes with each other. The establishment of isolation connotes the transformation of races into separate species and is therefore outside the scope of the present article. The point which should be made clear here is that a race becomes more and more a reality, and less and less an abstraction, as it approaches the species rank. Species attain a degree of existential concreteness which makes them independent actors in the drama of life. In terms of this histrionic analogy, races of a species may be likened only to members of a choir. The prime characteristic of a species is that individuals belonging to it are prevented from interbreeding with those of other species, but not with each other, by physiological isolating mechanisms. It is, therefore, legitimate to speak of pure species (contrasting them with hybrids between species). Yet, identical gene variants continue to occur in races that are well advanced toward the species rank, as well as in separate species.

The description of the racial composition of a species in terms of the variations in gene frequencies presupposes a careful genetic analysis of the material under study. Unquestionably, this is a slow and difficult task, especially where, as in man, the conditions for genetic work are unfavorable. A satisfactory insight into the nature and significance of the racial differences in man demands far more extensive and detailed information than is now available on the mode of inheritance of the characters causing interracial, as well as intraracial, variability; scientifically controlled data on the manifestation of the diverse genetic conditions in various environments; and a thorough knowledge of the incidence of the determining genes in the class, caste and race subdivisions of the mankind. Evidently, this task can be accomplished only at the expense of concerted efforts of many scientists and organizations in different parts of the world. Yet, the difficulty of the task is not a sufficient reason to cling to the outworn methods of racial study, the inadequacy of which is quite plain, and still less is it a reason for erecting far-reaching theories on the basis of admittedly faulty data. To do so would be a travesty on science. It is said that Menaechmus warned Alexander the Great that "There is no royal road to mathematics." There is no royal road to genetics either.

8.2   "Rh and the Races of Man" (1951)

William C. Boyd

The races of man used to be studied mainly by explorers and anthropologists, who journeyed to the far corners of the earth to examine them. Today they can also be studied by explorers who need look no farther than the test tubes in their laboratories. The investigators who are conducting this new kind of exploration are serologists, and their explorations are carried out on samples of blood. From this rich material they can obtain precise information on the genes of human populations in all parts of the world. Chiefly responsible for making such studies possible was the discovery of the Rh factor in the blood.

Fifty-one years ago a young Viennese pathologist named Karl Landsteiner found that human blood was not the same in all individuals—that it could be classified into several different types, now known as A, B, AB and O. It was many years before much attention was paid to this discovery; until the beginning of the First World War people were sometimes given transfusions without a check of their blood type, sometimes with fatal results.

But when in 1940 Landsteiner and Alexander S. Wiener discovered the Rh factor, the response was very different. By then the blood was much better understood, and so were the genetic mechanisms of the blood groups. Medical investigators almost immediately gave intensive study to the Rh factor, for it was soon recognized to be responsible for the fatal infants' disease called erythroblastosis fetalis, which may attack the child of an Rh-negative mother and an Rh-positive father. And serologists, realizing the value of the Rh factor as a genetic tool, promptly began to examine the blood of American Indians, Asiatics and various other human populations to see whether they differed in the frequency of the Rh genes.

More than eight Rh blood genes have now been identified. There is no general agreement yet on a system for naming them; the symbols used here will be those devised by Wiener. They designate the Rh genes as r, r$'$, r$''$, r^y, R^z, R^0, R^1 and R^2. The Rh-negative gene is r.

It was soon found that the proportion of Rh-negative individuals ranges from 13 to 17 per cent among peoples of European stock, meaning the white inhabitants of Europe, America and Australia. The question arose: Would other races be found to have a similar incidence of the Rh-negative type? The first studies done on American Indians showed they do not; in fact, pure American Indians are in the fortunate position of having no Rh-negative genes at all. No case of erythroblastosis has been found among Indians.

William C. Boyd, "Rh and the Races of Man," *Scientific American* 185, no. 5 (1951): 22–25.

Nearly all modern anthropologists believe that the American Indians originally came from Asia. If this is true, their blood types should show similarities to the types to be found in Asia, just as their physical appearance resembles in some respects that of the Mongoloids. Studies were made on the Rh types of the Chinese and Japanese to test the hypothesis, and they showed that these peoples have a very small percentage of Rh-negative individuals—so small that the Rh genes in the samples may have been contributed by just a few persons, thought to be "pure," who were actually of mixed European-Asiatic ancestry. Erythroblastosis also seems extremely rare or unknown among these peoples.

The Rh explorations were next taken up by workers in Australia. Taking advantage of the air transportation in the Pacific provided by the air forces during the Second World War, they had samples of blood from various Pacific peoples flown to their laboratories so that it arrived in suitable condition to be tested. They examined the bloods of Indonesians, Filipinos, Australian aborigines, Papuans, Maoris, Admiralty Islanders, Fijians, New Caledonians and Loyalty Islanders, and did not find a single sample of Rh-negative blood.

The next region to be explored was Africa, once called the "Dark Continent"—an appellation which is no longer quite so apt. Tests on American Negroes in New York City had suggested that Negroes have about half as much Rh-negative blood as do white Americans. This was confirmed by the studies of African Negroes.

If we plot all these data on a map of the world as pieces of a puzzle, a general picture now begins to emerge. We can divide the world into three great regions. The peoples of Asia and the Pacific have no Rh-negative genes. The Europeans and their descendants do have them, in about 16 per cent of the population. The inhabitants of the great continent of Africa have about seven per cent Rh-negative—half as much as possessed by the Europeans.

It is a matter of great interest that the people of Egypt, a gateway between Africa and Europe, stand between the Europeans and the Africans in their Rh inheritance. Two sampling studies show that about 10 per cent of the Egyptians have Rh-negative blood.

Where did the Europeans get their Rh-negative genes? Obviously this is a most difficult question to answer, but a gleam of light suddenly appeared when a South American investigator of Basque descent tested a group of Basques in South America. He found that 28 per cent of them had Rh-negative blood—far more than the average among Europeans. The Basques have long been known to be the remnants of a people older than the other inhabitants of Europe. They speak a language not known to be related to any other, and they have been in France and Spain for a very long time. They jealously kept their blood "pure"; a girl who married outside the Basque race found herself disowned by her parents. It seems reasonable to suppose, however, that although the Basques have managed rather successfully to maintain their own racial purity, they must have mixed with other Europeans to some extent during their long history. If the Basques had a high percentage of Rh-negative genes, in any mixture

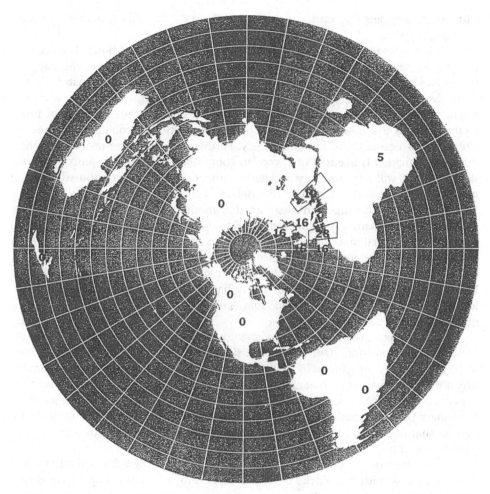

Figure 8.2.1

Rh-negative gene occurs in different percentages in the various populations of the world. Each number on this map indicates the percentage of the earliest known inhabitants of that region who have Rh-negative blood.

with other Europeans they must have contributed a certain frequency of these genes to the European stock.

If we postulate that it was the Basques who brought the Rh-negative gene to Europe, we have only pushed our difficulty back one step. Where did the Basques themselves get the gene? And why has the gene not been eliminated by natural selection? Assuming that there are two Rh genes in a population, one positive and one negative (a convenient oversimplification), selection due to deaths of children from erythroblastosis will mean that the only possible condition of equilibrium is one in which the two genes are of equal frequency. Furthermore, this is an unstable equilibrium. The situation is like that of a wooden cone balanced on its tip; push it ever so slightly and it will fall over. Any accidental variation in either direction will start a process which will lead to complete elimination of either the Rh-negative gene or the Rh-positive gene. This argument assumes that mothers of erythroblastotic children leave fewer descendants than the average. However, the British geneticist Ronald A. Fisher has questioned that assumption. He believes that a family which had lost a child would have a strong tendency to replace the child by having one or more extra children, and that in many cases Rh-negative women might have larger families than their Rh-positive sisters. Fisher has even suggested that this tendency to compensate for the loss of children by having large families may actually be increasing the Rh-negative frequency among the Basques.

Anthropologists have now begun to examine some of the rarer Rh genes in their study of populations. One of the most significant is the R^0 gene. This gene seems to be almost an identifying mark of the African peoples; among the Bantu of South Africa it is the most common of all the Rh types, whereas it occurs in only about two per cent of a European stock. (In speaking of Africa, of course, we mean Africa south of the Sahara Desert, because the people of North Africa and Egypt belong primarily to Europe, that is, the old Mediterranean civilization of which the Romans and Phoenicians formed a part.)

In the white population of England and the U.S. there are several rare Rh genes that occur with frequencies of only one per cent or so but are nevertheless important identifying factors. These genes, with few exceptions, are uniformly absent from the Asiatic and Pacific native populations. In fact, the genetic situation seems to be much simpler in the Pacific than here, for the peoples of that part of the world have only three Rh genes in any considerable amount, in contrast to the eight or more in our own population. The reason for this difference is, at present, a matter for speculation. Perhaps in the Pacific, some parts of which were peopled in relatively recent times, the missing genes never arrived; on the other hand, it may be that they have been eliminated by natural selection.

There is one rare Rh gene that does show up with unusual frequency among the Asiatics. The R^z gene, which occurs in less than one per cent of Europeans, is present in five per cent or more of the American Indians and the Australian aborigines. R^z may be called an Asiatic gene, in somewhat the same sense as R^0 may be termed an African gene, although it never reaches the same frequencies.

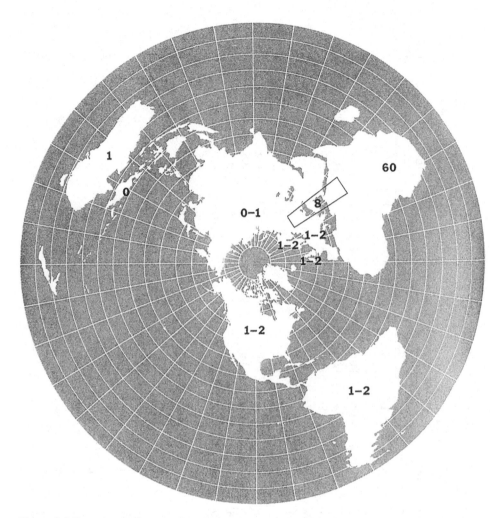

Figure 8.2.2
R° gene has a distribution of its own differing from that of the Rh-negative gene. As in figure 8.2.1, each number on this map indicates the percentage of the earliest known inhabitants of that region who have R° blood.

Table 8.2.1
Several Rh Genes Are Tabulated with Respect to Frequency

	Number of persons tested	Frequencies of Rh types (per cent)								
		rh cde	Rh₁ CDe	Rh₂ cDE	Rh₁Rh₂ CDe/cDE	Rh₀ cDe	rh' Cde	rh" cdE	rh'rh" Cde/cdE	Rh₁Rh₂ Cde/CDE
Basques	167	28.8	55.1	7.8	6.0	0.6	1.8	0.8	0	0
"Whites" (France)	501	17.0	51.7	13.6	13.0	3.6	0.4	0.6	0	0.6
Czechs (Prague)	181	16.0	50.3	11.6	11.6	1.1	0.6	0	0	0
"Whites" (Hollanders)	200	15.4	51.5	12.3	17.7	1.5	1.5	0.7	0	0
"Whites" (England)	1,038	15.3	54.8	14.7	11.6	2.3	0.6	0.7	0	
Sao Paulo (Brazil)	138	15.2	55.2	10.1	11.6	5.8	1.4	0.7		
"Whites" (Australia)	350	14.9	54.0	12.6	16.6	0.6	0.9	0.6	0	0
"Whites" (England)	927	14.8	54.9	12.2	13.6	2.5	0.7	1.3		0.1
"Whites" (U.S.A.)	7,317	14.7	53.5	15.0	12.9	2.2	1.1	0.6	0.01	
Spanish (Barcelona)	223	13.0	63.2	13.0	9.4	0.5	0	0.5	0	0.5
"Whites" (U.S.A.)	766	12.5	54.7	14.9	13.9	2.2	0.9	0.5	0	0.1
"Arabs" (Baghdad)	300	10.3	50.3	13.7	15.7	8.3	1.0	0.7	0	0
Puerto Ricans	179	10.1	39.1	19.6	14.0	15.1	1.7	0.5	0	
Negroes (U.S.A.)	223	8.1	20.2	22.4	5.4	41.2	2.7	0	0	
Negroes (U.S.A.)	135	7.4	23.7	16.3	4.4	45.9	1.5	0.7	0	0
Asiatic Indians (Moslems)	156	7.1	70.5	5.1	12.8	1.9	2.6	0	0	
South African Bantu	300	5.3	27.0	0	2.3	64.3	1.0	0	0	
Chinese	132	1.5	60.6	3.0	34.1	0.9	0	0	0	
Japanese	150	1.3	37.4	13.3	47.3	0	0	0	0	0.7
Japanese	180	0.6	51.7	8.3	39.4	0	0	0	0	
Indians (Mexico, Tuxpan)	95	0	48.1	9.5	38.1	1.1	0	0	0	3.1
Indians (Ramah, N. M.)	105	0	40.0	17.1	36.2	2.9	0.9	0	0	2.9
Indians (Ramah, N. M.)	305	0	28.5	20.0	41.0	0.7	3.0	0	0.7	6.2

Indians (Utah)	104	0	33.7	28.8	37.5	0	0	0	0	0
Indians (Brazil)	238	0	22.7	19.3	53.2	0.5	0	0.5	0	4.8
Indonesians	200	0	74.0	2.5	22.5	0	0	0	0.5	0
Filipinos	100	0	87.0	2.0	11.0	0	0	0	0	0
Australian Aborigines	100	0	53.0	21.0	15.0	4.0	1.0	0	0	6.0
Australian Aborigines	234	0	58.2	8.5	30.4	1.3	1.7	0	0	0
Papuans	100	0	93.0	0	4.0	0	0	0	0	3.0
Maoris	32	0	25.0	31.0	41.0	3.0	0	0	0	0
Admiralty Islanders	112	0	92.9	.9	6.2	0	0	0	0	0
Fuians	110	0	89.1	1.8	9.1	0	0	0	0	
New Caledonians	243	0	77.4	2.1	20.5	0	0	0	0	
Loyalty Islanders	103	0	77.7	2.9	19.4	0	0	0	0	
Siamese (Bangkok)	213	0	74.7	3.3	21.1	0.5	0	0	0	3.3

The terminology differs from that in text: rh in chart is r in text and Rh_0 is R^0.

The Rh factors are not the only new blood factors that have been found since the discovery of the classical blood groups. Actually eight independent gene systems have been discovered, largely by British workers. When reagents for these become generally available, it will become possible to determine several million different types of individual blood, and it is not beyond the bounds of possibility that we shall be able to identify a person by his blood almost as surely as by his face.

Attempts have been made to trace the history of the Rh types back further than their occurrence in *Homo sapiens*. If man and the chimpanzee, for instance, are common descendants of some remote common ancestral type, we might expect to get some information about the Rh genes by testing chimpanzees and other anthropoids. Wiener tested 10 chimpanzees and found that they all appeared to possess an Rh-negative gene. This would suggest that the Rh-negative factor may be older than the human race as we know it. It is possible, however, that this gene has arisen independently in chimpanzees and man. We know from studies on Egyptian mummies that the blood groups are certainly more than 5,000 years old.

The interpretation of these facts depends on one's particular theory of human evolution and one's views about anthropology in general. Geneticists who study populations are today convinced that probably four main agencies have operated to produce the geographical differentiations which account for our present human races. These mechanisms are mutation, selection, genetic drift or isolation, and mixture between races, possibly even between different species. We do not know whether selection is acting to increase or decrease the frequency of the Rh-negative gene or any of the other blood factors.

That genetic drift may account for the loss of genes is suggested by the example of the American Indian. Pureblooded American Indians have no type B blood, although in Asia, from which they are believed to have come, it is common. It is tempting to suppose that B has been lost in the Indians simply by a genetic accident or series of accidents.

To illustrate how this might happen, let us suppose that we have a relatively small population of about 100 or so coming from Asia to America. We then have two possibilities: First, there is a random chance, though a small one, that none of the 100 migrants happened to have B blood. Thus the new population would never have this gene. The second possibility is that an individual of group B did arrive in America but left no descendants or had children who happened to belong to group O, which would be quite possible if he had one gene for B and one for O. The University of Chicago geneticist Sewall Wright has shown mathematically that accidents of this sort can play an important role in the evolution of natural populations.

Could we account for the loss of the Rh-negative gene in a similar way? This seems much more doubtful, because the enormous population of Asia seems to have little, if any, of this gene, and yet we have no record of their having migrated to Asia from any other place; in fact, the probabilities are that man originated in Asia or Africa or some region combining the two. It would seem more likely that selection, rather than genetic accident ("drift"), may account for the absence of the Rh-negative gene in the Asiatics. If this is true, it leaves us with the problem: Why has selection not eliminated the gene from the Europeans? Possibly it is now doing so but has not yet been operating long enough to show much effect. It may be that selection against the Rh-negative gene is more intense in dense populations. Asia has long been densely populated, whereas the modern European is obviously separated by only a relatively short time from the Neolithic hunters who wandered in small hordes through the forests.

8.3 "On the Non-Existence of Human Races" (1962)

Frank B. Livingstone and Theodosius Dobzhansky

ON THE NON-EXISTENCE OF HUMAN RACES

Frank B. Livingstone

In this paper I would like to point out that there are excellent arguments for abandoning the concept of race with reference to the living populations of *Homo sapiens*. Although this may seem to be a rather unorthodox position among anthropologists, a growing minority of biologists in general are advocating a similar position with regard to such diverse organisms as grackles, martens, and butterflies (Brown 1957, Hagmeier 1958, Gillham 1956). Their arguments seem equally applicable to man. It should be pointed out that this position does not imply that there is no biological variability between the populations of organisms which comprise a species, but just that this variability does not conform to the discrete packages labelled races. The position can be stated in other words as: There are no races, there are only clines.

The term, race, has had a long history of anthropological usage and it can be generally defined as referring to a group of local or breeding populations within a species. Thus, it is a taxonomic term for sub-specific groupings greater than the local population. Most anthropologists today use a genetic definition of races as populations which differ in the frequency of some genes.

The term, race, or its newer synonym, geographical race, is used in a similar way with reference to biological species other than man. Where the term is used, it can be considered as approximately synonymous with the term, subspecies. In 1953 Wilson and Brown first suggested discarding the concept of subspecies since it did not accord with the facts. Their main argument was that the genetic variation among the local populations of a species was discordant.

Variation is concordant if the geographic variation of the genetic characters is correlated, so that a classification based on one character would reflect the variability in any other. Such a pattern of variation is almost never found among the local populations of a wide-ranging species, although it is usually found among related relatively allopatric species.

Thus, although it is possible to divide a group of related species into discrete units, namely the species, it is impossible to divide a single species into groups larger than the panmictic population. The causes of intraspecific biological variation

Frank B. Livingstone, and Theodosius Dobzhansky, "On the Non-Existence of Human Races," *Current Anthropology* 3, no. 3 (1962): 279–281.

are different from those of interspecific variation and to apply the term subspecies to any part of such variation not only is arbitrary or impossible but tends to obscure the explanation of this variation. If one genetic character is used, it is possible to divide a species into subspecies according to the variation in this character. If two characters are used, it may still be possible, but there will be some "problem populations," which, if you are an anthropologist, will be labelled composite or mixed. As the number of characters increases it becomes more nearly impossible to determine what the "actual races really are."

In addition to being a concept used to classify human variability, race has also been overworked as an explanation of this variability. When a particular blood group gene or hair form is found to be characteristic of the populations of a particular region, it is frequently "explained" as being a "racial" character. This type of explanation means, in other words, that this particular set of human populations possesses this character, while it is absent in the rest of humanity, because of the close common ancestry of the former. At times many characteristics which were thought to be racial have been found in many widely separated populations, so that the explanation in terms of race required the assumption of lengthy migrations. In this way race or common ancestry and migration have been used to explain much of the genetic variability among human populations. Unfortunately such explanations neither accord with our knowledge of the population structure and movements of hunters and gatherers, nor take into consideration the basic cause of biological variation, natural selection.

The incompatibility between race and natural selection has been recognized for a long time; so that if one's major aim was to discover the races of man, one has to disregard natural selection. Thus, non-adaptive characters were sought for and, in some instances, considered found. But the recognition of the role of natural selection has in the past ten years changed the course of research into human variability; or at least it has changed the thinking of the "aracial ultrapolymorphists."

If a central problem of physical anthropology is the explanation of the genetic variability among human population—and I think it is—then there are other methods of describing and explaining this variability which do not utilize the concept of race. This variability can also be described in terms of the concepts of cline and morphism (Huxley 1955). The variability in the frequency of any gene can be plotted in the same way that temperature is plotted on a weather map. Then one can attempt to explain this variability by the mathematical theory of population genetics. This is a very general theory and is capable of explaining all racial or gene frequency differences, although of course for any particular gene the exact magnitudes of factors, mutation, natural selection, gene drift, and gene flow, which control gene frequency differences are not known. All genes mutate, drift, flow, and for a given environment have fitnesses associated with their various genotypes. Hence differences in the frequency of any gene among a series of populations can be explained by these general factors which control gene frequency change. Gene frequency clines can result from many different types of interaction between the general factors which control gene frequencies. For example, a cline may be due to: 1) the recent advance of an advantageous gene; 2) gene flow between populations which inhabit environments with different equilibrium frequencies for the gene; or 3) a gradual change in the equilibrium value of the gene along the cline. The theoretical analysis of clines has barely begun but there seems to be no need for the concept of race in this analysis.

COMMENT

Theodosius Dobzhansky

I agree with Dr. Livingstone that if races have to be "discrete units," then there are no races, and if "race" is used as an "explanation" of the human variability, rather than vice versa, then that explanation is invalid. Races are genetically open systems while species are closed ones; therefore races can be discrete only under some exceptional circumstances. Races arise chiefly as a result of the ordering of the genetic variability by natural selection in conformity with the environmental conditions in different territories; therefore the variability precedes race and serves as a raw material for its formation.

The difficulties with the race concept arise chiefly from failure to realize that while race differences are objectively ascertainable biological phenomena, race is also a category of biological classification. Since human population (and those of other sexually-reproducing species) often, in fact usually, differ in the frequencies of one or more, usually several to many, genetic variables, they are by this test racially distinct. But it does not follow that any racially distinct populations must be given racial (or subspecific) labels. Discovery of races is a biological problem, naming races is a nomenclatorial problem. There is nothing arbitrary about whether race differences do or do not exist, but whether races should or should not be named, and if they should, how many should be recognized, is a matter of convenience and hence of judgment. Names are given to races, species, and other categories because names are convenient in writing and speaking about the populations or the organisms discovered or investigated.

If all racial variation formed clines (geographic gradients), if all clines were uniform, and if the clines in different characters (and genes) were absolutely independent and uncorrelated, then race differences would still exist. But racial names would not be conveniently applicable, and we would have to get along without them. But this is not the situation in the human species and in many animal and plant species. The clines are not uniform; they are steeper where natural, or social, impediments to travel and intermarriage interpose obstacles to gene exchange, and more gradual where the gene exchange is unobstructed. For the same reason, different variable characters, and the gene frequencies underlying them, are often correlated. This does not make the races any more or less "natural;" but it does make them more easily nameable.

The multiplication of racial or subspecific names has gone beyond the limits of convenience in the human and in some animal species. This was bound to provoke a reaction, and up to a point this was salutary. But if the reaction goes too far in its protest it breeds confusion. To say that mankind has no races plays into hands of race bigots, and this is least of all desirable when the "scientific" racism attempts to rear its ugly head (see *Current Anthropology*, October 1961).

REPLY

Frank B. Livingstone

Professor Dobzhansky appears to disagree with the major point of my article and puts forward several arguments against it. His final, and least important, argument that my position "plays into hands of racial bigots" is incompetent, irrelevant, and immaterial. The fact that some crank may make political hay of a biological fact, concept, or theory is no criterion of the validity of any of these in biological science. I also fail to comprehend how a position which denies the validity of a concept supports anyone using that concept.

With regard to the biological arguments, Dobzhansky begins in the first paragraph by stating "Races are ..." and then that "Races arise ..." Evidently Dobzhansky already knows what races are, but in his further discussion he employs two different concepts for the term, race. Races exist as biological entities which are discoverable, but we should only label some of these races. In this way Dobzhansky distinguishes between races and race differences. Racial differences do not necessarily mean differences in race. Dobzhansky states, and I would agree with him, that just about all human populations differ in the frequency of some gene. He then concludes that all human populations are racially distinct, which should imply that each is a separate race. I have come to much the same conclusion but think that the term, race, is therefore inapplicable. Race as a concept has been, and still is, used to classify or group together human populations. Since any grouping differs with the gene frequency used, the number of groupings is equal to the number of genes. Thus, one population could belong to several different races. These different kinds of races have been emphasized by Garn (1960), but I don't think they accord with the general use of the term race as a concept within the Linnean system of biological nomenclature. Any particular animal population cannot belong to several different genera, species, or any other taxonomic level within the Linnean system; such a usage is inconsistent with the assumptions of the system.

By definition, races are populations which differ in the frequency of some genes; but according to Dobzhansky all differences in gene frequency between populations are racial differences. Thus, the difference in the frequency of thalassemia between the town of Orosei in the lowlands of Sardinia and the town of Desulo in the highlands fifty kilometers away is a racial difference. With this definition, there is also a high sickle cell gene frequency race consisting of some Greeks, Italians, Turks, Arabs, Africans and Indians, and a low sickle cell race consisting of the rest of humanity. One could also speak of a high color-blind and a low color-blind race. Since Dobzhansky stresses that it may not be convenient to distinguish some racially different populations as races, these may be cases which he wouldn't want to distinguish. But someone else may find it convenient to distinguish them. However, the confusion arises from the fact that these similarities and differences are not "racial" in the sense that they are not due to closeness of common ancestry but to the operation of natural selection, a factor which I think Dobzhansky neglects.

At first, Dobzhansky agrees that natural selection is the major factor determining gene frequency differences. But later when Dobzhansky discusses the nature of genetic variability among human populations, his statements imply that gene exchange is the sole factor responsible for the shape of clines in gene frequency or, in other words, for the genetic differences between human populations. I attempted to show how natural selection and gene flow can interact in several ways to determine the shape of a cline. Although the study of clines in gene frequency among human populations has scarcely begun, surely the clines in the sickle cell gene are the result of more than gene flow. Indeed it is impossible to explain clines solely in terms of gene flow.

Finally, I think our disagreement hinges to a great extent on different views of the nature of scientific concepts and theories. To Dobzhansky races and race differences are things which exist "out there" as biological phenomena, but race is also a label by which we describe "out there." I disagree. No science can divorce its concepts, definitions, and theories so completely from its subject matter, so that Dobzhansky's dichotomy between biological and nomenclatorial problems is impossible. The names, labels, concepts, and definitions of a science are not simply conveniences, and their function is not just to serve as vehicles of communication. The concepts are also not just a set of labels by which we attempt to classify or divide up reality or our observations of it. Of course, there has to be a set of commonly accepted labels so that we can communicate, but the concepts of a science are also logically interconnected and form a coherent, consistent theory or system. The concepts of such a system are defined in terms of one another and certain primitive terms, and then the formal, mathematical, or logical properties of the system derived. For example, in the field of human biology, the concept of breeding population when considered as part of the mathematical theory of population genetics pertains to nothing in reality. But when combined with the concepts of gene frequency, random mating, etc., further concepts such as the Hardy-Weinberg Law or the principle of random gene drift can be logically derived. The latter are more or less the theorems of population genetics and are the logical outcomes of the more basic concepts or axioms. The science of population genetics then attempts to apply this theory to bodies of data and to attempt to determine which group of individuals in a particular area fits most closely the concept of breeding population, but the function of this concept or the theory is not to divide up or label reality, but to explain it. Of course, the concepts of the mathematical theory of population genetics have been developed from the data and findings of a particular sphere of reality and are approximations to these data. But they can also be considered solely as logical concepts and studied as a formal system with no reference to this reality. As Medawar (1960) has remarked, this theory has great generality in biology and in that science occupies a position analogous to Newton's Laws in physical science. An infinite population randomly mating without selection, mutation, or gene flow is analogous to Newton's body moving without friction at a constant velocity.

In applying the theory of population genetics to humanity, the species is divided into breeding populations although for any area or group of people this concept may be difficult to apply. It is likely that each breeding population will prove to be genetically unique, so that all will be racially distinct in Dobzhansky's terms. But this is not the general use of the concept of race in biology, and the concept has not in the past been associated with this theory of human diversity. Race has instead been considered as a concept of the Linnean system of classification within which it is applied to groups of populations within a species. To apply a concept of the Linnean system to a group of populations implies something about the evolutionary history of these populations, and it also implies that these populations are similar in whatever characters were used to classify them together because of close common ancestry. It is this implied explanation of whatever genetic variability is used to group populations into races which I consider to be false.

References Cited

Brown, W. L. Jr. 1957. Centrifugal speciation. *Quarterly Review of Biology* 32:247–277.

Garn, S. M. 1960. *Human Races*. C. C. Thomas, Springfield.

Gillham, N. W. 1956. Geographic variation and the subspecies concept in butterflies. *Systematic Zoology* 5:110–120.

Hagmeier, E. M. 1958. The inapplicability of the subspecies concept in the North American Marten. *Systematic Zoology* 7:1–7.

Huxley, J. S. 1955. Morphism and evolution. *Heredity* 9:1–52.

Medawar, P. B. 1960. *The Future of Man*. Methuen, London.

Wilson, E. O. and Brown, W. L. Jr. 1953. The subspecies concept and its taxonomic application. *Systematic Zoology* 2:97–111.

8.4   "The Genetic Basis of Human Races" (1968)

Bentley Glass

In the course of a recent conversation Ernst Mayr and I agreed that probably no two of you sitting next to each other in the audience at this session would differ by fewer than several hundred genes. On the other hand, as Professor Dobzhansky has pointed out, the differences which distinguish human populations from one another, whether they are so distinct that we call them races or not, are expressed by the geneticist in terms of the frequencies of different genes; that is to say, some of you here are of blood group A, some are of blood group B, some are of blood group O. Knowing the nature of this population, I could predict rather well what percentage of each blood type would be present in a sample. Facing an audience in Tokyo instead of Washington, I would predict different frequencies of A and B and O. The difference is not that in Japan there are no individuals of type A or type O and only individuals of type B; quite the contrary, all three types are present in such a population, just as they are present here. The difference is that in Japan there would be more individuals in the audience of type B, and consequently fewer of types A and O. There are practically no populations in the world in which all three blood types of the A-B-O system are not represented, although among the Indians of North and South America before they became intermixed with other peoples, there seem to have been no individuals of type B.

Very few such all-or-none characteristics of a genetic nature distinguish the different major races of mankind. In contradistinction to the four hundred or so genes by which you differ from your neighbor, there are probably not more than a dozen different genes that it would be easy to specify as occurring in one race but not in others.

Races are subdivisions of a species. There is no real distinction between races, in the anthropological or zoological sense, and subspecies. Races (or subspecies) always are separated from each other in space or time. In other words, contemporaneous races or subspecies always are separated from each other geographically. Different races of the same species differ adaptively, as a rule. They may or may not show some degree of genetic isolation, that is, preferential breeding with mates of their own race, or of lessened fertility of the interracial mating or of the hybrids so produced. The analysis of race and population structure as factors in evolution owes much to R. A. Fisher (1), Sewall Wright (2), and J. B. S. Haldane (3).

The formation of races, through geographic isolation and adaptation by means of natural selection, up to a stage of marked morphological difference and complete barriers to direct gene exchange among them, is dramatically illustrated in Figure

Bentley Glass, "The Genetic Basis of Human Races," in Margaret Mead, Theodosius Dobzhansky, Ethel Tobach, and Richard Light, eds., *Science and the Concept of Race* (New York: Columbia University Press, 1968), 88–93.

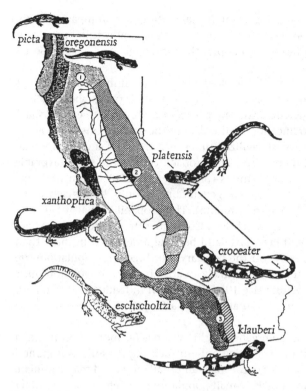

Figure 8.4.1
Subspecies of the slamander, *Ensatina eschscholtzi*, in California. (Courtesy of professor R. C. Stebbins.)

8.4.1 by a chain of races described by R. C. Stebbins (4) in the California salamander, *Ensatina eschscholtzi*, in which the distributions of the end members of the chain overlapped with no apparent interbreeding. Only an interruption of the ring by extinction of one or more intermediate races is therefore required to complete the process of speciation.*

The ring or chain of races—"circular overlap" of geographic races in Mayr's terminology—enables us to arrive at a fairly satisfactory definition of a species on a biological or genetic basis. A sexually reproducing species undergoing geographic differentiation is isolated from other related species by the lack of interchange between the several gene pools of the species. On the other hand, races belonging to the same species, although somewhat differentiated by selection, can still experience gene flow between their populations, at least indirectly, through intermediate races.

The races of *Homo sapiens*, whether we lump them into a few or split them into a great many, are clearly of geographic origin. There is no evidence whatsoever of genetic isolation between them. Offspring of interracial matings are unimpaired, if not superior, in vigor and fertility. Consequently, we may conclude that within our present species, the process of speciation has not yet approached the stage at which subspecies become full species.

The student of human genetics is therefore concerned mainly with an analysis of early stages of speciation, with changes in allelic frequencies brought about by mutation, selection, gene flow, or genetic drift. The effects of each of these evolutionary factors are strongly subject to modification by the sizes of the populations involved.

In the small populations of precivilized times, the effects of mutation would expectably be more diversifying than today. The probability that the same favorable mutation would arise in two small populations (breeding size, 250) within the same span of 25,000 years, or roughly 1000 human generations, is only 25 per cent if the gene has a mutation rate of 10^{-6} (one in a million). The probability that it becomes established in both populations is very much less. Consequently, in those early times favorable mutations would rarely, if ever, arise and become established in more than one population coincidentally. Today, on the contrary, our breeding populations are so large that any favorable mutation is quite likely to arise in every population quite frequently. In a population of breeding size 10^8 (100 million individuals; 200 million functioning gametes per generation), a mutation with a frequency of 10^{-6} will arise 200 times per generation.

Chance, in the form of random genetic drift, also has far more effect in small populations—in fact, it is significantly effective only in such populations. Abrupt and unpredictable shifts of allele frequencies occur in small populations and may cumulate over several generations in a "run of luck." Small populations therefore come to differ radically in their gene frequencies from the populations of their origin, and not only for the above reason but also because of the "founder effect." That is, whenever a new colony is established by a very few individuals, it cannot be fully and proportionately representative of the gene pool from which it is drawn. The large populations of today minimize these effects, except where culturally or religiously isolated groups of small size hold themselves strictly aloof from intermarriage with their neighbors (6).

Large populations are subject to a different kind of selection pressure than that operating on man's ancestors. In primitive times selection was probably largely for survival to adulthood. It involved such factors as lack of physical defects, good intelligence, and, especially, resistance to malnutrition and constitutional disease. With the crowding of man into villages and cities and the advent of agriculture, selection pressure must have lessened along these lines and increased along others, such as in resistance against infectious diseases. Today we have changed all that once again. Modern nutritional and medical standards have sharply lowered the death rate. Most babies in most countries now have at birth a life expectancy exceeding the reproductive period. Hence viability yields to fertility as the chief modern form of selection pressure. Crow (7) has shown that there are ample differences in fertility rates to serve as a new basis for selection.

Gene flow is probably the principal factor now reducing the differences between the gene pools of the various human races. Thus the gene pool of North American Negroes (socially defined) is now approximately 30 per cent derived from white ancestry (8, 9). In South America the amalgamation is considerably greater. The races are in fact disappearing, although the process will require thousands of years at present rates.

We should not deny that certain racial differences—physical, nutritive, immunological, and perhaps behavioral—are possibly adaptive, although this is difficult to prove in even a single case. Comparisons of relative superiority must be made on a common basis, but that very fact renders them invalid, since the superiority relates to adaptation to a particular set of conditions, to a particular environment, that is not the same for any two populations or races. We may be able to say that for a certain set of conditions one trait is superior (in viability or fertility) to another, but the superiority cannot be generalized. For a different set of conditions, if the traits are the "normal" differences of different races, superiority often will rest with the alternate one. The very complexity and changing nature of the environment assures that selection will itself foster a complex process of genetic adaptation.

Finally, it is clear that certain "normal" alleles, present in virtually all human races—as, for example, the A-B-O blood group alleles—are by no means exempt from selection, but interact most complexly (10). Thus it seems that in a population containing solely A and O blood types, the introduction of the typically Mongolian allele for B produces an improved balance of selective factors (11). It is not clear that the introduction of B into a population composed solely of type O would be equally advantageous. Pure B/O populations do not exist, although theoretically they might. Sickle-cell hemoglobin and thalassemia are detrimental enough to cause the death of homozygotes. However, in regions of endemic malaria—man's great killer in civilized times—both sickle-cell hemoglobin and thalassemia have become widely dispersed and have served to preserve populations threatened with extinction because normal hemoglobin is so tasty to the malaria plasmodium.

As the human races disappear, genetic differences will of course remain, but within a polyglot population, a composite gene pool. From the study of the phenomenon of race in man, so long as it exists, we can derive useful knowledge about the interplay of evolutionary factors and perhaps even foresee what genetic adaptations will be required in our changing world.

Note

* A brief note by Brown and Stebbins (5) indicates that there is some hybridization between the blotched and unblotched subspecies of *Ensatina* at the ends of the ring in Mill Canyon in southern California. That finding might seem to render the example inappropriate, but this is probably not so. Stebbins writes (personal communication) that further investigation by Brown, soon to be published, shows that hybrids are in fact rare between the extreme forms of the chain or ring. He states: "I feel at present that the Rassenkreis still can serve as

a good example of circular overlap because of the apparent low frequency of genetic breakdown in southern California."

References

1. R. A. Fisher, *The Genetical Theory of Natural Selection* (Oxford, Clarendon Press, 1930).

2. Sewall Wright, "Evolution in Mendelian Populations," *Genetics*, **16** (1931), 97–159; "Population Structure as a Factor in Evolution," in *Moderne Biologie: Festschrift zum 60 Geburtstag von Hans Nachtsheim* (Berlin, F. W. Peters, 1950), 275–87; and "The Genetical Structure of Populations," *Annals of Eugenics*, **15** (1951), 323–54.

3. J. B. S. Haldane, *The Causes of Evolution* (London, Longmans, Green and Co., 1932).

4. Robert C. Stebbins, "Intraspecific Sympatry in the Lungless Salamander *Ensatina eschscholtzi*," *Evolution*, **11** (1957), 265–70.

5. Charles W. Brown and Robert C. Stebbins, "Evidence for Hybridization between the Blotched and Unblotched Subspecies of the Salamander *Ensatina eschscholtzi*," *Evolution*, **18** (1965), 706–07.

6. Bentley Glass, Milton S. Sacks, Elsa F. Jahn, and Charles Hess, "Genetic Drift in a Religious Isolate: An Analysis of the Causes of Variation in Blood Group and Other Gene Frequencies in a Small Population," *American Naturalist*, **86** (1952), 145–59.

7. James F. Crow, "Mechanisms and Trends in Human Evolution," in Hudson Hoagland and R. W. Burhoe, eds., *Evolution and Man's Progress* (New York, Columbia University Press, 1962), 6–21.

8. Bentley Glass and C. C. Li, "The Dynamics of Racial Intermixture: An Analysis Based on the American Negro," *American Journal of Human Genetics*, **5** (1953), 1–20.

9. Arthur G. Steinberg, Rachel Stauffer, and Samuel H. Boyer, "Evidence for a *Gm^{ab}* Allele in the *Gm* System of American Negroes," *Nature*, **188** (1960), 169–70.

10. Alice M. Brues, "Selection and Polymorphism in the A-B-O Blood Groups," *American Journal of Physical Anthropology*, **12**, n.s. (1954), 559–97.

11. Bentley Glass, "On the Evidence of Random Genetic Drift in Human Populations," *American Journal of Physical Anthropology*, **14**, n.s. (1956), 541–56.

9 THE END OF RACE?

Introduction

Evelynn M. Hammonds and Rebecca M. Herzig

The tenor of American research on racial differences shifted after the Second World War. By the middle of the twentieth century, it had become increasingly difficult to sustain widespread scientific support for doctrines of racial inferiority. The 1950 UNESCO (United Nations Educational, Scientific and Cultural Organization) Statement on Race, drafted by an international committee of anthropologists, geneticists, biologists, sociologists and psychologists argued for man's essential unity and his shared mental, emotional and social capacities. Denying any causal connection between physical characteristics and social hierarchies, the statement boldly proclaimed that race "is not so much a biological phenomenon as a social myth."

Despite the committee's assertion that the statement epitomized contemporary scientific thought on race, not all investigators accepted the proposal. Several prominent geneticists and physical anthropologists challenged specific features of the committee's conclusions, most notably the declaration that, as a species, mankind displayed an innate drive towards cooperation and an "ethic of universal brotherhood." Their protests led to a redrafting of the statement in 1951, with a new committee. The second committee omitted the language of universal brotherhood and placed more emphasis on the significance of physical characteristics in distinguishing racial groups.

Although the UNESCO documents did not deny the existence of human races, they did publicly mark a trend towards critical reappraisals of the race concept. Over the next two decades, a growing number of American physical anthropologists, biologists, and psychologists would argue that the race concept was scientifically invalid and should be discarded. While they recognized that physical and genetic differences exist among human populations, these scientists claimed, as the population geneticist Frank Livingstone wrote in 1962 that "this variability does not conform to the discrete packages labeled races." Physical anthropologists, most prominently Ashley Montagu (one of the most outspoken critics of the race concept and the chair of the UNESCO Committee on Race), also stressed the arbitrary nature of racial classifications and argued that the long history of "admixture" between various human populations made it incredibly difficult to isolate stable groups with unique characteristics.

This postwar trend in anthropology and biology, which some scholars have labeled "the retreat of scientific racism," sprang from several factors. After the war, many American scientists publicly embraced anti-racist stances and critiqued racial classification as they attempted to distance their work from that of Nazi scientists and physicians. (Nazi racial science, which led to the extermination of millions of Jews, Roma, homosexuals, and other peoples determined to be "biologically inferior," had both informed and been informed by early twentieth century American work on racial differences.) The growth of population genetics and molecular biology lent credence to scientific opposition to racial typologies, as genetic analyses indicated a wider range of variation within races than between them. The rising status of Boasian anthropology, the emerging civil rights movement in the American South, and mounting anticolonial struggles in Africa, Asia, and South America further contributed to an environment in which scientific attacks on the concept of race flourished.

Yet, as mentioned, not all scientific investigators agreed with the conclusions reached by Montagu, Livingstone, and others. In response to such claims, a group of American anthropologists, geneticists, and biologists argued that races were in fact valid taxonomic units at the sub-species level, and that the race concept should be retained. Historians have characterized this debate, which raged most intensely in the 1950s and 1960s, as a battle between "splitters" (those who argued for the race concept as a legitimate and useful classificatory schema for observed physical differences among humans) and "lumpers" (those who acknowledge human physical variability but de-emphasized its significance and disputed the biological validity of racial classifications). Though it remained unacknowledged in scientists' texts, arguments between lumpers and splitters transpired as African Americans began to challenge Jim Crow laws and other forms of racial discrimination through boycotts and civil disobedience measures.

Today, social scientists and historians generally agree that the lumpers prevailed, and that race ceased to function as a valid scientific explanation for human differences in the post-civil rights era. However, this view overlooks the fact that debates over the status of race have persisted in multiple fields throughout the second half of the twentieth century and into the present. Despite the oft-heard assertion that "science has proven that races don't exist," there has never been a universal consensus on the definition and significance of race within and across the diverse fields of anatomy, biology, anthropology, public health and biomedicine. Discussions about race, its biological significance, and social importance have differed widely in their content and development according to context.

The primary documents included in this chapter offer a broad sampling of recent discussions about the utility and meaning of the race concept in anthropology, sociology, psychology, and biomedicine. A presentation given at the 1989 meeting of the American Association for the Advancement of Science by Canadian psychologist J. Phillipe Rushton (notorious for his claims that blacks display lower intelligence, higher sexual drive, and a more emotional nature than whites and "Orientals") inspired the AAPA Statement on Race, which was published in the *American Journal of Physical Anthropology* after seven years of debate and revision. The American Anthropological Association's statement, drafted, revised, and re-written over several years, arose in response to similar concerns among cultural anthropologists, who wished to preempt "spurious claim[s]" about racial differences and address "public confusion about the meaning of 'race.'"

If there is indeed "public confusion" about race, it would be hardly surprising, given the wide range of high-profile statements about the nature of difference appearing in scientific publications. Even as many professional social scientists sought to distance themselves from race as a biological category, the National Institutes of Health continued to fund the Office of Research on Minority Health (ORMH), an agency established in 1990 to promote clinical investigation of racial and ethnic health disparities. In 1995, forensic experts working at the Oklahoma City Alfred P. Murrah Federal Building bombing site in Oklahoma City identified skeletal remains by the presumed race of the victim. In 2005, the Food and Drug Administration approved the sale of BiDil, a combination drug for the treatment of heart failure in self-identified Black patients. Even as affirmative action, multicultural education, and other social programs designed to redress historic social and economic inequities experienced by racial minorities have come under attack, the differentiation, segmentation, and segregation of bodies along lines of race appear to be acquiring ever novel forms of social and financial investment. And, as always, these investments carry the palimpsest of their historical precursors. In his coverage of the zebrafish study included below, *New York Times* reporter Nicholas Wade drew on a language of acclimatization, pigmentation, and natural difference which, aside from the language of the gene, would have been perfectly at home in a nineteenth-century literary magazine: "The new gene ... may shed light on the evolutionary pressures to which Europeans were subjected as their ancestors, who were presumably dark skinned, moved into the northern latitudes some 40,000 years ago. Humans acquired dark skins in Africa about 1.5 million years ago to shield their newly hairless bodies from the sun. Its ultra-violet rays destroy folic acid, a shortage of which leads to birth defects."

The state of race relations in twenty-first century America remains unstable, and continues to shape scientific and medical debates on race in ways not yet thoroughly understood. As you read the documents below, consider how broader social and political transformations might operate in debates about the race concept in biology, anthropology, and public health. To what do authors refer when they use the term *race*? How do they distinguish between "scientific" and "common" meanings of the term? What assumptions about race, science, and society do these suggestions share with previous scientific and medical research on racial differences? Where do they depart from prevailing ideas and practices regarding race in biology, anthropology, medicine, and public health? Several authors propose replacing *race* with other terminology, variables, and objects of inquiry. Would you support these proposals? Why or why not?

Acknowledgment

Abigail Bass prepared an earlier draft of this introduction.

Bibliography

Barkan, Elazar. *The Retreat of Scientific Racism*. Cambridge: Cambridge University Press, 1992.

Brown, L. "Race, Ethnicity and Health: Can Genetics Explain Disparities?" *Perspectives in Biology and Medicine* 45, no. 2 (2002): 159–174.

9.1 "Statement on Race" (1950)

UNESCO

STATEMENT ON THE NATURE OF RACE AND RACE DIFFERENCES

The reasons for convening a second meeting of experts to discuss the concept of race were chiefly these:

Race is a question of interest to many different kinds of people, not only to the public at large, but to sociologists, anthropologists and biologists, especially those dealing with problems of genetics. At the first discussion on the problem of race, it was chiefly sociologists who gave their opinions and framed the 'Statement on race'. That statement had a good effect, but it did not carry the authority of just those groups within whose special province fall the biological problems of race, namely the physical anthropologists and geneticists. Secondly, the first statement did not, in all its details, carry conviction of these groups and, because of this, it was not supported by many authorities in these two fields.

In general, the chief conclusions of the first statement were sustained, but with differences in emphasis and with some important deletions.

There was no delay or hesitation or lack of unanimity in reaching the primary conclusion that there were no scientific grounds whatever for the racialist position regarding purity of race and the hierarchy of inferior and superior races to which this leads.

We agreed that all races were mixed and that intraracial variability in most biological characters was as great as, if not greater than, interracial variability.

We agreed that races had reached their present states by the operation of evolutionary factors by which different proportions of similar hereditary elements (genes) had become characteristic of different, partially separated groups. The source of these elements seemed to all of us to be the variability which arises by random mutation, and the isolating factors bringing about racial differentiation by preventing intermingling of groups with different mutations, chiefly geographical for the main groups such as African, European and Asiatic.

Man, we recognized, is distinguished as much by his culture as by his biology, and it was clear to all of us that many of the factors leading to the formation of minor races of men have been cultural. Anything that tends to prevent free exchange of genes amongst groups is a potential race-making factor and these partial barriers may be religious, social and linguistic, as well as geographical.

We were careful to avoid dogmatic definitions of race, since, as a product of evolutionary factors, it is a dynamic rather than a static concept. We were equally

UNESCO, "Statement on Race" (Issued July 18, 1950). *What Is Race?* Paris, UNESCO (1952), 76–80.

careful to avoid saying that, because races were all variable and many of them graded into each other, therefore races did not exist. The physical anthropologists and the man in the street both know that races exist; the former, from the scientifically recognizable and measurable congeries of traits which he uses in classifying the varieties of man; the latter from the immediate evidence of his senses when he sees an African, a European, an Asiatic and an American Indian together.

We had no difficulty in agreeing that no evidence of differences in innate mental ability between different racial groups has been adduced, but that here too intraracial variability is at least as great as interracial variability. We agreed that psychological traits could not be used in classifying races, nor could they serve as parts of racial descriptions.

We were fortunate in having as members of our conference several scientists who had made special studies of the results of intermarriage between members of different races. This meant that our conclusion that race mixture in general did not lead to disadvantageous results was based on actual experience as well as upon study of the literature. Many of our members thought it quite likely that hydridization of different races could lead to biologically advantageous results, although there was insufficient evidence to support any conclusion.

Since race, as a word, has become coloured by its misuse in connexion with national, linguistic and religious differences, and by its deliberate abuse by racialists, we tried to find a new word to express the same meaning of a biologically differentiated group. On this we did not succeed, but agreed to reserve race as the word to be used for anthropological classification of groups showing definite combinations of physical (including physiological) traits in characteristic proportions.

We also tried hard, but again we failed, to reach some general statement about the inborn nature of man with respect to his behaviour toward his fellows. It is obvious that members of a group show co-operative or associative behaviour towards each other, while members of different groups may show aggressive behaviour towards each other and both of these attitudes may occur within the same individual. We recognized that the understanding of the psychological origin of race prejudice was an important problem which called for further study.

Nevertheless, having regard to the limitations of our present knowledge, all of us believed that the biological differences found amongst human racial groups can in no case justify the views of racial inequality which have been based on ignorance and prejudice, and that all of the differences which we know can well be disregarded for all ethical human purposes.

L. C. Dunn *(rapporteur)*,
June 1951

1. Scientists are generally agreed that all men living today belong to a single species, *Homo sapiens*, and are derived from a common stock, even though there is some dispute as to when and how different human groups diverged from this common stock.

The concept of race is unanimously regarded by anthropologists as a classificatory device providing a zoological frame within which the various groups of mankind may be arranged and by means of which studies of evolutionary processes can be facilitated. In its anthropological sense, the word "race" should be reserved for groups of mankind possessing well-developed and primarily heritable physical differences from other groups. Many populations can be so classified but, because of the complexity of human history, there are also many populations which cannot easily be fitted into a racial classification.

2. Some of the physical differences between human groups are due to differences in hereditary constitution and some to differences in the environments in which they have been brought up. In most cases, both influences have been at work. The science of genetics suggests that the hereditary differences among populations of a single species are the results of the action of two sets of processes. On the one hand, the genetic composition of isolated populations is constantly but gradually being altered by natural selection and by occasional changes (mutations) in the material particles (genes) which control heredity. Populations are also affected by fortuitous changes in gene frequency and by marriage customs. On the other hand, crossing is constantly breaking down the differentiations so set up. The new mixed populations, in so far as they, in turn, become isolated, are subject to the same processes, and these may lead to further changes. Existing races are merely the result, considered at a particular moment in time, of the total effect of such processes on the human species. The hereditary characters to be used in the classification of human groups, the limits of their variation within these groups, and thus the extent of the classificatory sub-divisions adopted may legitimately differ according to the scientific purpose in view.

3. National, religious, geographical, linguistic and cultural groups do not necessarily co-incide with racial groups; and the cultural traits of such groups have no demonstrated connexion with racial traits. Americans are not a race, nor are Frenchmen, nor Germans; nor *ipso facto* is any other national group. Moslems and Jews are no more races than are Roman Catholics and Protestants; nor are people who live in Iceland or Britain or India, or who speak English or any other language, or who are culturally Turkish or Chinese and the like, thereby describable as races. The use of the term "race" in speaking of such groups may be a serious error, but it is one which is habitually committed.

4. Human races can be, and have been, classified in different ways by different anthropologists. Most of them agree in classifying the greater part of existing mankind into at least three large units, which may be called major groups (in French *grand-races*, in German *Hauptrassen*). Such a classification does not depend on any single physical character, nor does for example, skin colour by itself necessarily distinguish one major group from another. Furthermore, so far as it has been possible to analyse them, the differences in physical structure which distinguish one major group from another give no support to popular notions of any general "superiority" or "inferiority" which are sometimes implied in referring to these groups.

Broadly speaking, individuals belonging to different major groups of mankind are distinguishable by virtue of their physical characters, but individual members, or small groups belonging to different races within the same major group are usually not so distinguishable. Even the major groups grade into each other, and the physical traits by which they and the races within them are characterized overlap considerably. With respect to most, if not all, measurable characters, the differences among individuals belonging to the same race are greater than the differences that occur between the observed averages for two or more races within the same major group.

5. Most anthropologists do not include mental characteristics in their classification of human races. Studies within a single race have shown that both innate capacity and environmental opportunity determine the results of tests of intelligence and temperament, though their relative importance is disputed.

When intelligence tests, even non-verbal, are made on a group of non-literate people, their scores are usually lower than those of more civilized people. It has been recorded that different groups of the same race occupying similarly high levels of civilization may yield considerable differences in intelligence tests. When, however, the two groups have been brought up from childhood in similar environments, the differences are usually very slight. Moreover, there is good evidence that, given similar opportunities, the average performance (that is to say, the performance of the individual who is representative because he is surpassed by as many as he surpasses), and the variation round it, do not differ appreciably from one race to another.

Even those psychologists who claim to have found the greatest differences in intelligence between groups of different racial origin and have contended that they are hereditary, always report that some members of the group of inferior performance surpass not merely the lowest ranking member of the superior group but also the average of its members. In any case, it has never been possible to separate members of two groups on the basis of mental capacity, as they can often be separated on a basis of religion, skin colour, hair form or language. It is possible, though not proved, that some types of innate capacity for intellectual and emotional responses are commoner in one human group than in another, but it is certain that, within a single group, innate capacities vary as much as, if not more than, they do between different groups.

The study of the heredity of psychological characteristics is beset with difficulties. We know that certain mental diseases and defects are transmitted from one generation to the next, but we are less familiar with the part played by heredity in the mental life of normal individuals. The normal individual, irrespective of race, is essentially educable. It follows that his intellectual and moral life is largely conditioned by his training and by his physical and social environment.

It often happens that a national group may appear to be characterized by particular psychological attributes. The superficial view would be that this is due to race. Scientifically, however, we realize that any common psychological attribute is more likely to be due to a common historical and social background, and that such attributes may obscure the fact that, within different populations consisting of many human types, one will find approximately the same range of temperament and intelligence.

6. The scientific material available to us at present does not justify the conclusion that inherited genetic differences are a major factor in producing the differences between

the cultures and cultural achievements of different peoples or groups. It does indicate, on the contrary, that a major factor in explaining such differences is the cultural experience which each group has undergone.

7. There is no evidence for the existence of so-called "pure" races. Skeletal remains provide the basis of our limited knowledge about earlier races. In regard to race mixture, the evidence points to the fact that human hybridization has been going on for an indefinite but considerable time. Indeed, one of the processes of race formation and race extinction or absorption is by means of hybridization between races. As there is no reliable evidence that disadvantageous effects are produced thereby, no biological justification exists for prohibiting intermarriage between persons of different races.

8. We now have to consider the bearing of these statements on the problem of human equality. We wish to emphasize that equality of opportunity and equality in law in no way depend, as ethical principles, upon the assertion that human beings are in fact equal in endowment.

9. We have thought it worth while to set out in a formal manner what is at present scientifically established concerning individual and group differences:

a. In matters of race, the only characteristics which anthropologists have so far been able to use effectively as a basis for classification are physical (anatomical and physiological).

b. Available scientific knowledge provides no basis for believing that the groups of mankind differ in their innate capacity for intellectual and emotional development.

c. Some biological differences between human beings within a single race may be as great as, or greater than, the same biological differences between races.

d. Vast social changes have occurred that have not been connected in any way with changes in racial type. Historical and sociological studies thus support the view that genetic differences are of little significance in determining the social and cultural differences between different groups of men.

e. There is no evidence that race mixture produces disadvantageous results from a biological point of view. The social results of race mixture, whether for good or ill, can generally be traced to social factors.

Notes

Text drafted at Unesco House, Paris, on 8 June 1951, by:

Professor R. A. M. Borgman, Royal Tropical Institute, Amsterdam;

Professor Gunnar Dahlberg, Director, State Institute for Human Genetics and Race Biology, University of Uppsala;

Professor L. C. Dunn, Department of Zoology, Columbia University, New York;

Professor J. B. S. Haldane, Head, Department of Biometry, University College, London;

Professor M. F. Ashley Montagu, Chairman, Department of Anthropology, Rutgers University, New Brunswick, N.J.;

Dr. A. E. Mourant, Director, Blood Group Reference Laboratory, Lister Institute, London;

Professor Hans Nachtscheim, Director, Institut für Genetik, Freie Universität, Berlin;

Dr. Eugène Schreider, Directeur adjoint du Laboratoire d'Anthropologie Physique de l'Ecole des Hautes Etudes, Paris;

Professor Harry L. Shapiro, Chairman, Department of Anthropology, American Museum of Natural History, New York;

Dr. J. C. Trevor, Faculty of Archaeology and Anthropology, University of Cambridge;

Dr. Henri V. Vallois, Professeur au Musée d'Histoire Naturelle, Directeur du Musée de l'Homme, Paris;

Professor S. Zuckerman, Head, Department of Anatomy, Medical School, University of Birmingham;

Professor Th. Dobzhansky, Department of Zoology, Columbia University, New York;

Dr. Julian Huxley contributed to the final wording.

9.2   "Statement on Biological Aspects of Race" (1996)

American Association of Physical Anthropologists

Preamble

As scientists who study human evolution and variation, we believe that we have an obligation to share with other scientists and the general public our current understanding of the structure of human variation from a biological perspective.

Popular conceptualizations of race are derived from 19th and early 20th century scientific formulations. These old racial categories were based on externally visible traits, primarily skin color, features of the face, and the shape and size of the head and body, and the underlying skeleton. They were often imbued with nonbiological attributes, based on social constructions of race. These categories of race are rooted in the scientific traditions of the 19th century, and in even earlier philosophical traditions which presumed that immutable visible traits can predict the measure of all other traits in an individual or a population. Such notions have often been used to support racist doctrines. Yet old racial concepts persist as social conventions that foster institutional discrimination. The expression of prejudice may or may not undermine material well-being, but it does involve the mistreatment of people and thus it often is psychologically distressing and socially damaging. Scientists should try to keep the results of their research from being used in a biased way that would serve discriminatory ends.

Position

We offer the following points as revisions of the 1964 UNESCO statement on race:

1. All humans living today belong to a single species, *Homo sapiens*, and share a common descent.

Although there are differences of opinion regarding how and where different human groups diverged or fused to form new ones from a common ancestral group, all living populations in each of the earth's geographic areas have evolved from that ancestral group over the same amount of time.

Much of the biological variation among populations involves modest degrees of variation in the frequency of shared traits. Human populations have at times been isolated, but have never genetically diverged enough to produce any biological barriers to mating between members of different populations.

2. Biological differences between human beings reflect both hereditary factors and the influence of natural and social environments. In most cases, these differences are due to the interaction of both. The degree to which environment or heredity affects any particular trait varies greatly.

3. There is great genetic diversity within all human populations. Pure races, in the sense of genetically homogenous populations, do not exist in the human species today, nor is there any evidence that they have ever existed in the past.

4. There are obvious physical differences between populations living in different geographic areas of the world. Some of these differences are strongly inherited and others, such as body size and shape, are strongly influenced by nutrition, way of life, and other aspects of the environment. Genetic differences between populations commonly consist of differences in the frequencies of all inherited traits, including those that are environmentally malleable.

5. For centuries, scholars have sought to comprehend patterns in nature by classifying living things. The only living species in the human family, *Homo sapiens*, has become a highly diversified global array of populations. The geographic pattern of genetic variation within this array is complex, and presents no major discontinuity. Humanity cannot be classified into discrete geographic categories with absolute boundaries. Furthermore, the complexities of human history make it difficult to determine the position of certain groups in classifications. Multiplying subcategories cannot correct the inadequacies of these classifications.

Generally, the traits used to characterize a population are either independently inherited or show only varying degrees of association with one another within each population. Therefore, the combination of these traits in an individual very commonly deviates from the average combination in the population. This fact renders untenable the idea of discrete races made up chiefly of typical representatives.

6. In humankind as well as in other animals, the genetic composition of each population is subject over time to the modifying influence of diverse factors. These include natural selection, promoting adaptation of the population to the environment; mutations, involving modifications in genetic material; admixture, leading to genetic exchange between local populations, and randomly changing frequencies of genetic characteristics from one generation to another. The human features which have universal biological value for the survival of the species are not known to occur more frequently in one population than in any other. Therefore it is meaningless from the biological point of view to attribute a general inferiority or superiority to this or to that race.

7. The human species has a past rich in migration, in territorial expansions, and in contractions. As a consequence, we are adapted to many of the earth's environments in general, but to none in particular. For many millennia, human progress in any field has been based on culture and not on genetic improvement.

Mating between members of different human groups tends to diminish differences between groups, and has played a very important role in human history. Wherever different human populations have come in contact, such matings have taken place. Obstacles to such interaction have been social and cultural, not biological. The global process of urbanization, coupled with intercontinental migrations, has the potential to reduce the differences among all human populations.

8. Partly as a result of gene flow, the hereditary characteristics of human populations are in a state of perpetual flux. Distinctive local populations are continually coming into and passing out of existence. Such populations do not correspond to breeds of domestic animals, which have been produced by artificial selection over many generations for specific human purposes.

9. The biological consequences of mating depend only on the individual genetic makeup of the couple, and not on their racial classifications. Therefore, no biological justification exists for restricting intermarriage between persons of different racial classifications.

10. There is no necessary concordance between biological characteristics and culturally defined groups. On every continent, there are diverse populations that differ in language, economy, and culture. There is no national, religious, linguistic or cultural group or economic class that constitutes a race. However, human beings who speak the same language and share the same culture frequently select each other as mates, with the result that there is often some degree of correspondence between the distribution of physical traits on the one hand and that of linguistic and cultural traits on the other. But there is no causal linkage between these physical and behavioral traits, and therefore it is not justifiable to attribute cultural characteristics to genetic inheritance.

11. Physical, cultural and social environments influence the behavioral differences among individuals in society. Although heredity influences the behavioral variability of individuals within a given population, it does not affect the ability of any such population to function in a given social setting. The genetic capacity for intellectual development is one of the biological traits of our species essential for its survival. This genetic capacity is known to differ among individuals. The peoples of the world today appear to possess equal biological potential for assimilating any human culture. Racist political doctrines find no foundation in scientific knowledge concerning modern or past human populations.

9.3 "Statement on Race" (1999)

American Anthropological Association

The following statement was adopted by the AAA Executive Board on May 17, 1998, acting on a draft prepared by a committee of representative American anthropologists. It does not reflect a consensus of all members of the AAA, as individuals vary in their approaches to the study of "race." We believe that it represents generally the contemporary thinking and scholarly positions of a majority of anthropologists.

In the US both scholars and the general public have been conditioned to viewing human races as natural and separate divisions within the human species based on visible physical differences. With the vast expansion of scientific knowledge in this century, however, it has become clear that human populations are not unambiguous, clearly demarcated, biologically distinct groups. Evidence from the analysis of genetics (eg, DNA) indicates that there is greater variation within racial groups than between them. This means that most physical variation, about 94%, lies within so-called racial groups. Conventional geographic "racial" groupings differ from one another only in about 6% of their genes. In neighboring populations there is much overlapping of genes and their phenotypic (physical) expressions. Throughout history whenever different groups have come into contact, they have interbred. The continued sharing of genetic materials has maintained all of humankind as a single species.

Physical variations in any given trait tend to occur gradually rather than abruptly over geographic areas. And because physical traits are inherited independently of one another, knowing the range of one trait does not predict the presence of others. For example, skin color varies largely from light in the temperate areas in the north to dark in the tropical areas in the south; its intensity is not related to nose shape or hair texture. Dark skin may be associated with frizzy or kinky hair or curly or wavy or straight hair, all of which are found among different indigenous peoples in tropical regions. These facts render any attempt to establish lines of division among biological populations both arbitrary and subjective.

Historical research has shown that the idea of race has always carried more meanings than mere physical differences; indeed, physical variations in the human species have no meaning except the social ones that humans put on them. Today scholars in many fields argue that race as it is understood in the USA was a social mechanism invented during the 18th century to refer to those populations brought together in colonial America: the English and other European settlers, the conquered Indian peoples, and those peoples of Africa brought in to provide slave labor.

American Anthropological Association, "Statement on Race," *American Anthropologist* 100, no. 3 (1999): 712–713.

From its inception, this modern concept of race was modeled after an ancient theorem of the Great Chain of Being which posited natural categories on a hierarchy established by God or nature. Thus race was a mode of classification linked specifically to peoples in the colonial situation. It subsumed a growing ideology of inequality devised to rationalize European attitudes and treatment of the conquered and enslaved peoples. Proponents of slavery in particular during the 19th century used race to justify the retention of slavery. The ideology magnified the differences among Europeans, Africans and Indians, established a rigid hierarchy of socially exclusive categories underscored and bolstered unequal rank and status differences, and provided the rationalization that the inequality was natural or God-given. The different physical traits of African-Americans and Indians became markers or symbols of their status differences.

As they were constructing US society, leaders among European-Americans fabricated the cultural/behavioral characteristics associated with each race, linking superior traits with Europeans and negative and inferior ones to blacks and Indians. Numerous arbitrary and fictitious beliefs about the different peoples were institutionalized and deeply embedded in American thought.

Early in the 19th century the growing fields of science began to reflect the public consciousness about human differences. Differences among the racial categories were projected to their greatest extreme when the argument was posed that Africans, Indians and Europeans were separate species, with Africans the least human and closer taxonomically to apes.

Ultimately race as an ideology about human differences was subsequently spread to other areas of the world. It became a strategy for dividing, ranking and controlling colonized people used by colonial powers everywhere. But it was not limited to the colonial situation. In the latter part of the 19th century it was employed by Europeans to rank one another and to justify social, economic and political inequalities among their peoples. During World War II, the Nazis under Adolf Hitler enjoined the expanded ideology of race and racial differences and took them to a logical end: the extermination of 11 million people of "inferior races" (eg, Jews, Gypsies, Africans, homosexuals and so forth) and other unspeakable brutalities of the Holocaust.

Race thus evolved as a world view, a body of prejudgments that distorts our ideas about human differences and group behavior. Racial beliefs constitute myths about the diversity in the human species and about the abilities and behavior of people homogenized into racial categories. The myths fused behavior and physical features together in the public mind, impeding our comprehension of both biological variations and cultural behavior, implying that both are genetically determined. Racial myths bear no relationship to the reality of human capabilities or behavior. Scientists today find that reliance on such folk beliefs about human differences in research has led to countless errors.

At the end of the 20th century, we now understand that human cultural behavior is learned, conditioned into infants beginning at birth, and always subject to modification. No human is born with a built-in culture or language. Our temperaments, dispositions and personalities, regardless of genetic propensities, are developed within sets of meanings and values that we call "culture." Studies of infant and early childhood learning and behavior attest to the reality of our cultures in forming who we are.

It is a basic tenet of anthropological knowledge that all normal human beings have the capacity to learn any cultural behavior. The American experience with immigrants from hundreds of different language and cultural backgrounds who have acquired some version of American culture traits and behavior is the clearest evidence of this fact. Moreover, people of all physical variations have learned different cultural behaviors and continue to do so as modern transportation moves millions of immigrants around the world.

How people have been accepted and treated within the context of a given society or culture has a direct impact on how they perform in that society. The racial world view was invented to assign some groups to perpetual low status, while others were permitted access to privilege, power and wealth. The tragedy in the US has been that the policies and practices stemming from this world view succeeded all too well in constructing unequal populations among Europeans, Native Americans and peoples of African descent. Given what we know about the capacity of normal humans to achieve and function within any culture, we conclude that present-day inequalities between so-called racial groups are not consequences of their biological inheritance but products of historical and contemporary social, economic, educational and political circumstances.

9.4 "Categorization of Humans in Biomedical Research: Genes, Race, and Disease" (2002)

Neil Risch, Esteban Burchard, Elad Ziv, and Hua Tang

A major discussion has arisen recently regarding optimal strategies for categorizing humans, especially in the United States, for the purpose of biomedical research, both etiologic and pharmaceutical. Clearly it is important to know whether particular individuals within the population are more susceptible to particular diseases or most likely to benefit from certain therapeutic interventions. The focus of the dialogue has been the relative merit of the concept of "race" or "ethnicity," especially from the genetic perspective. For example, a recent editorial in the *New England Journal of Medicine* [1] claimed that "race is biologically meaningless" and warned that "instruction in medical genetics should emphasize the fallacy of race as a scientific concept and the dangers inherent in practicing race-based medicine." In support of this perspective, a recent article in *Nature Genetics* [2] purported to find that "commonly used ethnic labels are both insufficient and inaccurate representations of inferred genetic clusters." Furthermore, a supporting editorial in the same issue [3] concluded that "population clusters identified by genotype analysis seem to be more informative than those identified by skin color or self-declaration of 'race.'" These conclusions seem consistent with the claim that "there is no biological basis for 'race'" [3] and that "the myth of major genetic differences across 'races' is nonetheless worth dismissing with genetic evidence" [4]. Of course, the use of the term "major" leaves the door open for possible differences but *a priori* limits any potential significance of such differences.

In our view, much of this discussion does not derive from an objective scientific perspective. This is understandable, given both historic and current inequities based on perceived racial or ethnic identities, both in the US and around the world, and the resulting sensitivities in such debates. Nonetheless, we demonstrate here that from both an objective and scientific (genetic and epidemiologic) perspective there is great validity in racial/ethnic self-categorizations, both from the research and public policy points of view.

Definition of Risk Factors: Human Categorization

The human population is not homogeneous in terms of risk of disease. Indeed, it is probably the case that every human being has a uniquely defined risk, based on his/her inherited (genetic) constitution plus non-genetic or environmental characteristics acquired during life. It is the goal of etiological epidemiological research to characterize such risks, both on an individual as well as population level, for the effective planning

Neil Risch, Esteban Burchard, Elad Ziv, and Hua Tang, "Categorization of Humans in Biomedical Research: Genes, Race, and Disease," *Genome Biology* 3, no. 7 (2002): 1–12.

of prevention and/or treatment strategies. This public health perspective applies not only to disease, but to variation in normal traits (for example, for quantitative variables that are risk factors for disease such as blood pressure) as well as treatment response and adverse effects of pharmacologic agents.

The term "risk factor" is widely used in epidemiology to define a characteristic associated either directly or indirectly with risk of disease. Some risk factors are fixed at birth (for example, sex or ethnicity) while others are acquired during life (for example, exposure to tobacco smoke or other environmental toxins). It is often assumed that risk factors fixed at birth are non-modifiable while those acquired after birth are modifiable and thus amenable to intervention strategies. But a multifactorial model of risk requires the interplay of multiple inherited and non-inherited factors in producing a particular risk profile. Identification of inherited factors can both aid in the discovery of their environmental counterparts as well as provide a rational strategy for identifying, *a priori*, the most vulnerable members of our population, on whom prevention strategies can be focused. These concepts apply not only to disease prevention but to disease treatment as well, given that providing timely and efficacious treatment to individuals benefits both patients and health-care providers.

The ultimate goal of characterizing each individual's unique risk would require knowledge of every causal factor and the quantitative relationship of all possible combinations of such factors. In most cases, causal variables are not known, however, so epidemiologists resort to other means of categorizing people by the use of surrogate variables. Examples of such variables include gender, occupation, geographic location, socioeconomic status and dietary intake of a given food. It is understood that these variables do not themselves reflect a direct causal relationship with disease but rather are correlated with such a causal variable or variables.

Each of these classification systems masks within it inherent risk heterogeneity that could be further resolved if the specific agent(s) were identified. For example, geographic gradients in disease rates are well known, but it is not the geographic location, *per se*, that is causally related but rather some underlying correlated causal factor(s) such as temperature, humidity, rainfall, sunlight or presence of ground toxins. Even geographic categorizations, for example those based on latitude, mask heterogeneity within strata. Vancouver has a similar latitude to Winnipeg, but individuals born and raised in these two locations have very different climatologic experiences.

Although risk factor associations do not usually imply direct causal links, they do provide a starting point for further investigation. For example, there are sex differences in the rate of a variety of disorders. Sometimes these differences are related to endogenous (and hence non- or poorly-modifiable) differences between men and women (for example, the differential rates of breast cancer in men and women), while other examples are due to behavioral (presumably modifiable) differences (for example, differential lung cancer rates in men and women due to different smoking experiences). When direct causal factors are identified, risk estimates on both an individual and population basis can be made much more precise. Before such identification, however, the use of cruder surrogate factors can still provide valuable input for prevention and treatment decisions, even while acknowledging the latent heterogeneity within strata defined by such variables.

Rationale for the Genetic Categorization of Humans

The discussion above provides an objective perspective from which to examine the question of genetic categorization of humans for biomedical research purposes. As for non-genetic risk factors, the ultimate goal of genetic research is to identify those specific genes and gene variants that influence the risk of disease, a quantitative outcome of interest, or response to a particular drug. Once all such genes are identified, every individual's unique risk can be assessed according to some quantitative model (if the number of genetic factors involved is large, however, problems will arise from the large number of possible combinations). In this case, categorization can occur at the level of the individual, irrespective of his/her racial, ethnic or geographic origins, moving us "closer to the ultimate goal of individualized therapy" [3]. But unless financial considerations in genetic testing become moot, there may still be practical issues concerning which genes to test in which individuals, and whether all genes should be tested in everyone.

To date, few genes underlying susceptibility to common diseases or influencing drug response have been identified. A question then arises, as it does when considering nongenetic factors, as to whether humans can or should be categorized genetically according to a surrogate scheme in the absence of known, specific gene effects. Wilson et al. [2] argue forcefully that genetic structure exists in the human population, that it can easily be identified even with a relatively modest number of marker loci, and that such structure is highly predictive of drug response: "We conclude that it is not only feasible but a clinical priority to assess genetic structure as a routine part of drug evaluation." [2]. This conclusion was based on their identification of four genetic "clusters" within a diverse sample of humans, and significant differences in the frequencies across these clusters for functional allelic variants in drug metabolizing enzymes.

Human Evolution

Probably the best way to examine the issue of genetic sub-grouping is through the lens of human evolution. If the human population mated at random, there would be no issue of genetic subgrouping because the chance of any individual carrying a specific gene variant would be evenly distributed around the world. For a variety of reasons, however, including geography, sociology and culture, humans have not and do not currently mate randomly, either on a global level or within countries such as the US. A clearer picture of human evolution has emerged from numerous studies over the past decade using a variety of genetic markers and involving indigenous populations from around the world. In summary, populations outside Africa derive from one or more migration events out of Africa within the last 100,000 years [5–11]. The greatest

328 9 The End of Race?

genetic variation occurs within Africans, with variation outside Africa representing either a subset of African diversity or newly arisen variants. Genetic differentiation between individuals depends on the degree and duration of separation of their ancestors. Geographic isolation and in-breeding (endogamy) due to social and/or cultural forces over extended time periods create and enhance genetic differentiation, while migration and inter-mating reduce it.

With this as background, it is not surprising that numerous human population genetic studies have come to the identical conclusion—that genetic differentiation is greatest when defined on a continental basis. The results are the same irrespective of the type of genetic markers employed, be they classical systems [5], restriction fragment length polymorphisms (RFLPs) [6], microsatellites [7–11], or single nucleotide polymorphisms (SNPs) [12]. For example, studying 14 indigenous populations from 5 continents with 30 microsatellite loci, Bowcock et al. [7] observed that the 14 populations clustered into the five continental groups, as depicted in Figure 9.4.1. The African branch included three sub-Saharan populations, CAR pygmies, Zaire pygmies, and the Lisongo; the Caucasian branch included Northern Europeans and Northern Italians; the Pacific Islander branch included Melanesians, New Guineans and Australians; the East Asian branch included Chinese, Japanese and Cambodians; and the Native American branch included Mayans from Mexico and the Surui and Karitiana from the Amazon basin. The identical diagram has since been derived by others, using a similar or greater number of microsatellite markers and individuals [8, 9]. More recently, a survey of 3,899 SNPs in 313 genes based on US populations (Caucasians, African-Americans, Asians and Hispanics) once again provided distinct and non-overlapping clustering of the Caucasian, African-American and Asian samples [12]: "The results confirmed the integrity of the self-described ancestry of these individuals". Hispanics, who represent a recently admixed group between Native American, Caucasian and African, did not form a distinct subgroup, but clustered variously with the other groups. A previous cluster analysis based on a much smaller number of SNPs led to a similar conclusion: "A tree relating 144 individuals from 12 human groups of Africa, Asia, Europe and Oceania, inferred from an average of 75 DNA polymorphisms/individual, is remarkable in that most individuals cluster with other members of their regional group" [13]. Effectively, these population genetic studies have recapitulated the classical definition of races based on continental ancestry—namely African, Caucasian (Europe and Middle East), Asian, Pacific Islander (for example, Australian, New Guinean and Melanesian), and Native American.

The terms race, ethnicity and ancestry are often used interchangeably, but some have also drawn distinctions. For the purpose of this article, we define racial groups on the basis of the primary continent of origin, as discussed above (with some modifications described below). Ethnicity is a self-defined construct that may be based on geographic, social, cultural and religious grounds. It has potential meaning from the genetic perspective, provided it defines an endogamous group that can be differentiated from other such groups. Ancestry refers to the race/ethnicity of an individual's ancestors, whatever the individual's current affiliation. From the genetic perspective,

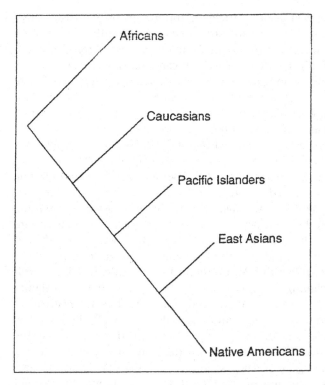

Figure 9.4.1
The evolutionary tree of human races. Population genetic studies of world populations support the categorization into five major groups, as shown. See text for further details.

the important concept is mating patterns, and the degree to which racially or ethnically defined groups remain endogamous.

The continental definitions of race and ancestry need some modification, because it is clear that migrations have blurred the strict continental boundaries. For example, individuals currently living in South Africa, although currently Africans, have very different ancestry, race and ethnicity depending on the ancestry of their forbears (for example from Europe or Asia) and the degree to which they have remained endogamous. For our purposes here, on the basis of numerons population genetic surveys, we categorize Africans as those with primary ancestry in sub-Saharan Africa; this group includes African Americans and Afro-Caribbeans. Caucasians include those with ancestry in Europe and West Asia, including the Indian subcontinent and Middle East; North Africans typically also are included in this group as their ancestry derives largely from the Middle East rather than sub-Saharan Africa. "Asians" are those from eastern Asia including China, Indochina, Japan, the Philippines and Siberia. By contrast, Pacific Islanders are those with indigenous ancestry from Australia, Papua New Guinea,

Melanesia and Micronesia, as well as other Pacific Island groups further east. Native Americans are those that have indigenous ancestry in North and South America. Populations that exist at the boundaries of these continental divisions are sometimes the most difficult to categorize simply. For example, east African groups, such as Ethiopians and Somalis, have great genetic resemblance to Caucasians and are clearly intermediate between sub-Saharan Africans and Caucasians [5]. The existence of such intermediate groups should not, however, overshadow the fact that the greatest genetic structure that exists in the human population occurs at the racial level.

Most recently, Wilson et al. [2] studied 354 individuals from 8 populations deriving from Africa (Bantus, Afro-Caribbeans and Ethiopians), Europe/Mideast (Norwegians, Ashkenazi Jews and Armenians), Asia (Chinese) and Pacific Islands (Papua New Guineans). Their study was based on cluster analysis using 39 microsatellite loci. Consistent with previous studies, they obtained evidence of four clusters representing the major continental (racial) divisions described above as African, Caucasian, Asian, and Pacific Islander. The one population in their analysis that was seemingly not clearly classified on continental grounds was the Ethiopians, who clustered more into the Caucasian group. But it is known that African populations with close contact with Middle East populations, including Ethiopians and North Africans, have had significant admixture from Middle Eastern (Caucasian) groups, and are thus more closely related to Caucasians [14]. Furthermore, the analysis by Wilson et al. [2] did not detect subgroups within the four major racial clusters (for example, it did not separate the Norwegians, Ashkenazi Jews and Armenians among the Caucasian cluster), despite known genetic differences among them. The reason is clearly that these differences are not as great as those between races and are insufficient, with the amount of data provided, to distinguish these subgroups.

Are Racial Differences Merely Cosmetic?

Two arguments against racial categorization as defined above are firstly that race has no biological basis [1, 3], and secondly that there are racial differences but they are merely cosmetic, reflecting superficial characteristics such as skin color and facial features that involve a very small number of genetic loci that were selected historically; these superficial differences do not reflect any additional genetic distinctiveness [2]. A response to the first of these points depends on the definition of "biological." If biological is defined as genetic then, as detailed above, a decade or more of population genetics research has documented genetic, and therefore biological, differentiation among the races. This conclusion was most recently reinforced by the analysis of Wilson et al. [2]. If biological is defined by susceptibility to, and natural history of, a chronic disease, then again numerous studies over past decades have documented biological differences among the races. In this context, it is difficult to imagine that such differences are not meaningful. Indeed, it is difficult to conceive of a definition of "biological" that does not lead to racial differentiation, except perhaps one as extreme as speciation.

A forceful presentation of the second point—that racial differences are merely cosmetic—was given recently in an editorial in the *New England Journal of Medicine* [1]: "Such research mistakenly assumes an inherent biological difference between black-skinned and white-skinned people. It falls into error by attributing a complex phys-iological or clinical phenomenon to arbitrary aspects of external appearance. It is implausible that the few genes that account for such outward characteristics could be meaningfully linked to multigenic diseases such as diabetes mellitus or to the intri-cacies of the therapeutic effect of a drug." The logical flaw in this argument is the assumption that the blacks and whites in the referenced study differ only in skin pig-ment. Racial categorizations have never been based on skin pigment, but on indige-nous continent of origin. For example, none of the population genetic studies cited above, including the study of Wilson et al. [2], used skin pigment of the study subjects, or genetic loci related to skin pigment, as predictive variables. Yet the various racial groups were easily distinguishable on the basis of even a modest number of random genetic markers; furthermore, categorization is extremely resistant to variation accord-ing to the type of markers used (for example, RFLPs, microsatellites or SNPs).

Genetic differentiation among the races has also led to some variation in pig-mentation across races, but considerable variation within races remains, and there is substantial overlap for this feature. For example, it would be difficult to distinguish most Caucasians and Asians on the basis of skin pigment alone, yet they are easily dis-tinguished by genetic markers. The author of the above statement [1] is in error to as-sume that the only genetic differences between races, which may differ on average in pigmentation, are for the genes that determine pigmentation.

Common versus Rare Alleles

Despite the evidence for genetic differentiation among the five major races, as defined above, numerous studies have shown that local populations retain a great deal of ge-netic variation. Analysis of variance has led to estimates of 10% for the proportion of variance due to average differences between races, and 75% of the variance due to ge-netic variation within populations. Comparable estimates have been obtained for clas-sical blood markers [15–16], microsatellites [17], and SNPs [12]. Unfortunately, these analysis of variance estimates have also led to misunderstandings or misinterpreta-tions. Because of the large amount of variation observed within races versus between races, some commentators have denied genetic differentiation between the races; for example, "Genetic data ... show that any two individuals within a particular popula-tion are as different genetically as any two people selected from any two populations in the world." [18]. This assertion is both counter-intuitive and factually incorrect [12, 13]. If it were true, it would be impossible to create discrete clusters of humans (that end up corresponding to the major races), for example as was done by Wilson et al. [2], with even as few as 20 randomly chosen genetic markers. Two Caucasians are more similar to each other genetically than a Caucasian and an Asian.

In these variance assessments, it is also important to consider the frequency of the allelic variants examined. These studies are based primarily on common alleles, and may not reflect the degree of differentiation between races for rare alleles. This is an important concern because alleles underlying disease susceptibility, especially deleterious diseases, may be less frequent than randomly selected alleles. Similarly, it has also been shown that among different classes of SNPs, those that lead to non-conservative amino-acid substitutions (which most frequently are associated with clinical outcomes) occur least often, and when they do occur they tend to have lower allele frequencies than non-coding or synonymous coding changes [12, 19–20].

It is likely that genetic differentiation among races is enhanced for disease-predisposing alleles because such alleles tend to be in the lower frequency range. It is well known that rarer alleles are subject to greater fluctuation in frequency due to genetic drift than common alleles. Indeed, for most Mendelian diseases, even higher frequency alleles are found only in specific races (for example, cystic fibrosis and hemochromatosis in Caucasians). Furthermore, recent SNP surveys of the different races have shown that lower frequency variants are much more likely to be specific to a single race or shared by only two races than are common variants [12, 19–20]. In one study, only 21% of 3,899 SNPs were found to be pan-ethnic, and some race-specific SNPs were found to have a frequency greater than 25% [12].

Admixture and Genetic Categorization in the United States

Most population genetic studies that focus on human evolution and the relatedness of people, including the ones cited above, utilize indigenous groups from the various continents. These groups would not necessarily adequately depict the US population, for example, where admixture between races has occurred over many centuries. Nonetheless, during the same period of time, as well as currently, mating patterns are far from random. The tendency toward endogamy is reflected within the 2000 US Census [21], which allowed individuals to report themselves to be of a single race or of mixed race. Six racial categories were provided (White; Black or African American; American Indian and Alaska Native; Asian; Native Hawaiian and other Pacific Islander; Some other race). In response to this question, 97.6% of subjects reported themselves to be of one race, while 2.4% reported themselves to be of more than one race; 75% reported themselves as White, 12.3% as Black or African American, 3.6% as Asian, 1% as American Indian or Alaska Native, 0.1% as Hawaiian or Pacific Islander; and 5.5% of other race. Of the 5.5% who reported themselves as 'other race', most (97%) also reported themselves to be Hispanic. According to these numbers, if mating were at random with respect to these racial categories, 42% of individuals would result from 'mixed' matings and hence derive from more than one race, as opposed to the 2.4% reported. These figures highlight the strong deviation from random mating in the US.

What are the implications of these census results and the admixture that has occurred in the US population for genetic categorization in biomedical research studies in the US? Gene flow from non-Caucasians into the US Caucasian population has been modest. On the other hand, gene flow from Caucasians into African Americans has been greater; several studies have estimated the proportion of Caucasian admixture in African Americans to be approximately 17%, ranging regionally from about 12% to 23% [22]. Thus, despite the admixture, African Americans remain a largely African group, reflecting primarily their African origins from a genetic perspective. Asians and Pacific Islanders have been less influenced by admixture and again closely represent their indigenous origins. The same is true for Native Americans, although some degree of Caucasian admixture has occurred in this group as well [23].

The most complex group is made up of those who self-identify as Hispanic/Latino. The US Census did not consider this group as a separate race, although 42% of respondents who considered themselves Hispanic checked the category "other race" for the racial question, while 48% checked "White." Hispanics are typically a mix of Native American, Caucasian and African/African American, with the relative proportions varying regionally. Southwest Hispanics, who are primarily Mexican-American, appear to be largely Caucasian and Native American; recent admixture estimates are 39% Native American, 58% Caucasian and 3% African [24]. By contrast, East Coast Hispanics are largely Caribbean in origin, and have a greater proportion African admixture [25]. Thus, depending on geography, self-identified Hispanics could aggregate genetically with Caucasians, Native Americans, African Americans or form their own cluster.

The persistence of genetic differentiation among these US racial groups (as defined by the US Census) has also been verified recently in a study of nearly 4,000 SNPs in 313 genes [12]. These authors found distinct clusters for Caucasian Americans, African Americans and Asian Americans; the Hispanic Americans did not form a separate cluster but were either grouped with Caucasians or not easily classified. Although the US Census results suggest the large majority of individuals can be categorized into a single ancestral group, there remain individuals of mixed ancestry who will not be easily categorized by any simple system of finite, discrete categories. On the other hand, such individuals can be particularly informative in epidemiologic studies focused on differentiating genetic versus environmental sources for racial/ethnic difference, as we describe further below.

Genetic Clustering versus Self-Reported Ancestry

A major conclusion from the study of Wilson et al. [2], reiterated in accompanying editorials, is that "Clusters identified by genotyping...are far more robust than those identified using geographic and ethnic labels" [26]. But closer examination of the study and other data actually leads to the opposite conclusion: namely, that self-defined race, ethnicity or ancestry are actually more genetically informative than clusters based on analysis of random genetic markers.

In their analysis, Wilson et al. [2] found greater variation in allele frequencies for drug metabolizing enzymes based on four "genetically" defined clusters than in

Table 9.4.1

Allele Frequency Differentiation of Drug Metabolizing Enzymes on the Basis of "Genetic Clusters" versus "Racial Groups," from the Data of Wilson et al. [2]

| Locus | Genetic clusters | | | | | Racial groups | | | |
| | Allele frequencies | | | | | Allele frequencies | | | |
	C	A	B	D	σ^2	African	Cauca-sian	East Asian	σ^2
CYP1A2	0.60	0.66	0.69	0.59	0.0023	0.58	0.68	0.67	0.0030
GSTM1	0.31	0.47	0.53	0.45	0.0087	0.33	0.49	0.52	0.0104
CYP2C19	0.27	0.09	0.37	0.25	0.0134	0.22	0.08	0.354	0.0182
DIA4	0.19	0.22	0.11	0.53	0.0340	0.21	0.21	0.32	0.0040
NAT2	0.46	0.74	0.17	0.33	0.0582	0.58	0.74	0.15	0.0931
CYP2D6	0.70	0.53	0.39	0.42	0.0197	0.70	0.49	0.37	0.0279

Genetic clusters: C, primarily African; A, primarily Caucasian; B, primarily Pacific Islander; D, primarily East Asian. Racial groups: East Asian denotes Chinese plus Papua New Guinean.

three "ethnically" defined clusters. The ethnic clusters included Caucasians (Norwegians, Ashkenazi Jews and Armenians), Africans (Bantus, Afro-Caribbeans and Ethiopians), and Asians (Chinese and Papua New Guineans). The inclusion of Papua New Guineans as Asians would be considered highly controversial by most population geneticists, as all prior studies of this group show them to cluster with Pacific Islanders [7, 8], and as we discussed above, population genetic studies have shown Pacific Islanders to be distinct from Asians [6–9]. Furthermore, the racial categories of the US census would also not merge Chinese with New Guineans. Nonetheless, examination of the variance in allele frequencies for the six drug-metabolizing enzymes across the four "genetically defined" clusters versus the three "racial" groups does not reveal greater differentiation of the former (Table 9.4.1). In fact, for 5 of the 6 loci, the variance is greater among the three "racial" groups, although for most loci the variance is very similar for the two categorization schemes. This is not surprising, as the racial categories aligned nearly perfectly with genotype clusters. The only exception was for the Ethiopians, who (as we discussed above) are known to genetically resemble Caucasians, probably as a result of considerable Caucasian admixture [14]. On the other hand, neither of these ethnic categorizations (of New Guineans and Ethiopians) would have much impact on studies in the US, as these groups represent only a tiny fraction of the US population, even among Pacific Islanders and African Americans, respectively.

Indeed, in another respect, the results of Wilson et al. [2] demonstrate the superiority of ethnic labels over genetic clustering. Consider the group they labeled Caucasian, consisting of Norwegians, Ashkenazi Jews and Armenians. Their genetic cluster analysis lumped these three populations together into a single (Caucasian) cluster. Yet numerous genetic studies of these groups have shown them to differ in allele frequencies for a variety of loci. For example, the hemochromatosis gene mutation C282Y has a frequency of less than 1% in Armenians and Ashkenazi Jews but of 8% in Norwegians [27]. Thus, in this case, self-defined ethnicity provides greater discriminatory power than the single genotype cluster obtained by Wilson et al. [2].

This conclusion obtains not just for the ethnically homogeneous but for admixed subjects as well. A study by Williams et al. [28] considered the degree of Caucasian admixture in a Pima Indian population. These authors compared self-report (the number of Caucasian and Pima grandparents) with a genetic estimate of admixture based on 18 conventional blood markers. Using type 2 diabetes mellitus (high frequency in Pimas, low frequency in Caucasians) as the outcome, disease frequency correlated more strongly with self-reported admixture than with genetically estimated admixture.

As seen in the study of Wilson et al. [2], genetic cluster analysis is only powerful in separating out individuals whose ancestors diverged many millennia ago, leading to substantial genetic differences. It is much less capable of differentiating more recently separated groups, whose genetic differentiation is smaller. This conclusion may also be a result of the small number of genetic markers employed, however, and finer resolution might be possible with a much larger number of loci.

How Many Loci Are Needed for Clustering?

A natural question arises as to the number of loci required to categorize individuals into ancestrally defined clusters. The answer depends on the degree of genetic differentiation of the populations in question. Two groups with ancient separation and no migration will require far fewer markers than groups that have separated more recently or have been influenced by recent migrations and/or admixture.

A simple quantification of this question is possible, as described in Box 9.4.1. The number of biallelic loci necessary for given misclassification rates is given in Table 9.4.2; when δ_{av} is high, 20 or fewer loci are adequate for accurate classification, but even for low δ_{av} values 200 markers are more than adequate. How do these numbers relate to the ability to differentiate genetically various racial/ethnic groups? Recent large-scale surveys of 257 SNPs [29] and 744 short tandem repeat polymorphisms (STRPs, or microsatellites) [30] provide an answer to this question, based on the distribution of δ values observed in these surveys for various population comparisons. Both studies included Caucasian Americans (CA), Asian Americans (AS) and African Americans (AA); Dean et al. [29] also included Native Americans (NA), while Smith et al. [30] included Hispanic Americans (HA). Table 9.4.3 provides the median δ values for all markers, for the top 50th percentile of δ values, and for the top 20th percentile of δ values.

In conjunction with Table 9.4.2, we can estimate that about 120 unselected SNPs or 20 highly selected SNPs can distinguish group CA from NA, AA from AS and AA from NA. A few hundred random SNPs are required to separate CA from AA, CA from AS and AS from NA, or about 40 highly selected loci. STRP loci are more powerful

Box 9.4.1

How Many Loci Are Needed for Clustering?

Suppose there are two populations, P1 and P2, and n biallelic loci with alleles A_i and B_i, respectively. Assume that the A allele is always more frequent in P1 than in P2. Let the frequency of A_i in P1 and P2 be $p_i + \delta_i/2$ and $p_i - \delta_i/2$, respectively. Thus, the average frequency of A_i in P1 and P2 is p_i and the absolute difference in frequency between P1 and P2 is δ_i. For a total of n loci and $2n$ alleles, we then construct a statistic T representing the total proportion of A alleles for an individual across all n loci. Let T_1 represent the T value for a randomly selected individual from P1, and similarly T_2 for an individual from P2. The statistic T is used to determine population membership. The degree of overlap in the distributions of T_1 and T_2 is directly related to the misclassification rate. For randomly selected loci, it is reasonable to assume the p_i values are symmetrically distributed around $\frac{1}{2}$ and the average value p_{av} of p_i across n loci is $\frac{1}{2}$. In that case, we can employ a threshold of $\frac{1}{2}$ for T, so that for a randomly selected individual we place them in P1 if $T > \frac{1}{2}$ and in P2 if $T < \frac{1}{2}$. An upper bound on the genetic misclassification' rate mis can be obtained by the following formula (assuming equal representation of P1 and P2)

$$mis = \tfrac{1}{2}\Pr\!\left(T_1 \leq \tfrac{1}{2}\right) + \tfrac{1}{2}\Pr\!\left(T_2 \geq \tfrac{1}{2}\right) \tag{1}$$

As a first approximation, we invoke the central limit theorem and assume that T_1 and T_2 are normally distributed. The mean value of T_1 and T_2 are simply $p_{av} + \frac{1}{2}\delta_{av}$ and $p_{av} - \frac{1}{2}\delta_{av}$, respectively, where p_{av} and δ_{av} are the average values of p_i and δ_i across n loci. Similarly, a close upper bound on the variance of T_1 and T_2 is given by $\sigma^2 = (1 - \delta_{av}^2)/8n$. Formula (1) above then translates into

$$mis = \frac{1}{2}\Phi\!\left(\frac{\tfrac{1}{2} - p_{av} - \tfrac{1}{2}\delta_{av}}{\sigma}\right) + \frac{1}{2}\Phi\!\left(\frac{-\tfrac{1}{2} + p_{av} - \tfrac{1}{2}\delta_{av}}{\sigma}\right)$$

where ϕ is the cumulative normal distribution function. Assuming $p_{av} = \frac{1}{2}$ gives

$$mis = \Phi\!\left(-\frac{\delta_{av}}{2\sigma}\right) \tag{2}$$

To determine the number of loci required for a given misclassification rate mis, we invert formula (2) above, substituting $\sigma^2 = (1 - \delta_{av}^2)/8n$ to obtain

$$n = \frac{[\Phi^{-1}(mis)]^2(1 - \delta_{av}^2)}{2\delta_{av}^2} \tag{3}$$

The number of biallelic loci necessary for a misclassification rate of 10^{-3}, 10^{-4} or 10^{-5} is given in table 9.4.2 as a function of δ_{av}. When δ_{av} is high (>0.5), 20 or fewer loci are adequate for accurate classification; but even for δ_{av} values as low as 0.2, 200 markers are more than adequate.

Table 9.4.2
The Number of Markers Required for Clustering As a Function of the Misclassification Rate
(Calculated As Shown in Box 9.4.1)

δ value	Misclassification rate		
	10^{-3}	10^{-4}	10^{-5}
0.6	9	13	17
0.5	15	21	28
0.4	25	37	48
0.3	49	71	92
0.2	115	166	218
0.1	474	687	901

and have higher effective δ values because they have multiple alleles. Table 9.4.3 reveals that fewer than 100 random STRPs, or about 30 highly selected loci, can distinguish the major racial groups. As expected, differentiating Caucasians and Hispanic Americans, who are admixed but mostly of Caucasian ancestry, is more difficult and requires a few hundred random STRPs or about 50 highly selected loci. These results also indicate that many hundreds of markers or more would be required to accurately differentiate more closely related groups, for example populations within the same racial category.

Gene-Environment Correlation and Confounding—The Real Problem

From an epidemiologic perspective, the use of "genetic" clusters, as suggested by Wilson et al. [2], instead of self-reported ethnicity will not alleviate but rather will actually create and/or exacerbate problems associated with genetic inferences based on racial differences. The true complication is due to the fact that racial and ethnic groups differ from each other on a variety of social, cultural, behavioral and environmental variables as well as gene frequencies, leading to confounding between genetic and environmental risk factors in an ethnically heterogeneous study. For example, with respect to treatment response, "An individual's response to a drug depends on a host of factors, including overall health, lifestyle, support system, education and socioeconomic status—all of which are difficult to control for and likely to be affected, at least in the United States, by a person's 'race'" [3].

Specifically, let us consider the practical implications of the "race-neutral approach" [3] advocated by Wilson et al. [2]. As an example, we revisit a recent study of the efficacy of inhibitors of angiotensin-converting enzyme (ACE) in 1,200 white versus 800 black patients with congestive heart failure [31] that generated a great deal of controversy [1, 32]. In that study, the authors showed that black patients on the ACE inhibitor Enalapril showed no reduction in hospitalization compared with those on placebo, whereas white patients showed a strong, statistically significant difference

Table 9.4.3

Median δ Values for Different Racial/Ethnic Group Comparisons, from Data of Dean et al. [29] and Smith et al. [30]

Groups	All markers		Top 50% of markers		Top 20% of markers	
	SNP	STRP	SNP	STRP	SNP	STRP
CA-AA	0.17	0.28	0.24	0.36	0.37	0.44
CA-AS	0.17	0.36	0.27	0.47	0.37	0.59
CA-NA	0.21		0.29		0.40	
CA-HA		0.18		0.27		0.34
AA-AS	0.20	0.42	0.34	0.52	0.46	0.59
AA-NA	0.20		0.36		0.48	
AA-HA		0.27		0.36		0.43
AS-NA	0.16		0.29		0.42	
AS-HA		0.32		0.42		0.50

Groups are as follows: CA, Caucasians; AA, African Americans; AS, East Asians; NA, Native Americans; HA, Hispanic Americans.

between treatment versus placebo arms. Let us suppose that instead of using racial labels, the authors had performed genotype cluster analysis on their combined sample. They would have obtained two clusters—cluster A containing approximately 1,200 subjects, and cluster B, containing approximately 800 subjects. They would then demonstrate that cluster A treated subjects show a dramatic response to Enalapril compared to placebo subjects, while cluster B subjects show no such response. The direct inference from this analysis would be that the difference in responsiveness between individuals in cluster A and cluster B is genetic—that is, due to a frequency difference in one or more alleles between the two groups. But the problem should be obvious: cluster A is composed of the Caucasian subjects and cluster B the African Americans.

Although a genetic difference in treatment responsiveness between these two groups is inferred, the conclusion is completely confounded with the myriad other ways these two groups might differ from each other; hence the culprit may not be genetic at all.

A racial difference in the frequency of some phenotype of interest (disease, or drug response) or quantitative trait is but a first clue in the search for etiologic causal factors. As we have illustrated, without such racial/ethnic labels, these underlying factors cannot be adequately investigated. Although some investigators might quickly jump to a genetic explanation for an ethnic difference, this is rarely the case with epidemiologists, who have a broad view of the complex nature of most human traits [33]. Indeed, epidemiologists employ several different approaches to disentangling genetic from environmental causes of ethnic differences, including migrant studies and stratified analyses.

The rationale underlying migrant studies is to compare the frequency of a trait (such as disease) between members of the same ethnic group (who are assumed to be genetically homogeneous) who are residing in different environments. For example, breast cancer rates in Asian (Chinese and Japanese) women are vastly lower than the rates among US Caucasians. However, the breast cancer rates of Chinese and Japanese women living in Hawaii and the San Francisco Bay Area are comparable to those of US Caucasians [34]. These results suggest an environmental source of the racial difference. Asians are also known to have much lower rates of multiple sclerosis than European Caucasians [35]. But within a single country, namely Canada, this racial difference persists [36], increasing support for (but not proving) a genetic explanation.

The best approach to resolve confounding is through matched, adjusted or stratified analyses, but this depends on having the confounding variables (or their surrogates). Such analyses can be performed in a racially heterogeneous sample, but it is potentially more powerful when performed within a single group. The reason is that the correlation between confounding variables (such as genes and environment) may be stronger in a heterogeneous study population than in a more homogeneous one. The ability to disentangle the effects of confounding variables is greater when their sample correlation is low.

A simple example is provided in Figure 9.4.2. Here we assume two populations (for example, races), groups A and B. As shown in the figure, where both environmental and genetic factors differ between the populations, it is impossible to determine which is the functional cause of the racial difference if the genetic and environmental effects are completely correlated in frequency within the two groups. More importantly, if the relative frequency in the two groups of the environmental factor was not measured, analysis stratified on the genetic differences yields the correct interpretation that the genetic difference does not contribute to the racial difference if the genetic and environmental factors are not correlated. But if they are fully correlated, analysis stratified on the genetic factor alone would lead to the incorrect conclusion that it is the cause of the racial difference.

Epidemiologists often perform analyses of racial differences stratified on numerous environmental variables, such as socio-economic status, access to health care, education, and so on. The persistence of racial differences after accounting for these covariates raises the index of suspicion that genetic differences may be involved. For example, Karter et al. [37] recently demonstrated persistence of racial differences in diabetes complications in a health maintenance organization after controlling for numerous potential confounders including measures of socioeconomic status, education, and health-care access and utilization. Such evidence is indirect, however, as other unmeasured factors may still be responsible [38]. Ultimate proof depends on identifying a specific gene effect within each population, with an allele frequency difference between populations. One such example involves the lower risk of type 1 diabetes in US Hispanic versus Caucasian children. The HLA allele DR3 is predisposing to type 1 diabetes in both populations but has a lower frequency in Hispanics than Caucasians [39].

(a) Confounding

		Group A	Group B
Frequency of genotype:	G1	0.50	0.10
	G2	0.50	0.90
Frequency of environment:	E1	0.50	0.10
	E2	0.50	0.90
Prevalence of D		0.055	0.019

(b) Stratified analysis

	Group A		Group B	
	G and E independent	G and E correlated	G and E independent	G and E correlated
Stratify on G:				
P(D|G1)	0.055	0.10	0.019	0.10
P(D|G2)	0.055	0.01	0.019	0.01
Stratify on E:				
P(D|E1)	0.10	0.10	0.10	0.10
P(D|E2)	0.01	0.01	0.01	0.01

Figure 9.4.2
An example of confounding and a stratified analysis of environmental and genetic factors. Here we assume two populations (for example, races), groups A and B. G1 and G2 represent dichotomous genotype classes at a candidate gene locus (here one of the classes represents two genotypes for simplification, as would be the case for a dominant model), and E1 and E2 represent two strata of an environmental factor. (a) We assume that the probability (P) of trait D depends only on E, so that the risk of D given E1 is 10%, versus 1% given E2. In group A, the frequency of G1, G2, E1, and E2 are each 50%, whereas in group B the prevalence is 1.9%; hence, a racial difference exists in the prevalence of D. (b) We next consider the prevalence of D within strata defined by G and E. First, we assume G and E are frequency-independent within each group. In this case, the frequency difference in D between groups A and D persists within strata defined by G, but not within strata defined by E. Thus, the environmental factor E can completely explain the racial difference between groups A and B, but the genetic factor does not. Next consider the case where G and E are completely correlated in frequency within groups. In this case, analysis stratified on G or E eliminates the prevalence difference between groups A and B, and it is impossible to determine which is the functional cause of the racial difference. More important, consider the situation where factor E was not measured. Then for the first scenario (G and E independent within group), analysis stratified on G yields the correct interpretation that G does not contribute to the racial difference; for the second scenario (G and E fully correlated), however, analysis stratified on G would lead to the incorrect conclusion that G is the cause of the racial difference. P(D/G1) denotes the probability of disease given an individual has genotype G1, and similarly for G2, E1, and E2.

Another approach often taken by genetic epidemiologists is to consider the prevalence of disease or drug response (D) in individuals who are admixed between groups A and B—for example, in individuals who are 100%A, 75%A–25%B, 50%A–50%B, 25%A–75%B and 100%B (corresponding to 4, 3, 2, 1 and 0 grandparents that are group A, respectively). A continuous cline in the frequency of D with genome proportion that is group A is taken as suggestive evidence of genetic factors explaining the prevalence difference between groups A and B. An example of this type of analysis is the decreasing trend in type 2 diabetes in Pima Indians with degree of Caucasian admixture [23]. Analyses stratified on environmental factors can again strengthen the argument. But the same caveat applies here as described above. If an unmeasured environmental variable (such as socio-economic status) covaries in the same fashion as the proportion group A, the racial difference could be due to this unmeasured variable. At best, one could argue that the racial difference is not explained by any of the measured covariates.

Identical Treatment Is Not Equal Treatment

Both for genetic and non-genetic reasons, we believe that racial and ethnic groups should not be assumed to be equivalent, either in terms of disease risk or drug response. A "race-neutral" or "color-blind" approach to biomedical research is neither equitable nor advantageous, and would not lead to a reduction of disparities in disease risk or treatment efficacy between groups. Whether African Americans, Hispanics, Native Americans, Pacific Islanders or Asians respond equally to a particular drug is an empirical question that can only be addressed by studying these groups individually. Differences in treatment response or disease prevalence between racial/ethnic groups need to be studied carefully; naive inferences about genetic causation without evidence should be avoided. At the same time, gratuitous dismissal of a genetic interpretation without evidence for doing so is also unjustified.

We strongly support the search for candidate genes that contribute both to disease susceptibility and treatment response, both within and across racial/ethnic groups. Identification of such genes can help provide more precise individualized risk estimates. Environmental variables that influence risk and interact with genetic variables also require identification. Only if consideration of all these variables leaves no residual difference in risk between racial/ethnic groups is it justified to ignore race and ethnicity.

Is there any advantage to using random genetic markers and genetically defined clusters over self-reported ancestry in attempting to assess risk? In the case where a sample was collected without ancestry information, or for individuals on whom ancestral background is missing—for example, adoptees—such data could substitute for self-identified ancestry. On the other hand, we see considerable disadvantage in avoiding self-reported ancestry in favor of a "color-blind" approach of genetically defined clusters. Any study in the US that randomly samples subjects without regard to ancestry will obtain, on average, 75% Caucasians, 12% African Americans, 4%

Asians, 12% Hispanics (broadly defined) and few Pacific Islanders, although these frequencies would vary regionally. Thus, results from such studies would be largely derived from the Caucasian majority, with obtained parameter estimates that might not apply to the groups with minority representation. It might be possible in such studies to subsequently identify racial/ethnic differences, either based on self-reported ancestry of the study subjects or by genetic cluster analysis. However, the low frequency of the non-Caucasian groups in the total sample would lead to reduced power to detect and investigate any racial/ethnic differences that might be present. The best way to avoid this power issue is to specifically over-sample the lower frequency racial/ethnic groups to obtain larger sample sizes. Obviously, the only way to effectively and economically do so is based on self-identified ancestry, as genotype information from the population at large is not available and not likely to become so. Fortunately, the National Institutes of Health have instituted policies to encourage the inclusion of US ethnic minorities, and we strongly support continuation and expansion of this policy.

Finally, we believe that identifying genetic differences between races and ethnic groups, be they for random genetic markers, genes that lead to disease susceptibility or variation in drug response, is scientifically appropriate. What is not scientific is a value system attached to any such findings. Great abuse has occurred in the past with notions of "genetic superiority" of one particular group over another. The notion of superiority is not scientific, only political, and can only be used for political purposes.

As we enter this new millennium with an advancing arsenal of molecular genetic tools and strategies, the view of genes as immutable is too simplistic. Every race and even ethnic group within the races has its own collection of clinical priorities based on differing prevalence of diseases. It is a reflection of the diversity of our species—genetic, cultural and sociological. Taking advantage of this diversity in the scientific study of disease to gain understanding helps all of those afflicted. We need to value our diversity rather than fear it. Ignoring our differences, even if with the best of intentions, will ultimately lead to the disservice of those who are in the minority.

Acknowledgments

We are grateful to Andrew Karter and Catherine Schaefer for many helpful comments and discussion on an earlier version of this manuscript. N.R. was supported by NIH grant GM057672, E.B. and E.Z. by the Sandler Family Supporting Foundation, and H.T. by a Howard Hughes Medical Institute fellowship.

References

1. Schwartz R. S.: Racial profiling in medical research. *N Engl J Med* 2001, 344:1392–1393.

2. Wilson J. F., Weale M. E., Smith A. C., Gratrix F., Fletcher B., Thomas M. F., Bradman N., Goldstein D. B.: Population genetic structure of variable drug response. *Nat Genet* 2001, 29:265–269.

3. Editorial: Genes, drugs and race. *Nat Genet* 2001, 29:239–240.

4. Owens K., King, M.-C.: Genomic views of human history. *Science* 1999, 286:451–453.

5. Cavalli-Sforza L. L., Piazza A., Menozzi P., Mountain J.: Reconstruction of human evolution; bringing together genetic, archaeological, and linguistic data. *Proc Natl Acad Sci USA* 1988, 85:6002–6006.

6. Bowcock A. M., Kidd J. R., Mountain J. L., Hebert J. M., Carotenuto L., Kidd K. K., Cavalli-Sforza L. L.: Drift, admixture, and selection in human evolution: a study with DNA polymorphisms. *Proc Natl Acad Sci USA* 1991, 88:839–843.

7. Bowcock A. M., Ruiz-Linares A., Tomfohrde J., Minch E., Kidd J. R., Cavalli-Sforza L. L.: High resolution of human evolutionary trees with polymorphic microsatellites. *Nature* 1994, 368:455–457.

8. Perez-Lezaun A., Calafell F., Mateu E., Comas D., Ruiz-Pacheco R., Bertranpetit J.: Microsatellite variation with the differentiation of modern humans. *Hum Genet* 1997, 99:1–7.

9. Calafell F., Shuster A., Speed W. C., Kidd J. R., Kidd K. K.: Short tandem repeat polymorphism evolution in humans. *Eur J Hum Genet* 1998, 6:38–49.

10. Tishkoff S. A., Dietzsch E., Speed W., Pakstis A. J., Kidd J. R., Cheung K., Bonne-Tamir B., Santachiara-Benerecetti A. S., Moral P., Krings M., et al.: Global patterns of linkage disequilibrium at the CD4 locus and modern human origins. *Science* 1996, 271:1380–1387.

11. Jorde L. B., Rogers A. R., Bamshad M., Watkins W. S., Krakowiak P., Sung S., Kere J., Harpending H. C.: Microsatellite diversity and the demographic history of modern humans. *Proc Natl Acad Sci USA* 1997, 94:3100–3103.

12. Stephens J. C., Schneider J. A., Tanguay D. A., Choi J., Acharya T., Stanley S. E., Jiang R., Messer C. J., Chew A., Han J. H., et al.: Haplotype variation and linkage disequilibrium in 313 human genes. *Science* 2001, 293:489–493.

13. Mountain J. L., Cavalli-Sforza L. L.: Multilocus genotypes, a tree of individuals, and human evolutionary history. *Am J Hum Genet* 1997, 61:705–718.

14. Cavalli-Sforza L. L., Menozzi P., Piazza A.: *The History and Geography of Human Genes.* Princeton University Press: Princeton New Jersey; 1994.

15. Lewontin R. C.: The apportionment of human diversity. *Evol Biol* 1972, 6:381–398.

16. Latter B. D. H.: Genetic differences within and between populations of the major human subgroups. *Amer Nat* 1980, 116:220–237.

17. Barbujani G., Magagni A., Minch E., Cavalli-Sforza L. L.: An apportionment of human DNA diversity. *Proc Natl Acad Sci USA* 1997, 94:4516–4519.

18. Editorial: Census, race and science. *Nat Genet* 2000, 24:97–98.

19. Cargill M., Altshuler D., Ireland J., Sklar P., Ardlie K., Patil N., Shaw N., Lane C. R., Lim E. P., Kalyanaraman N., et al.: Characterization of single-nucleotide polymorphisms in coding regions of human genes. *Nat Genet* 1999, 22:231–238.

20. Halushka M. K., Fan J. B., Bentley K., Hsie L., Shen N., Weder A., Cooper R., Lipshutz R., Chakravarti A.: Patterns of single-nucleotide polymorphisms in candidate genes for blood-pressure homeostasis. *Nat Genet* 1999, 22:239–247.

21. *Overview of Race and Hispanic Origin: Census 2000 Brief.* United States Census 2000, US Census Bureau, US Department of Commerce. [http://www.census.gov/population/www/cen2000/briefs.html]

22. Parra E. J., Marcini A., Akey J., Martinson J., Batzer M. A., Cooper R., Forrester T., Allison D. B., Deka R., Ferrell R. E., Shriver M. D.: Estimating African American admixture proportions by use of population-specific alleles. *Am J Hum Genet* 1998, 63:1839–1851.

23. Knowler W. C., Williams R. C., Pettitt D. J., Steinberg A. G.: Gm3;5,13,14 and type 2 diabetes mellitus: an association in American Indians with genetic admixture. *Am J Hum Genet* 1988, 43:520–526.

24. Tseng M., Williams R. C., Maurer K. R., Schanfield M. S., Knowler W. C., Everhart J. E.: Genetic admixture and gallbladder disease in Mexican Americans. *Am J Phys Anthropol* 1998, 106:361–371.

25. Hanis C. L., Newett-Emmett D., Bertin T. K., Schull W. J.: Origins of U.S. Hispanics. Implications for diabetes. *Diabetes Care* 1991, 14:618–627.

26. McLeod H. L.: Pharmacogenetics: more than skin deep. *Nat Genet* 2001, 29:247–248.

27. Merryweather-Clarke A. T., Pointon J. J., Jouanolle A. M., Rochette J., Robson K. J.: Geography of HFE C282Y and H63D mutations. *Genet Test* 2000, 42:183–198.

28. Williams R. C., Long J. C., Hanson R. L., Sievers M. L., Knowler W. C.: Individual estimates of European genetic admixture associated with lower body-mass index, plasma glucose, and prevalence of type 2 diabetes in Pima Indians. *Am J Hum Genet* 2000, 66:527–538.

29. Dean M., Stephens J. C., Winkler C., Lomb D. A., Ramsburg M., Boaze R., Stewart C., Charbonneau L., Goldman D., Albaugh B. J., et al.: Polymorphic admixture typing in human ethnic populations. *Am J Hum Genet* 1994, 55:788–808.

30. Smith M. W., Lautenberger J. A., Shin H. D., Chretien J.-P., Shrestha S., Gilbert D. A., O'Brien S. J.: Markers for mapping by admixture linkage disequilibrium in African American and Hispanic populations. *Am J Hum Genet* 2001, 69:1080–1094.

31. Exner D. V., Dries D. L., Domanski M. J., Cohn J. N.: Lesser response to angiotensin-converting-enzyme inhibitor therapy in black as compared with white patients with left ventricular dysfunction. *N Engl J Med* 2001, 344:1351–1357.

32. Wood A. J. J.: Racial differences in the response to drugs—pointers to genetic differences. *N Engl Med* 2001, 344:1393–1396.

33. Lin S. S., Kelsey J. L.: Use of race and ethnicity in epidemiologic research: concepts, methodological issues, and suggestions for research. *Epidem Revs* 2000, 22:187–202.

34. Ziegler R. G., Hoover R. N., Pike M. C., Hildesheim A., Nomura A. M., West D. W., Wu-Williams A. H., Koloner L. N., Horn-Ross P. L., Rosenthal J. F., et al.: Migration patterns and breast cancer risk in Asian-American women. *J Natl Cancer Inst* 1993, 85:1819–1827.

35. Compston A.: Distribution of multiple sclerosis. In *McAlpine's Multiple Sclerosis*. Edited by Compston A., Ebers G., Lassman H., McDonald I., Matthews B., Wekerle H. Churchill Livingston: London; 1991:63–100.

36. Ebers G. C., Sadovnick A. D.: Epidemiology. In *Multiple Sclerosis*. Edited by Paty D. W., Ebers G. C. Philadelphia: FA Davis Company; 1998:5–28.

37. Karter A. J., Ferrara A., Liu J. Y., Moffet H. H., Ackerson L. M., Selby J. V.: Ethnic disparities in diabetic complications in an insured population. *JAMA* 2002, 287:2519–2527.

38. Kaufman J. S., Cooper R. S.: Commentary: considerations for use of racial/ethnic classificaiton in etiologic research. *Am J Epidemiol* 2001, 154:291–298.

39. Cruickshanks K. J., Jobim L. F., Lawler-Heavner J., Neville T. G., Gay E. C., Chase H. P., Klingensmith G., Todd J. A., Hamman R. F.: Ethnic differences in human leukocyte antigen markers of susceptibility to IDDM. *Diabetes Care* 1994, 17:132–137.

9.5 The Importance of Collecting Data and Doing Social Scientific Research on Race (2003)

American Sociological Association

The question of whether to collect statistics that allow the comparison of differences among racial and ethnic groups in the census, public surveys, and administrative databases is not an abstract one. Some scholarly and civic leaders believe that measuring these differences promotes social divisions and fuels a mistaken perception that race is a biological concept. California voters are likely to face a referendum in 2004 to prohibit the collection of racial data by most state government agencies. As the leading voice for 13,000 academic and practicing sociologists, the ASA takes the position that calls to end the collection of data using racial categories are ill advised, although racial categories do not necessarily reflect biological or genetic categories. The failure to gather data on this socially significant category would preserve the status quo and hamper progress toward understanding and addressing inequalities in primary social institutions. The ASA statement highlights significant research findings on the role and consequences of race relations in social institutions such as schools, labor markets, neighborhoods, and health care scholarship that would not have been possible without data on racial categories.

The longstanding debate over racial classification in the United States is certain to generate greater public interest as our population becomes more diverse. The ASA hopes to continue to play a meaningful role in that important dialogue.

The following statement was adopted by the elected Council of the American Sociological Association (ASA) on August 9, 2002, acting on a document prepared by a Task Force of ASA members. Council believes that this statement summarizes the views of sociologists with expertise in matters related to race in America.

Executive Summary

Race is a complex, sensitive, and controversial topic in scientific discourse and in public policy. Views on race and the racial classification system used to measure it have become polarized. At the heart of the debate in the United States are several fundamental questions: What are the causes and consequences of racial inequality? Should we continue to use racial classification to assess the role and consequences of race? And, perhaps most significantly, under what conditions does the classification of people by race promote racial division, and when does it aid the pursuit of justice and equality?

American Sociological Association, *The Importance of Collecting Data and Doing Social Scientific Research on Race* (Washington, D.C.: American Sociological Association, 2003).

The answers to these questions are important to scientific inquiry, but they are not merely academic. Some scholarly and civic leaders have proposed that the government stop collecting data on race altogether. Respected voices from the fields of human molecular biology and physical anthropology (supported by research from the Human Genome Project) assert that the concept of race has no validity in their respective fields. Growing numbers of humanist scholars, social anthropologists, and political commentators have joined the chorus in urging the nation to rid itself of the concept of race.

However, a large body of social science research documents the role and consequences of race in primary social institutions and environments, including the criminal justice, education and health systems, job markets, and where people live. These studies illustrate how racial hierarchies are embedded in daily life, from racial profiling in law enforcement, to "red-lining" communities of color in mortgage lending, to sharp disparities in the health of members of different population groups. Policymakers, in fact, have recognized the importance of research into the causes of racial disparities. For example, the 2000 Minority Health and Health Disparities Research and Education Act directed the National Institutes of Health to support continued research on health gaps between racial groups, with the ultimate goal of eliminating such disparities. Moreover, growth among some racial and ethnic groups (notably, Asians and Hispanics) and the diversification of the nation's racial and ethnic composition underscore the need for expanded research on the health and socio-economic status of these groups.

Sociologists have long examined how race—a social concept that changes over time—has been used to place people in categories. Some scientists and policymakers now contend that research using the concept of race perpetuates the negative consequences of thinking in racial terms. Others argue that measuring differential experiences, treatment, and outcomes across racial categories is necessary to track disparities and to inform policymaking to achieve greater social justice.

The American Sociological Association (ASA), an association of some 13,000 U.S. and international sociologists, finds greater merit in the latter point of view. Sociological scholarship on "race" provides scientific evidence in the current scientific and civic debate over the social consequences of the existing categorizations and perceptions of race; allows scholars to document how race shapes social ranking, access to resources, and life experiences; and advances understanding of this important dimension of social life, which in turn advances social justice. Refusing to acknowledge the fact of racial classification, feelings, and actions, and refusing to measure their consequences will not eliminate racial inequalities. At best, it will preserve the status quo.

The following statement sets forth the basis for ASA's position, and illustrates the importance of data on race to further scientific investigation and informed public discourse. ASA fully recognizes the global nature of the debate over race, racial classification, and the role of race in societies; this statement focuses attention on the treatment of race in the United States and the scholarly and public interest in continuing to measure it.

Racial Classifications As the Basis for Scientific Inquiry

Race is a complex, sensitive, and controversial topic in scientific discourse and in public policy. Views on race and the racial classification system used to measure it have become polarized. In popular discourse, racial groups are viewed as physically distinguishable populations that share a common geographically based ancestry. "Race" shapes the way that some people relate to each other, based on their belief that it reflects physical, intellectual, moral, or spiritual superiority or inferiority. However, biological research now suggests that the substantial overlap among any and all biological categories of race undermines the utility of the concept for scientific work in this field.

How, then, can it be the subject of valid scientific investigation at the social level? The answer is that social and economic life is organized, in part, around race as a social construct. When a concept is central to societal organization, examining how, when, and why people in that society use the concept is vital to understanding the organization and consequences of social relationships.

Sociological analysis of the family provides an analogue. We know that families take many forms; for example, they can be nuclear or extended, patrilineal or matrilineal. Some family categories correspond to biological categories; others do not. Moreover, boundaries of family membership vary, depending on a range of individual and institutional factors. Yet regardless of whether families correspond to biological definitions, social scientists study families and use membership in family categories in their study of other phenomena, such as well-being. Similarly, racial statuses, although not representing biological differences, are of sociological interest in their form, their changes, and their consequences.

The Social Concept of Race

Individuals and social institutions evaluate, rank, and ascribe behaviors to individuals on the basis of their presumed race. The concept of race in the United States—and the inevitable corresponding taxonomic system to categorize people by race—has changed, as economic, political, and historical contexts have changed (19). Sociologists are interested in explaining how and why social definitions of race persist and change. They also seek to explain the nature of power relationships between and among racial groups, and to understand more fully the nature of belief systems about race—the dimensions of how people use the concept and apply it in different circumstances.

Social Reality and Racial Classification

The way we define racial groups that comprise "the American mosaic" has also changed, most recently as immigrants from Asia, Latin America, and the Caribbean have entered the country in large numbers. One response to these demographic shifts has been the effort (sometimes contentious) to modify or add categories to the government's official statistical policy on race and ethnicity, which governs data collection in the census, other federal surveys, and administrative functions. Historically, changes in racial categories used for administrative purposes and self-identification have occurred within the context of a polarized biracialism of Black and White; other immigrants to the United States, including those from Asia, Latin America, and the Caribbean, have been "racialized" or ranked in between these two categories (26).

Although racial categories are legitimate subjects of empirical sociological investigation, it is important to recognize the danger of contributing to the popular conception of race as biological. Yet refusing to employ racial categories for administrative purposes and for social research does not eliminate their use in daily life, both by individuals and within social and economic institutions. In France, information on race is seldom collected officially, but evidence of systematic racial discrimination remains (31, 10). The 1988 Eurobarometer revealed that, of the 12 European countries included in the study, France was second (after Belgium) in both anti-immigrant prejudice and racial prejudice (29). Brazil's experience also is illustrative: The nation's then-ruling military junta barred the collection of racial data in the 1970 census, asserting that race was not a meaningful concept for social measurement. The resulting information void, coupled with government censorship, diminished public discussion of racial issues, but it did not substantially reduce racial inequalities. When racial data were collected again in the 1980 census, they revealed lower socio-economic status for those with darker skin (38).

The Consequences of Race and Race Relations in Social Institutions

Although race is a social construct (in other words, a social invention that changes as political, economic, and historical contexts change), it has real consequences across a wide range of social and economic institutions. Those who favor ignoring race as an explicit administrative matter, in the hope that it will cease to exist as a social concept, ignore the weight of a vast body of sociological research that shows that racial hierarchies are embedded in the routine practices of social groups and institutions.

Primary areas of sociological investigation include the consequences of racial classification as:

- A sorting mechanism for mating, marriage and adoption.
- A stratifying practice for providing or denying access to resources.
- An organizing device for mobilization to maintain or challenge systems of racial stratification.
- A basis for scientifically investigating proximate causes.

Race As a Sorting Mechanism for Mating, Marriage, and Adoption
Historically, race has been a primary sorting mechanism for marriage (as well as friendship and dating). Until anti-miscegenation laws were outlawed in the United States in

1967, many states prohibited interracial marriage. Since then, intermarriage rates have more than doubled to 2.2 percent of all marriages, according to the latest census information (14, 28). When Whites (the largest racial group in the United States) intermarry, they are most likely to marry Native Americans/American Indians and least likely to marry African Americans. Projections to the year 2010 suggest that intermarriage and, consequently, the universe of people identifying with two or more races is likely to increase, although most marriages still occur within socially designated racial groupings (7).

Race As a Stratifying Practice

Race serves as a basis for the distribution of social privileges and resources. Among the many arenas in which this occurs is education. On the one hand, education can be a mechanism for reducing differences across members of racial categories. On the other hand, through "tracking" and segregation, the primary and secondary educational system has played a major role in reproducing race and class inequalities. Tracking socializes and prepares students for different education and career paths. School districts continue to stratify by race and class through two-track systems (general and college prep/advanced) or systems in which all students take the same courses, but at different levels of ability. African Americans, Hispanics, American Indians, and students from low socioeconomic backgrounds, regardless of ability levels, are over-represented in lower level classes and in schools with fewer Advanced Placement classes, materials, and instructional resources (11, 13, 20, 23).

Race As an Organizing Device for Mobilization to Maintain or Challenge Systems of Racial Stratification

Understanding how social movements develop in racially stratified societies requires scholarship on the use of race in strategies of mobilization. Racial stratification has clear beneficiaries and clear victims, and both have organized on racial terms to challenge or preserve systems of racial stratification. For example, the apartheid regime in South Africa used race to maintain supremacy and privilege for Whites in nearly all aspects of economic and political life for much of the 20th century. Blacks and others seeking to overthrow the system often were able to mobilize opposition by appealing to its victims, the Black population. The American civil rights movement was similarly successful in mobilizing resistance to segregation, but it also provoked some White citizens into organizing their own power base (for example, by forming White Citizens' Councils) to maintain power and privilege (2, 24).

Race and Ethnicity As a Basis for the Scientific Investigation of Proximate Causes and Critical Interactions

Data on race often serve as an investigative key to discovering the fundamental causes of racially different outcomes and the "vicious cycle" of factors affecting these outcomes. Moreover, because race routinely interacts with other primary categories of social life, such as gender and social class, continued examination of these bases of fundamental social interaction and social cleavage is required. In the health arena, hypertension levels are much higher for African Americans than other groups. Sociological investigation suggests that discrimination and unequal allocation of society's resources might expose members of this racial group to higher levels of stress, a proximate cause of hypertension (40). Similarly, rates of prostate cancer are much higher for some groups of men than others. Likewise, breast cancer is higher for some groups of women than others. While the proximate causes may appear to be biological, research shows that environmental and socio-economic factors disproportionately place at greater risk members of socially subordinated racial and ethnic groups. For example, African Americans' and Hispanics' concentration in polluted and dangerous neighborhoods result in feelings of depression and powerlessness that, in turn, diminish the ability to improve these neighborhoods (35, 40, 41). Systematic investigation is necessary to uncover and distinguish what social forces, including race, contribute to disparate outcomes.

Research Highlights: Race and Ethnicity As Factors in Social Institutions

The following examples highlight significant research findings that illustrate the persistent role of race in primary social institutions in the United States, including the job market, neighborhoods, and the health care system. This scientific investigation would not have been possible without data on race.

Job Market

Sociological research shows that race is substantially related to workplace recruitment, hiring, firing, and promotions. Ostensibly neutral practices can advantage some racial groups and adversely affect others. For example, the majority of workers obtain their jobs through informal networks rather than through open recruitment and hiring practices. Business-as-usual recruitment and hiring practices include recruiting at predominantly White schools, advertising only in suburban newspapers, and employing relatives and friends of current workers. Young, White job seekers benefit from family connections, studies show. In contrast, a recent study revealed that word-of-mouth recruitment through family and friendship networks limited job opportunities for African Americans in the construction trades. Government downsizing provides another example of a "race neutral" practice with racially disparate consequences: Research shows that because African Americans have successfully established employment niches in the civil service, government workforce reductions displace disproportionate numbers of African American-and increasingly, Hispanic-employees. These and other social processes, such as conscious and unconscious prejudices of those with power in the workplace, affecting the labor market largely explain the persistent two-to-one ratio of Black to White unemployment (4, 5, 9, 15, 32, 39, 42, 43).

Neighborhood Segregation

For all of its racial diversity, the highly segregated residential racial composition is a defining characteristic of American cities and suburbs. Whites and African Americans tend to live in substantially homogenous communities, as do many Asians and Hispanics. The segregation rates of Blacks have declined slightly, while the rates of Asians and Hispanics have increased. Sociological research shows that the "hyper-segregation" between Blacks and Whites, for example, is a consequence of both public and private policies, as well as individual attitudes and group practices.

Sociological research has been key to understanding the interaction between these policies, attitudes, and practices. For example, according to attitude surveys, by the 1990s, a majority of Whites were willing to live next door to African Americans, but their comfort level fell as the proportion of African Americans in the neighborhood increased. Real estate and mortgage-industry practices also contribute to neighborhood segregation, as well as racially disparate homeownership rates (which, in turn, contribute to the enormous wealth gap between racial groups). Despite fair housing laws, audit studies show, industry practices continue to steer African American homebuyers away from White neighborhoods, deny African Americans information about available loans, and offer inferior property insurance.

Segregation profoundly affects quality of life. African American neighborhoods (even relatively affluent ones) are less likely than White neighborhoods to have high quality services, schools, transportation, medical care, a mix of retail establishments, and other amenities. Low capital investment, relative lack of political influence, and limited social networks contribute to these disparities (1, 6, 8, 9, 17, 21, 22, 25, 30, 35, 36, 37, 42, 44).

Health
Research clearly documents significant, persistent differences in life expectancy, mortality, incidence of disease, and causes of death between racial groups. For example, African Americans have higher death rates than Whites for eight of the ten leading causes of death. While Asian-Pacific Islander babies have the lowest mortality rates of all broad racial categories, infant mortality for Native Hawaiians is nearly three times higher than for Japanese Americans. Genetics accounts for some health differences, but social and economic factors, uneven treatment, public health policy, and health and coping behaviors play a large role in these unequal health outcomes.

Socio-economic circumstances are the strongest predictors of both life span and freedom from disease and disability. Unequal life expectancy and mortality reflect racial disparities in income and incidence of poverty, education and, to some degree, marital status. Many studies have found that these characteristics and related environmental factors such as over-crowded housing, inaccessibility of medical care, poor sanitation, and pollution adversely impact life expectancy and both overall and cause-specific mortality for groups that have disproportionately high death rates.

Race differences in health insurance coverage largely reflect differences in key socio-economic characteristics. Hispanics are least likely to be employed in jobs that provide health insurance and relatively fewer Asian Americans are insured because they are more likely to be in small low-profit businesses that make it hard to pay for health insurance. Access to affordable medical care also affects health outcomes. Sociological research shows that highly segregated African American neighborhoods are less likely to have health care facilities such as hospitals and clinics, and have the highest ratio of patients to physicians. In addition, public policies such as privatization of medicine and lower Medicaid and Medicare funding have had unintended racial consequences; studies show a further reduction of medical services in African American neighborhoods as a result of these actions.

Even when health care services are available, members of different racial groups often do not receive comparable treatment. For example, African Americans are less likely to receive the most commonly performed diagnostic procedures, such as cardiovascular and orthopedic procedures. Institutional discrimination, including racial stereotyping by medical professionals, and systemic barriers, such as language difficulties for newer immigrants (the majority of whom are from Asia and Latin America), partly explain differential treatment patterns, stalling health improvements for some racial groups.

All of these factors interact to produce poorer health outcomes, indicating that racial stratification remains an important explanation for health disparities (3, 12, 16, 18, 21, 27, 33, 34, 40, 41).

Summary: The Importance of Sociological Research on Race

A central focus of sociological research is systematic attention to the causes and consequences of social inequalities. As long as Americans routinely sort each other into racial categories and act on the basis of those attributions, research on the role of race and race relations in the United States falls squarely within this scientific agenda. Racial profiling in law enforcement activities, "redlining" of predominantly minority neighborhoods in the mortgage and insurance industries, differential medical treatment, and tracking in schools, exemplify social practices that should be studied. Studying race as a social phenomenon makes for better science and more informed policy debate. As the United States becomes more diverse, the need for public agencies to continue to collect data on racial categories will become even more important. Sociologists are well qualified to study the impact of "race"—and all the ramifications of racial categorization—on people's lives and social institutions. The continuation of the collection and scholarly analysis of data serves both science and the public interest. For all of these reasons, the American Sociological Association supports collecting data and doing research on race.

Task Force Members

Diane Brown (Council)
Urban Health Program
Wayne State University

Manuel de la Puente
Population Division
U.S. Bureau of the Census

Bette J. Dickerson
Department of Sociology
American University

Troy Duster, Task Force Chair
Department of Sociology
New York University and
University of California-
Berkeley

Charles Hirschman
Department of Sociology
University of Washington

Deborah K. King
Department of Sociology
Dartmouth College

Sharon M. Lee
Department of Sociology
Portland State University

Felice J. Levine
Past Executive Officer
American Sociological
Association

Suzanne Model
Department of Sociology
University of
Massachusetts

Michael Omi
Department of Ethnic
Studies
University of California-
Berkeley

Willie Pearson, Jr.
School of History, Science
and Technology, Ivan
Allen College Georgia
Institute of Technology

C. Matthew Snipp
Department of Sociology
Stanford University

Roberta M. Spalter-Roth
Staff Liaison American
Sociological Association

Edward Telles
Department of Sociology
University of California-Los
Angeles

Hernan Vera
Department of Sociology
University of Florida

Lynn Weber
Women's Studies Program
University of South Carolina

David Wellman
Community Studies
Department
University of California-
Santa Cruz

David R. Williams
Institute for Social Research
University of Michigan

J. Milton Yinger
Professor Emeritus
Oberlin College

Note

1. The federal government defines race categories for statistical policy purposes, program administrative reporting, and civil rights compliance, and sets forth minimum categories for the collection and reporting of data on race. The current standards, adopted in October 1997, include five race categories: American Indian or Alaska Native; Asian; Black or African American; Native Hawaiian or Other Pacific Islander; and White. Respondents to federal data collection activities must be offered the option of selecting one or more racial designations. Hispanics or Latinos, whom current standards define as an ethnic group, can be of any race. However, before the government promulgated standard race categories in 1977, some U.S. censuses designated Hispanic groups as race categories (e.g., the 1930 census listed Mexicans as a separate race).

References

1. Alba, Richard D., John R. Logan, and Brian J. Stults. 2000. "The Changing Neighborhood Contexts of the Immigrant Metropolis." *Social Forces* 79:587–621.

2. Bloom, Jack M. 1987. *Class, Race and the Civil Rights Movement.* Bloomington, IN: Indiana University Press.

3. Bobo, Lawrence D. 2001. "Racial Attitudes and Relations at the Close of the Twentieth Century," pp. 264–201, in *America Becoming: Racial Trends and Their Consequences*, vol. 2, edited by Neil J. Smelser, William J. Wilson, and Faith Mitchell. Washington, DC: National Research Council.

4. Bobo, Lawrence D., Devon Johnson, and Susan Suh. 2002. "Racial Attitudes and Power in the Workplace: Do the Haves Differ from the Have-Nots?" pp. 491–522, in *Prismatic Metropolis: Inequality in Los Angeles*, edited by Lawrence D. Bobo, Melvin J. Oliver, James H. Johnson, Jr., and Abel Valenzuela Jr. New York, NY: Russell Sage Foundation.

5. DiTomaso, Nancy. 2000. "Why Anti-Discrimination Policies Are Not Enough: The Legacies and Consequences of Affirmative Inclusion—For Whites." Presented at the 95th annual meeting of the American Sociological Association, August 16, Anaheim, CA.

6. Drier, Peter, John Mollenkopf, and Todd Swanstrom. 2001. *Place Matters: Metropolitics in the 21st Century*. Lawrence, KS: University of Kansas Press.

7. Edmonston, Barry, Sharon M. Lee, and Jeffrey Passel. (in press). "Recent Trends in Intermarriage and Immigration, and Their Effects on the Future Racial Composition of the U.S. Population," in *The New Race Question*, edited by Joel Perlmann and Mary C. Waters. New York, NY: Russell Sage Foundation.

8. Farley, Reynolds. 1996. *The New American Reality: Who We Are, How We Got Here, Where We Are Going?* New York, NY: Russell Sage Foundation.

9. Farley, Reynolds, Sheldon, Danzinger, and Harry Holzer. 2001. *Detroit Divided*. New York, NY: Russell Sage Foundation.

10. Galap, Jean. 1991. "Phenotypes et Discrimination des Noirs en France: Question de Methode." *Intercultures* 14 (Juillet): 21–35.

11. Hallinan, Maureen T. 2001. "Sociological Perspectives on Black-White Inequalities in American Schooling." *Sociology of Education* (Extra Issue 2001): 50–70.

12. Hayward, Mark D, Eileen M. Crimmins, Toni P. Miles, and Yu Yang. 2000. "Socioeconomic Status and the Racial Gap in Chronic Health Conditions." *American Sociological Review* 65: 910–930.

13. Heubert, Jay P., and Robert M. Hauser, eds. 1999. *High Stakes: Testing for Tracking, Promotion, and Graduation*. Washington, DC: National Research Council.

14. Jones, Nicholas A., and Amy Symens Smith. 2001. *The Two or More Races Population 2000: Census 2000 Brief*. U.S. Bureau of the Census (November). Retrieved June 19, 2002, (http:www.census.gov/population/www/cen2000/briefs.html).

15. Kirshenman, Joleen, and Kathryn M. Neckerman. 1992. "We'd Love to Hire Them, But...: The Meaning of Race for Employers," pp. 203–234, in *The Urban Underclass*, edited by C. Jencks and P. Peterson. Washington, DC: The Brookings Institution.

16. Klinenberg, Eric. 2002. Heat Wave: *A Social Autopsy of Disaster in Chicago*. Chicago, IL: University of Chicago Press.

17. LaVeist, Thomas. 1992. "The Political Empowerment and Health Status of African Americans: Mapping a New Territory." *American Journal of Sociology* 97: 1080–1095.

18. LaViest, Thomas A., C. Diala, and N. C. Jarrett. 2000. "Social Status and Perceived Discrimination: Who Experiences Discrimination in the Health Care System and Why?" pp. 194–208, in *Minority Health in America*, edited by Carol J. R. Hogue, Martha A. Hargraves, and Karen Scott-Collins. Baltimore, MD: Johns Hopkins University Press.

19. Lee, Sharon M. 1993. "Racial Classifications in the U.S. Census: 1890–1990." *Ethnic and Racial Studies* 16: 75–94.

20. Lucas, Samuel Roundfield. 1999. *Tracking Inequality: Stratification and Mobility in American High Schools*. New York, NY: Teachers College Press.

21. Massey, Douglas S. 2001. "Residential Segregation and Neighborhood Conditions in U.S. Metropolitan Areas," pp. 391–434, in *America Becoming: Racial Trends and Their Consequences*, vol. 1, edited by Neil J. Smelser, William J. Wilson, and Faith Mitchell. Washington, DC: National Research Council.

22. Massey, Douglas S., and Nancy Denton. 1993. *American Apartheid: Segregation and the Making of the Underclass*. Cambridge, MA: Harvard University Press.

23. Mikelson, Roslyn A. 2002. "What Constitutes Racial Discrimination in Education? A Social Science Perspective." Prepared for workshop on Measuring Racial Disparities and Discrimination in Elementary and Secondary Education, National Research Council Committee on the National Statistics Center for Education, July 2002. For a grant from the Ford Foundation and National Science Foundation.

24. Morris, Aldon D. 1986. *The Origins of the Civil Rights Movement: Black Communities Organizing for Change*. New York, NY: The Free Press.

25. Oliver Melvin L., and Thomas J. Shapiro. 1995. *Black Wealth? White Wealth?: A New Perspective on Racial Inequality*. New York, NY: Routledge.

26. Omi, Michael. 2001. "The Changing Meaning of Race," pp. 243–263, in *America Becoming: Racial Trends and Their Consequences*, edited by Neil J. Smelser, William J. Wilson, and Faith Mitchell. Washington, DC: National Academy Press.

27. Quadagno, Jill. 2000. "Promoting Civil Rights through the Welfare State: How Medicare Integrated Southern Hospitals." *Social Problems* 47: 68–89.

28. Qian, Zhenchao. 1997. "Breaking the Racial Barriers: Variations in Interracial Marriage Between 1980 and 1990." *Demography* 34: 263–276.

29. Quillian, Lincoln. 1995. Prejudice as a Response to Perceived Group Threat: Population Composition and Anti-Immigrant and Racial Prejudice in Europe. *American Sociological Review* 60: 586–611.

30. Rankin, Bruce H., and James M. Quane. 2000. "Neighborhood Poverty and Social Isolation of Inner-City African American Families." *Social Forces* 79: 139–164.

31. Raveau, F., B. Kilborne, L. Frere, J. M. Lorin, and G. Trempe. 1976. *"Perception Sociale de la Couleur et Discrimination."* Cahiers d'Anthropologie 4: 23–42.

32. Reskin, Barbara F. 1998. *The Realities of Affirmative Action in Employment.* Washington, DC: The American Sociological Association.

33. Rogers, Richard, Robert Hummer, Charles B. Nam, Kimberly Peters. 1996. "Demographic, Socioeconomic, and Behavioral Factors Affecting Ethnic Mortality by Cause." *Social Forces* 74: 1419–1438.

34. Ross, Catherine E. and John Mirowsky. 2001. "Neighborhood Disadvantage, Disorder, and Health." *Journal of Health and Social Behavior* 42: 258–276.

35. Sampson, Robert J., Gregory D. Squires, and Min Zhou. 2001. *How Neighborhoods Matter: The Value of Investing at the Local Level.* Washington, DC: The American Sociological Association.

36. Schuman, Howard, Charlotte Steeh, Lawrence Bobo, and Maria Kryson. 1997. *Racial Attitudes in America.* 2 ed. Cambridge MA: Harvard University Press.

37. Squires, Gregory D., and Sally O'Connor. 2001. *Color and Money: Politics and Prospects for Community Reinvestment in Urban America.* Albany, NY: SUNY Press.

38. Telles, Edward. 2002. "Racial Ambiguity among the Brazilian Population." *Ethnic and Racial Studies* 25: 415–441.

39. Waldinger, Roger. 1996. *Still the Promised City? African-Americans and New Immigrants in Postindustrial New York.* Cambridge, MA: Harvard University Press.

40. Williams, David R. 2001. "Racial Variations in Adult Health Status: Patterns, Paradoxes, and Prospects," pp. 371–410 2 in *America Becoming: Racial Trends and Their Consequences,* vol. 2, edited by Neil J. Smelser, William J. Wilson, and Faith Mitchell. Washington, DC: National Research Council.

41. Williams, David R. and Chiquita Collins. (in press). "Racial Residential Segregation: A Fundamental Cause of Racial Disparities in Health." *Public Health Reports.*

42. Wilson, William J. 1996. *When Work Disappears: The World of the New Urban Poor.* New York, NY: Alfred A. Knopf, Inc.

43. Woo, Deborah. 2000. *Glass Ceilings and Asian Americans: The New Face of Workplace Barriers.* Walnut Creek, CA: AltaMira Press.

44. Yinger, John. 1995. *Closed Doors, Opportunities Lost.* New York, NY: Russell Sage Foundation.

9.6   "Zebrafish Researchers Hook Gene for Human Skin Color" (2005)

Michael Balter

People come in many different hues, from black to brown to white and shades in between. The chief determinant of skin color is the pigment melanin, which protects against ultraviolet rays and is found in cellular organelles called melanosomes. But the genetics behind this spectrum of skin colors have remained enigmatic. Now, on page 1782 of this week's issue of *Science*, an international team reports the identification of a zebrafish pigmentation gene and its human counterpart, which apparently accounts for a significant part of the difference between African and European skin tones. One variant of the gene seems to have undergone strong natural selection for lighter skin in Europeans.

The new work is raising goose bumps among skin-color researchers. "Entirely original and groundbreaking," says molecular biologist Richard Sturm of the University of Queensland in Brisbane, Australia. Anthropologist Nina Jablonski of the California Academy of Sciences in San Francisco, California, notes that the paper "provides very strong support for positive selection" of light skin in Europeans. Researchers have not been sure whether European pale skin is the result of some selective advantage or due to a relaxation of selection for dark skin, after the ancestors of modern Europeans migrated out of Africa into less sunny climes.

Yet the authors agree that the new gene, SLC24A5, is far from the whole story: Although at least 93% of Africans and East Asians share the same allele, East Asians are usually light skinned too. This means that variation in other genes, a handful of which have been previously identified, also affects skin color.

The *Science* paper is the culmination of a decade of work, says team leader Keith Cheng, a geneticist at Pennsylvania State University College of Medicine in Hershey. He and his colleagues were using the zebrafish as a model organism to search for cancer genes and became curious about a zebrafish mutation called *golden*, which lightens the fish's normally dark, melanin-rich stripes. Cheng's team identified the mutated gene and found that the zebrafish version shared about 69% of its sequence with the human gene SLC24A5, which is thought to be involved in ion exchange across cellular membranes—an important process in melanosome formation. And when Cheng and his co-workers injected human SLC24A5 messenger RNA (an intermediary molecule in protein synthesis) into *golden* zebrafish embryos, wild-type pigmentation pattern was restored.

Researchers say the ability of human SLC24A5 to "rescue" the mutant zebrafish is strong evidence that the gene has a similar function in fish and humans. "The

Michael Balter, "Zebrafish Researchers Hook Gene for Human Skin Color," *Science* 310, no. 5755 (16 December 2005): 1754–1755.

zebrafish data are extremely compelling," says human geneticist Neil Risch of the University of California, San Francisco.

The team then searched for genetic variants among humans. Data from the HapMap database of human genetic diversity (*Science*, 28 October, p. 601) showed that SLC24A5 has two primary alleles, which vary by one amino acid. Nearly all Africans and East Asians have an allele with alanine in a key locus, whereas 98% of Europeans have threonine at that locus. These marked frequency differences combined with the pattern of variation in nearby genes suggest that the threonine variant has been the target of a recent selective sweep among the ancestors of modern Europeans, Cheng's team concluded.

Finally, the team measured the pigmentation levels of 203 African Americans and 105 African Caribbeans—groups that represent an admixture of African and European ancestry—and compared their SLC24A5 genotypes. Subjects homozygous for the threonine allele tended to be lightest skinned, those homozygous for the alanine allele were darkest, and heterozygotes were in between, as shown by the degree of reflectance of their skin. The team concludes that between 25% and 38% of the skin-color difference between Europeans and Africans can be attributed to SLC24A5 variants. The experiments provide "a beautiful example of the critical role that model organism genetics continues to play for understanding human gene function," says geneticist Gregory Barsh of Stanford University in California.

The new work doesn't solve the question of why fair skin might have been favored among Europeans. However, it is consistent with a long-standing but unproven hypothesis that light skin allows more absorption of sunshine and so produces more vitamin D, a trait that would be favored at less sunny European latitudes.

Barsh adds that the paper "indicates how the genetics of skin-color variation is quite different from, and should not be confused with, the concept of race." Rather, he says, "one of the most obvious characteristics that distinguishes among different humans is nothing more than a simple change in activity of a protein expressed in pigment cells." Jablonski agrees: "Skin color does not equal race, period."

Note

See figure 9.6.1 in color insert.

9.7 "SLC24A5, a Putative Cation Exchanger, Affects Pigmentation in Zebrafish and Humans" (2005)

Rebecca L. Lamason, Manzoor-Ali P. K. Mohideen, Jason R. Mest, Andrew C. Wong, Heather L. Norton, Michele C. Aros, Michael J. Jurynec, Xianyun Mao, Vanessa R. Humphreville, Jasper E. Humbert, Soniya Sinha, Jessica L. Moore, Pudur Jagadeeswaran, Wei Zhao, Gang Ning, Izabela Makalowska, Paul M. McKeigue, David O'Donnell, Rick Kittles, Esteban J. Parra, Nancy J. Mangini, David J. Grunwald, Mark D. Shriver, Victor A. Canfield, and Keith C. Cheng

Pigment color and pattern are important for camouflage and the communication of visual cues. In vertebrates, body coloration is a function of specialized pigment cells derived from the neural crest (1). The melanocytes of birds and mammals (homologous to melanophores in other vertebrates) produce the insoluble polymeric pigment melanin. Melanin plays an important role in the protection of DNA from ultraviolet radiation (2) and the enhancement of visual acuity by controlling light scatter (3). Melanin pigmentation abnormalities have been associated with inflammation and cancer, as well as visual, endocrine, auditory, and platelet defects (4).

Despite the cloning of many human albinism genes and the knowledge of over 100 genes that affect coat color in mice, the genetic origin of the striking variations in human skin color is one of the remaining puzzles in biology (5). Because the primary ultrastructural differences between melanocytes of dark-skinned Africans and lighter-skinned Europeans include changes in melanosome number, size, and density (6, 7), we reasoned that animal models with similar differences may contribute to our understanding of human skin color. Here we present evidence that the human ortholog of a gene associated with a pigment mutation in zebrafish, SLC24A5, plays a role in human skin pigmentation.

The Zebrafish *Golden* Phenotype

The study of pigmentation variants (5, 8) has led to the identification of most of the known genes that affect pigmentation and has contributed to our understanding of basic genetic principles in peas, fruit flies, corn, mice, and other classical model systems. The first recessive mutation studied in zebrafish (*Danio rerio*), golden (*gol*[b1]), causes

Rebecca L. Lamason, Manzoor-Ali P. K. Mohideen, Jason R. Mest, Andrew C. Wong, Heather L. Norton, Michele C. Aros, Michael J. Jurynec, Xianyun Mao, Vanessa R. Humphreville, Jasper E. Humbert, Soniya Sinha, Jessica L. Moore, Pudur Jagadeeswaran, Wei Zhao, Gang Ning, Izabela Makalowska, Paul M. McKeigue, David O'Donnell, Rick Kittles, Esteban J. Parra, Nancy J. Mangini, David J. Grunwald, Mark D. Shriver, Victor A. Canfield, and Keith C. Cheng, "SLC24A5, a Putative Cation Exchanger, Affects Pigmentation in Zebrafish and Humans," *Science* 310, no. 5755 (16 December 2005): 1782–1786.

hypopigmentation of skin melanophores (Fig. 9.7.1; [all figures printed in color insert]) and retinal pigment epithelium (Fig. 9.7.2) (9). Despite its common use for the calibration of germ-line mutagenesis (10), the *golden* gene remained unidentified.

The *golden* phenotype is characterized by delayed and reduced development of melanin pigmentation. At approximately 48 hours postfertilization (hpf), melanin pigmentation is evident in the melanophores and retinal pigment epithelium (RPE) of wild-type embryos (Fig. 9.7.2A) but is not apparent in *golden* embryos (Fig. 9.7.2B). By 72 hpf, *golden* melanophores and RPE begin to develop pigmentation (Fig. 9.7.2, F and G) that is lighter than that of wild type (Fig. 9.7.2, D and E). In adult zebrafish, the melanophore-rich dark stripes are considerably lighter in *golden* compared with wild-type animals (Fig. 9.7.1, A and B). In regions of the ventral stripes where melanophore density is low enough to distinguish individual cells, it is apparent that the melanophores of *golden* adults are less melanin-rich than those in wild-type fish (Fig. 9.7.1, A and B).

Transmission electron microscopy was used to determine the cellular basis of *golden* hypopigmentation in skin melanophores and RPE of ~55-hpf wild-type and *golden* zebrafish. Wild-type melanophores contained numerous, uniformly dense, round-to-oval melanosomes (Fig. 9.7.1, C and E). The melanophores of *golden* fish were thinner and contained fewer melanosomes (Fig. 9.7.1D). In addition, *golden* melanosomes were smaller, less electron-dense, and irregularly shaped (Fig. 9.7.1F). Comparable differences between wild-type and *golden* melanosomes were present in the RPE (fig. S1, A and B [figs. S1–S6 not reprinted here]).

Dysmorphic melanosomes have also been reported in mouse models of Hermansky-Pudlak syndrome (HPS) (11, 12). Because HPS is characterized by defects in platelet-dense granules and lysosomes as well as melanosomes, we examined whether the *golden* mutation also affects thrombocyte function in the zebrafish. A comparison of *golden* and wild-type larvae in a laser-induced arterial thrombosis assay (13) revealed no significant difference in clotting time (35 versus 30 s). The *golden* phenotype thus appears to be restricted to melanin pigment cells in zebrafish.

The Zebrafish *Golden* Gene Is SLC24A5/NCKX5

Similarities between zebrafish *golden* and light-skinned human melanosomes suggested that the positional cloning of *golden* might lead to the identification of a phylogenetically conserved class of genes that regulate melanosome morphogenesis. Positional cloning, morpholino knockdown, DNA and RNA rescue, and expression analysis were used to identify the gene underlying the *golden* phenotype. Linkage analysis of 1126 homozygous *gol*[b1] embryos (representing 2252 meioses) revealed a single crossover between *golden* and microsatellite marker z13836 on chromosome 18. This map distance of 0.044 centimorgans (cM) [95% confidence interval (CI), 0.01 to 0.16 cM] corresponds to a physical distance of about 33 kilobases (kb) (using 1 cM = 740 kb) (14). Marker z9484 was also tightly linked to *golden* but informative in fewer individuals; no recombinants between z9484 and *golden* were identified in 468 embryos (95% CI, distance < 0.32 cM). Polymerase chain reaction (PCR) analysis of a γ-radiation-induced deletion allele, gol^{b13} (*15*), showed a loss of markers z10264, z9404, z928, and z13836, but not z9484 (fig. S2A). Screening of a zebrafish genomic library (*16*) led to the identification of a clone (PAC215f11) containing both z13836 and z9484 within an ~85-kb insert. Microinjection of PAC215f11 into *golden* embryos produced mosaic rescue of wild-type pigmentation in embryonic melanophores and RPE (Fig. 9.7.2, H and I), indicating the presence of a functional *golden* gene within this clone.

Shotgun sequencing, contig assembly, and gene prediction revealed two partial and three complete genes within PAC215f11 (fig. S2B): the 3' end of a thrombospondin-repeat–containing gene (FLJ13710), a putative potassium-dependent sodium/calcium exchanger (SLC24A5), myelin expression factor 2 (MYEF2), a cortexin homolog (CTXN2), and the 5' end of a sodium/potassium/chloride cotransporter gene (SLC12A1). We screened each candidate gene using morpholino anti-sense oligonucleotides directed against either the initiation codon (*17*) or splice donor junctions (*18*). Only embryos injected with a morpholino targeted to SLC24A5 (either of two splice-junction morpholinos or one start codon morpholino) successfully phenocopied *golden* (Fig. 9.7.2C). In rescue experiments, injection of full-length, wild-type SLC24A5 transcript into homozygous gol^{b1} embryos led to the partial restoration of wild-type pigmentation in both melanophores and RPE (Fig. 9.7.2, J and K). Taken together, these results confirm the identity of *golden* as SLC24A5.

To identify the mutation in the gol^{b1} allele, we compared complementary DNA (cDNA) and genomic sequence from wild-type and gol^{b1} embryos. A C $\rightarrow$ A nucleotide transversion that converts Tyr^{208} to a stop codon was found in gol^{b1} cDNA clones (GenBank accession number AY682554) and verified by sequencing gol^{b1} genomic DNA (fig. S3C). Conceptual translation of the mutant sequence predicts the truncation of the *golden* polypeptide to about 40% of its normal size, with loss of the central hydrophilic loop and the C-terminal cluster of potential transmembrane domains.

In wild-type embryos, the RNA expression pattern of SLC24A5 (Fig. 9.7.3A) resembled that of the melanin biosynthesis marker *dct* (Fig. 9.7.3B), consistent with expression of SLC24A5 in melanophores and RPE. In contrast, SLC24A5 expression was nearly undetectable in *golden* embryos (Fig. 9.7.3C), the expected result of non-sense-mediated mRNA decay (*19*). The extent of protein deletion associated with the gol^{b1} mutation, together with its low expression, suggests that gol^{b1} is a null mutation. The persistence of melanosome morphogenesis, despite likely absence of function, suggests that *golden* plays a modulatory rather than essential role in the formation of the melanosome. The pattern of *dct* expression seen in *golden* embryos (Fig. 9.7.3D) resembles that of wild-type embryos, indicating that the *golden* mutation does not affect the generation or migration of melanophores.

Conservation of *Golden* Gene Structure and Function in Vertebrate Evolution

Comparison of *golden* cDNA (accession number AY538713) to genomic (accession number AY581204) sequences shows that the wild-type gene contains nine exons (fig.

S2C) encoding 513 amino acids (fig. S3A). BLAST searches revealed that the protein is most similar to potassium-dependent sodium/calcium exchangers (encoded by the NCKX gene family), with highest similarity (68 to 69% amino acid identity) to murine SLC24A5 (accession number BAC40800) and human SLC24A5 (accession number NP_995322) (fig. S3B). The zebrafish *golden* gene shares less similarity with other human NCKX genes (35 to 41% identity to SLC24A1 to SLC24A4) or sodium/calcium exchanger (NCX) genes (26 to 29% identity to SLC8A1 to SLC8A3). Shared intron/exon structure and gene order (SLC24A5, MYEF2, CTXN2, and SLC12A5) between fish and mammals further supports the conclusion that the zebrafish *golden* gene and SLC24A5 are orthologs. The high sequence similarity among the orthologous sequences from fish and mammals (fig. S3A) suggested that function may also be conserved. The ability of human SLC24A5 mRNA to rescue melanin pigmentation when injected into *golden* zebrafish embryos (Fig. 9.7.2, L and M, and fig. S4) demonstrated functional conservation of the mammalian and fish polypeptides across vertebrate evolution.

Tissue-Specific Expression of SLC24A5

Quantitative reverse transcriptase PCR (RT-PCR) was used to examine SLC24A5 expression in normal mouse tissues and in the B16 melanoma cell line (Fig. 9.7.3E). SLC24A5 expression varied 1000-fold between tissues, with concentrations in skin and eye at least 10-fold higher than in other tissues. The mouse melanoma showed ~100-fold greater expression of SLC24A5 compared with normal skin and eye. These results suggest that mammalian SLC24A5, like zebrafish *golden*, appears to be highly expressed in melanin-producing cells.

Model for the Role of SLC24A5 in Pigmentation

SLC24A5 shares with other members of the protein family a potential hydrophobic signal sequence near the amino terminus and 11 hydrophobic segments, forming two groups of potential trans-membrane segments separated by a central cytoplasmic domain. This structure is consistent with membrane localization, although the specific topology of these proteins remains controversial (20). Elucidation of the specific role of this exchanger in melanosome morphogenesis requires knowledge of its subcellular localization and transport properties. Although previously characterized members of the NCKX and NCX families have been shown to be plasma membrane proteins (21), the melanosomal phenotype of *golden* suggested the possibility that the SLC24A5 protein resides in the melanosome membrane. To distinguish between these alternatives, confocal microscopy was used to localize green fluorescent protein (GFP)—and hemagglutinin (HA)-tagged derivatives of zebrafish SLC24A5 in MNT1, a constitutively pigmented human melanoma cell line (22). Both SLC24A5 fusion proteins displayed an intracellular pattern of localization (Fig. 9.7.4, A and B), which is distinct from that of a known plasma membrane control (Fig. 9.7.4C). The HA-tagged protein showed phenotypic rescue of the *golden* phenotype (Fig. 9.7.4D), indicating that tag addition did not abrogate its function. Taken together, these results indicate that the SLC24A5 protein functions in intracellular, membrane-bound structures, consistent with melanosomes and/or their precursors.

Several observations suggest a model for the involvement of SLC24A5 in organellar calcium uptake (Fig. 9.7.4E). First, the intracellular localization of the SLC24A5 protein suggests that it affects organellar, rather than cytoplasmic, calcium concentrations, in contrast with other members of the NCX and NCKX families. Second, the accumulation of calcium in mammalian melanosomes appears to occur in a transmembrane pH gradient–dependent manner (*23*). Third, several subunits of the vacuolar proton adenosine triphosphatase (V-ATPase) and at least two intracellular sodium/proton exchangers have also been localized to melanosomes (*24, 25*). In the model, active transport of protons by the V-ATPase is coupled to SLC24A5-mediated calcium transport via a sodium/proton exchanger. The melanosomal phenotype of the zebrafish *golden* mutant suggests that the calcium accumulation predicted by the model plays a role in melanosome morphogenesis and melanogenesis. The observations that processing of the melanosomal scaffolding protein pmel17 is mediated by a furin-like protease (*26*) and that furin activity is calcium-dependent (*27*) are consistent with this view. The role of pH in melanogenesis has been studied far more extensively than that of calcium, with alterations in pH affecting both the maturation of tyrosinase and its catalytic activity (*25, 28*). The interdependence of proton and calcium gradients in the model may thus provide a second mechanism, in addition to calcium-dependent melanosome morphogenesis, by which the activity of SLC24A5 might affect melanin pigmentation.

Role of SLC24A5 in Human Pigmentation

To evaluate the potential impact of SLC24A5 on the evolution of human skin pigmentation, we looked for polymorphisms within the gene. We noted that the G and A alleles of the single nucleotide polymorphism (SNP) rs1426654 encoded alanine or threonine, respectively, at amino acid 111 in the third exon of SLC24A5. This was the only coding SNP within SLC24A5 in the International Haplotype Map (HapMap) release 16c.1 (*29*). Sequence comparisons indicate the presence of alanine at the corresponding position in all other known members of the SLC24 (NCKX) gene family (fig. S5). The SNP rs1426654 had been previously shown to rank second (after the FY null allele at the Duffy antigen locus) in a tabulation of 3011 ancestry-informative markers (*30*). The allele frequency for the *Thr*[111] variant ranged from 98.7 to 100% among several European-American population samples, whereas the ancestral alanine allele (*Ala*[111]) had a frequency of 93 to 100% in African, Indigenous American, and East Asian population samples (fig. S6) (*29, 30*). The difference in allele frequencies between the European and African populations at rs1426654 ranks within the top 0.01% of SNP markers in the HapMap database (*29*), consistent with the possibility that this SNP has been a target of natural or sexual selection.

A striking reduction in heterozygosity near SLC24A5 in the European HapMap sample (Fig. 9.7.5A) constitutes additional evidence for selection. The 150-kb region on chromosome 15 that includes SLC24A5, MYEF2, CTNX2, and part of SLC12A1 has an average heterozygosity of only 0.0072 in the European sample, which is considerably lower than that of the non-European HapMap samples (0.175 to 0.226). This region, which contains several additional SNPs with high-frequency differences between populations, was the largest contiguous autosomal region of low heterozygosity in the European (CEU) population sample (Fig. 9.7.5B). This pattern of variation is consistent with the occurrence of a selective sweep in this genomic region in a population ancestral to Europeans. For comparison, diminished heterozygosity is seen in a 22-kb region encompassing the 3′ half of MATP (SLC45A2) in European samples, and more detailed analysis of this genomic region shows evidence for a selective sweep (31). However, the gene for agouti signaling protein (ASIP), which is known to be involved in pigmentation differences (32), shows no such evidence.

The availability of samples from two recently admixed populations, an African-American and an African-Caribbean population, allowed us to determine whether the rs1426654 polymorphism in SLC24A5 correlates with skin pigmentation levels, as measured by reflectometry (33). Regression analysis using ancestry and SLC24A5 genotype as independent variables revealed an impact of SLC24A5 on skin pigmentation (Fig. 9.7.6). Despite considerable overlap in skin pigmentation between genotypic groups, regression lines for individuals with GG versus AG and GG versus AA genotypes were separated by about 7 and 9.5 melanin units, respectively (Fig. 9.7.6A). These differences are more evident in plots of skin pigmentation separated by genotype (Fig. 9.7.6B). SLC24A5 genotype contributed an estimated 7.5, 9.5, or 11.2 melanin units to the differences in melanin pigmentation among African-Americans and African-Caribbeans in the dominant, unconstrained (additive effect plus dominance deviation), or additive models, respectively.

The computer program ADMIXMAP provides a test of gene effect that corrects for potential biases caused by uncertainty in the estimation of admixture from marker data (34). Score tests for association of melanin index with the SLC24A5 polymorphism were significant in both African-American ($P = 3 \times 10^{-6}$) and African-Caribbean population subsamples ($P = 2 \times 10^{-4}$). The effect of SLC24A5 on melanin index is between 7.6 and 11.4 melanin units (95% confidence limits). The data suggest that the skin-lightening effect of the A (Thr) allele is partially dominant to the G (Ala) allele. Based on the average pigmentation difference between European-Americans and African-Americans of about 30 melanin units (33), our results suggest that SLC24A5 explains between 25 and 38% of the European-African difference in skin melanin index.

Relative Contributions of SLC24A5 and Other Genes to Human Pigment Variation

Our estimates of the effect of SLC24A5 on pigmentation are consistent with previous work indicating that multiple genes must be invoked to explain the skin pigmentation differences between Europeans and Africans (*5*, *35*). Significant effects of several previously known pigmentation genes have been demonstrated, including those of MATP (*36*), ASIP (*32*), TYR (*33*), and OCA2 (*33*), but the magnitude of the contribution has been determined only for ASIP, which accounts for ≤4 melanin units (*32*). MATP may have a larger effect (*37*), but it can be concluded that much of the remaining difference in skin pigmentation remains to be explained.

Variation of skin, eye, and hair color in Europeans, in whom a haplotype containing the derived Thr^{111} allele predominates, indicates that other genes contribute to pigmentation within this population. For example, variants in MC1R have been linked to red hair and very light skin [reviewed in (*37*)], whereas OCA2 or a gene closely linked to it is involved in eye color (*7*, *38*). The lightening caused by the derived allele of SLC24A5 may be permissive for the effect of other genes on eye or hair color in Europeans.

Because Africans and East Asians share the ancestral Ala^{111} allele of rs1426654, this polymorphism cannot be responsible for the marked difference in skin pigmentation between these groups. Although we cannot rule out a contribution from other polymorphisms within this gene, the high heterozygosity in this region argues against a selective sweep in a population ancestral to East Asians. It will be interesting to determine whether the polymorphisms responsible for determining the lighter skin color of East Asians are unique to these populations or shared with Europeans.

The Importance of Model Systems in Human Gene Discovery

Our identification of the role of SLC24A5 in human pigmentation began with the positional cloning of a mutation in zebrafish. Typically, the search for genes associated with specific phenotypes in humans results in multiple potential candidates. Our results suggest that distinguishing the functional genes from multiple candidates may require a combination of phylogenetic analysis, nonmammalian functional genomics, and human genetics. Such cross-disciplinary approaches thus appear to be an effective way to mine societal benefit from our investment in the human genome.

References and Notes

1. R. N. Kelsh, *Pigment Cell Res.* 17, 26 (2004).

2. N. P. M. Smit et al., *Photochem. Photobiol.* 74, 424 (2001).

3. A. T. Hewitt, R. Adler, in *Retina*, S. J. Ryan, Ed. (The C. V. Mosby Company, St. Louis, MO, 1989), p. 57.

4. R. A. King, V. J. Hearing, D. J. Creel, W. S. Oetting, in *The Metabolic and Molecular Bases of Inherited Disease*, C. R. Scriver, A. L. Beaudet, W. S. Sly, D. Valle, Eds. (McGraw-Hill, St. Louis, MO, 2001), p. 5587.

5. G. S. Barsh, *PLoS Biol.* 1, 19 (2003).

6. J. L. Bolognia, S. J. Orlow, in *Dermatology*, J. L. Bolognia, J. L. Jorizzo, R. P. Rapini, Eds. (Mosby, London, 2003), p. 935.

7. R. A. Sturm, T. N. Frudakia, *Trends Genet.* 20, 327 (2004).

8. S. Fukamachi, A. Shimada, A. Shima, *Nat. Genet.* 28, 381 (2001).

9. G. Streisinger, C. Walker, N. Dower, D. Knauber, F. Singer, *Nature* 291, 293 (1981).

10. M. C. Mullins, M. Hammerschmidt, P. Haffter, C. Nusslein-Volhard, *Curr. Biol.* 4, 189 (1994).

11. T. Nguyen et al., *J. Invest. Dermatol.* 119, 1156 (2002).

12. T. Nguyen, M. Wei, *J. Invest. Dermatol.* 122, 452 (2004).

13. M. Gregory, P. Jagadeeswaran, *Blood Cells Mol. Dis.* 28, 418 (2002).

14. N. Shimoda et al., *Genomics* 58, 219 (1999).

15. C. Walker, G. Streisinger, *Genetics* 103, 125 (1983).

16. C. T. Amemiya, L. I. Zon, *Genomics* 58, 211 (1999).

17. A. Nasevicius, S. C. Ekker, *Nat. Genet.* 26, 216 (2000).

18. B. W. Draper, P. A. Morcos, C. B. Kimmel, *Genesis* 30, 154 (2001).

19. L. E. Maquat, *Curr. Biol.* 12, R196 (2002).

20. K. D. Philipson, D. A. Nicoll, *Annu. Rev. Physiol.* 62, 111 (2000).

21. X. Cai, J. Lytton, *Mol. Biol. Evol.* 21, 1692 (2004).

22. M. Cuomo et al., *J. Invest. Dermatol.* 96, 446 (1991).

23. R. Salceda, G. Sanchez-Chavez, *Cell Calcium* 27, 223 (2000).

24. V. Basrur et al., *Proteome Res.* 2, 69 (2003).

25. D. R. Smith, D. T. Spaulding, H. M. Glenn, B. B. Fuller, *Exp. Cell Res.* 298, 521 (2004).

26. J. F. Berson et al., *J. Cell Biol.* 161, 521 (2003).

27. G. Thomas, *Nat. Rev. Mol. Cell Biol.* 3, 53 (2002).

28. H. Watabe et al., *J. Biol. Chem.* 279, 7971 (2004).

29. The International HapMap Consortium, *Nature* 426, 789 (2003); available at www .hapmap.org, release 16b, 31 May 2005.

30. M. W. Smith et al., *Am. J. Hum. Genet.* 74, 1001 (2004).

31. M. Soejima et al., *Mol. Biol. Evol.*, doi: 10.1093/molbev/msj018 (2005).

32. C. Bonilla et al., *Hum. Genet.* 116, 402 (2005).

33. M. D. Shriver et al., *Hum. Genet.* 112, 386 (2003).

34. C. J. Hoggart et al., *Am. J. Hum. Genet.* 72, 1492 (2003).

35. G. A. Harrison, G. T. Owen, *Ann. Hum. Genet.* 28, 27 (1964).

36. J. Graf, R. Hodgson, A. van Daal, *Hum. Mutat.* 25, 278 (2005).

37. K. Makova, H. Norton, *Peptides* 26, 1901 (2005).

38. T. Frudakis et al., *Genetics* 165, 2071 (2003).

39. We thank P. Hubley for excellent management of our zebrafish facility; B. Blasiole, W. Boehmler, A. Sidor, and J. Gershenson for experimental assistance; R. Levenson, B. Kennedy, G. Chase, J. Carlson, and E. Puffenberger for helpful discussions; L. Rush for help on the cover design; and V. Hearing and M. Marks for MNT1 cells. Supported by funding from the Jake Gittlen Memorial Golf Tournament (K.C.), NSF (grant MCB9604923 to K.C.), NIH (grants CA73935, HD40179, and RR017441 to K.C.; HD37572 to D.G.; HG002154 to M.S.; EY11308 to N.M.; and HL077910 to P.J.), the Pennsylvania Tobacco Settlement Fund (K.C.), and the Natural Sciences and Engineering Research Council of Canada (E.P.). This work is dedicated to the open, trusting, and generous atmosphere fostered by the late George Streisinger.

Figure 9.6.1 Human rainbow. A newly discovered gene partly explains the light skin of Europeans, but not East Asians, as compared to Africans.

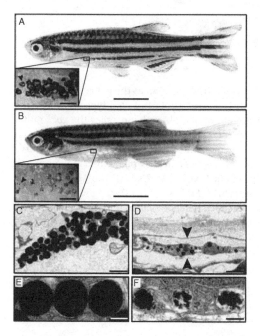

Figure 9.7.1 Phenotype of *golden* zebrafish. Lateral views of adult wild-type (A) and *golden* (B) zebrafish. Insets show melanophores (arrowheads). Scale bars, 5 mm (inset, 0.5 mm). *gol*[b1] mutants have melanophores that are, on average, smaller, more pale, and transparent. Transmission electron micrographs of skin melanophore from 55-hpf wild-type (C and E) and *gol*[b1] (D and F) larvae. *gol*[b1] skin melanophores (arrowheads show edges) are thinner and contain fewer melanosomes than do those of wild type. Melanosomes of *gol*[b1] larvae are fewer in number, smaller, less-pigmented, and irregular compared with wild type. Scale bars in (C) and (D), 1000 nm; in (E) and (F), 200nm.

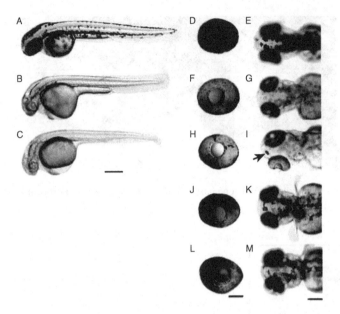

Figure 9.7.2 Rescue and morpholino knockdown establish SLC24A5 as the golden gene. Lateral views of 48-hpf (A) wild-type and (B) *gol^b1* zebrafish larvae. (C) 48-hpf wild-type larva injected with morpholino targeted to the traditional start site of SLC24A5 phenocopies the *gol^b1* mutation. Lateral view of eye (D) and dorsal view of head (E) of 72-hpf wild-type embryos. (F and G) *gol^b1* pigmentation pattern at 72 hpf, showing lightly pigmented cells. (H and I) 72 hpf *gol^b1* larva injected with PAC215f11 show mosaic rescue; arrow identifies a heavily pigmented melanophore. (J and K) 72-hpf larva injected with full-length zebrafish SLC24A5 RNA. (L and M) 72-hpf *gol^b1* larvae injected with full-length human European (Thr^111) SLC24A5 RNA. Rescue with the ancestral human allele (Ala^111) is shown in fig. S4 [not included here]. Rescue in RNA-injected embryos is more apparent in melanophores (K) and (M) than in RPE. Scale bars in [(A) to (C)], 300 μm; in (D), (F), (H), (J), and (L), 100 μm; in (E), (G), (I), (K) and (M), 200 μm.

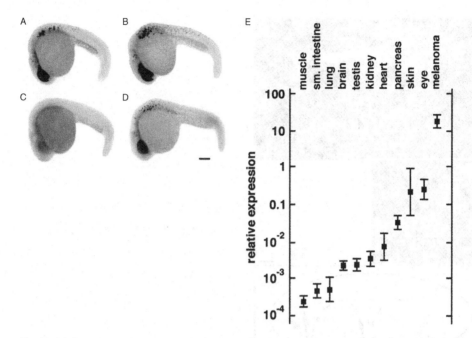

Figure 9.7.3 Expression of SLC24A5 in zebrafish embryos and adult mouse tissues. The expression of SLC24A5 (A) and DCT (B) in melanophores and RPE of a 24-hpf wild-type zebrafish larva show similar patterns. (C) *gol*[b1] larvae lack detectable SLC24A5 expression. (D) DCT expression in 24-hpf *gol*[b1] larva is similar to that in wild type. Scale bar, 200 μm. (E) Quantitative RT-PCR analysis of SLC24A5 expression in mouse tissues and B16 melanoma. Expression was normalized using the ration between SLC24A5 and the control transcript, RNA polymerase II (POLR2E).

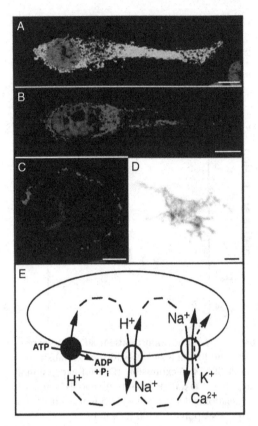

Figure 9.7.4 Subcellular localization of SLC24A5. Human MNT1 cells transfected with (A) GFP-tagged zebrafish SLC24A5 (green) and (B) HA-tagged SLC24A5 (red) clearly show intracellular expression. (C) HA-tagged D3 dopamine receptor localizes to the plasma membrane in MNT1 cells (red). 4′,6′-diamidino-2-phenylindole (DAPI) counterstain was used to visualize nuclei. Scale bars in (A) and (B), 10μm; in (C), 5 μm. (D) Rescue of dark pigmentation in a melanophore of a *golden* embryo by HA-tagged SLC24A5. These dark cells appear in *golden* embryos injected with the HA-tagged construct, but not in mock-injected embryos. Scale bar, 10 μm. (E) Model for calcium accumulation in melanosomes. Protons are actively transported into the melanosome by the V-ATPase (left). The proton electrochemical potential gradient drives sodium uptake via the sodium (Na$^+$)/proton (H$^+$) exchanger (center). Sodium efflux is coupled to calcium uptake by the SLC24A5 polypeptide (right). If potassium (dashed arrow) is cotransported with calcium, it must either accumulate within the melanosome or exit by means of additional transporters (not depicted). P$_i$, inorganic phosphate; ADP, adenosine diphosphate.

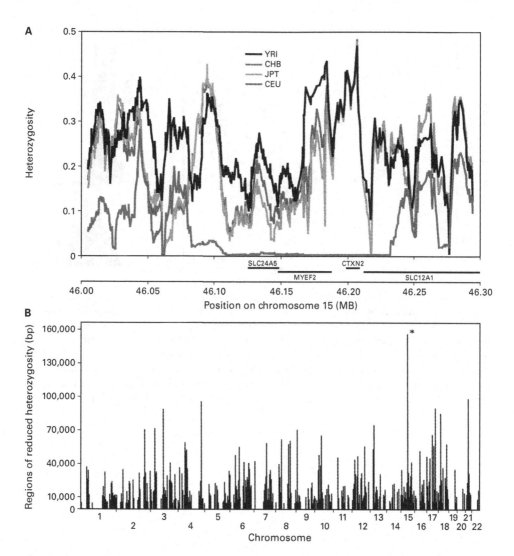

Figure 9.7.5 Region of decreased heterozygosity in Europeans on chromosome 15 near SLC24A5. (A) Heterozygosity for four HapMap populations plotted as averages over 10-kb intervals. YRI, Yoruba from Ibadan, Nigeria; CHB, Han Chinese from Beijing; JPT, Japanese from Tokyo; CEU, CEPH (Foundation Jean Dausset-Centre d'Etude du Polymorphism Humain) population of northern and western European ancestry from Utah. The data are from HapMap release 18 (phase II). (B) Distribution in genome of extended regions with low heterozygosity in the CEU sample. Only regions larger than 5 kb in which all SNPs have minor allele frequencies ≤0.05 and which contained at least one SNP with a population frequency difference between CEU and YRI of greater than 0.75 were plotted. Regions were divided at gaps between genotyped SNPs exceeding 10 kb. The data are from HapMap release 16c.1. An asterisk marks the region containing SLC24A5 within 15q21.

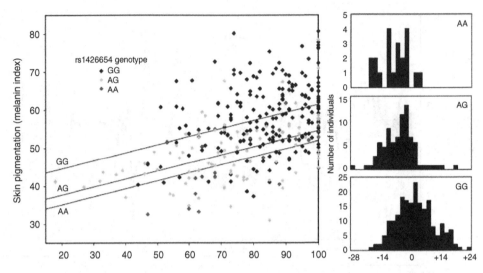

Figure 9.7.6 Effect of SLC24A5 genotype on pigmentation in admixed populations. (A) Variation of measured pigmentation with estimated ancestry and SLC24A5 genotype. Each point represents a single individual. Lines show regressions, constrained to have equal slopes, for each of the three genotypes. (B) Histograms showing the distribution of pigmentation after adjustment for ancestry for each genotype. Values shown are the difference between the measured melanin index and the calculated GG regression line ($y = 0.2113x + 30.91$). The corresponding uncorrected histograms are shown in fig. S7 [not included here]. Mean and SD (in parentheses) are given as follows for GG, 0 (8.5), $n = 202$ individuals; for AG, –7.0 (7.4), $n = 85$; for AA, –9.6 (6.4), $n = 21$.

Index

A-B-O system of blood types, 302, 305
Acclimatization of races, 113–115
Acromegaly, 225–226
Adamson, Edward William, 161
Adaptation, 281–286
Adaptive race, 7
Addison's disease, 226, 233
Admixture
  and ancestry, 279, 309, 320
  Caucasian, 42, 255, 330, 341, 359
  and genetic categorization, 332–333, 365
  mulatto, 252
  "normal," 143
  and racial identity, 237, 238, 267
Adoption, 349–350
Adrenal glands, 203–204, 218–219
African Americans. *See* Blacks
Alleles, 331–332, 358, 364–365
Almanac, 15
American Anthropological Association statement on race, 310, 322–324
American Anti-Slavery Society, 15
American Association of Physical Anthropologists statement on race, 310, 319–321
American Indians. *See* Native Americans
American Journal of Physical Anthropology, 310
American Sociological Association statement on race
  classification and social reality, 348–349
  classifications as the basis for scientific inquiry, 348
  consequences of race and race relations in social institutions, 349
  executive summary, 346–347
  race and ethnicity as basis for scientific investigation of proximate causes and critical interactions, 350–351

race as an organizing device for mobilization to maintain or challenge systems racial stratification, 350
race as a sorting mechanism, 349–350
race as a stratifying practice, 350
research highlights, 351–353
social concept of race, 348
Anatomical Record, 148
Anatomy. *See also* Blood; Glands
  and athletics, 185–194
  brain, 68–69, 205
  discussions on, 16–17
  ear, 42–51
  Frederick Douglass on, 37–40
  hair, 58, 266–270
  measurement research, 147–148
  muscular, 193
  neuromuscular, 193
  nose, 52–57
  pelvic, 59
  respiratory system, 69–70, 73–74, 205
  skeletal, 190–193, 225–226
  skin, 168–177
  sweat gland, 161–167
  Thomas Jefferson on differences in, 24–25
  and vulnerability to disease, 67–86
Ancestry, self-reported, 333–335
Anderson, George, 186
Animals, laws regarding, 22–23
Anthropology, physical. *See* Physical anthropology
Anti-Chinese Riot, 15
Arena, The, 107
Army, U.S. *See* Military
Aron, Hans, 161
Aryan race, 5
Asians and Pacific Islanders, 4, 15, 64, 220, 245, 254, 288
  alleles and skin color in, 358–359
  and health care, 352–353

Printed in the United States
By Bookmasters